D1274701

RADIOACTIVE
WASTE MANAGEMENT

RADIOACTIVE WASTE MANAGEMENT

ROBERT E. BERLIN
Department of Mechanical Engineering, Manhattan College

CATHERINE C. STANTON
Catherine C. Stanton & Associates, Inc.

WILEY

A Wiley-Interscience Publication

JOHN WILEY & SONS

New York • Chichester • Brisbane • Toronto • Singapore

Library of Congress Cataloging-in-Publication Data

Berlin, Robert E.
    Radioactive waste management.

    "A Wiley-Interscience publication."
    Bibliography: p.
    Includes index.
    1. Radioactive waste disposal—United States.
2. Radioactive waste sites—United States.   I. Stanton,
Catherine C.   II. Title.
TD812.2.B47   1990   363.7'28   88-17359
ISBN 0-471-85792-0

For our respective spouses,
Ruth and Harry,
for their tolerance and understanding

FOREWORD

Generally speaking, whenever the subject of radioactive waste comes up, it invariably is associated with controversy. The public's understanding of radioactive waste is frequently clouded by a lack of clear, concise, and timely information. All too often radioactive waste is thought to be a single entity, when in reality it takes various forms, with varying degrees of hazard. For example, the waste form could be liquid, solid, or gaseous. As to intensity, it could be low-level, intermediate-level, transuranic, mill tailings, or high-level waste. As a matter of fact, when surveying the field of waste, a spectrum emerges. At the low end of the spectrum there is the ubiquitous municipal solid waste—normal household garbage, followed by industrial waste, then hazardous chemical waste, low-level radioactive waste, intermediate-level radioactive waste, transuranic waste, and finally high-level radioactive waste.

While all these waste streams differ greatly as to quantity, form, and degree of hazard, they have one thing in common. It is difficult to site facilities for their storage, treatment, and disposal. Nevertheless, it is imperative that all these wastes be properly managed, for if they are not, the consequence may very well be disastrous. Dr. Berlin's and Mrs. Stanton's book takes on the task of addressing the radioactive waste portion of the spectrum. It is a much needed guide for the beginner and practitioner alike in that it includes in one place a comprehensive treatment of radioactive waste management.

A statement frequently made regarding radioactive waste is that the solution to the problem is not so much one of technical feasibility, but rather social or political—and there is a good deal of truth in that thought. The recently enacted fiscal year 1988 budget resolution included a provision designating Yucca Mountain in the state of Nevada as the site of this country's first high level radioactive waste site. This particular legislative initiative was enacted in large measure as a result of Congressional concern that the then existing process was making slow progress and was exceedingly expensive. The nation's high level radioactive waste needs to be placed in final disposal in a reasonable amount of time. For transuranic waste, the Department of Energy's Waste Isolation Pilot Plant, near Carlsbad, New Mexico, fills the bill. There are three existing low level radioactive waste facilities now in operation with additional capacity due to come on line in the early 1990s. And finally, the Department of Energy is proceeding with the required remedial action program for abandoned uranium mill tailings piles.

The credibility of the entire radioactive waste management program rests, in large measure, on the public's confidence that the effort is proceeding in a technically sound manner—this book makes a significant contribution toward that goal.

SHELDON MEYERS

Director, Office of Radiation Program
U.S. Environmental Protection Agency
Washington, DC
March 1988

PREFACE

This book is intended to provide a comprehensive and current review of the field of radioactive waste management to be used by practitioners in the field and by students of the subject. The motivation for the book was the lack of a unified body of information relating to the historical, regulatory, environmental, and technological aspects of radioactive waste management. The information presented herein is intended to remedy that lack by showing the continuity of evolution of the field, with its attendant problems and successes, leading to the current national program, and the waste management concepts and approaches emerging for future use.

The text material is organized to take the reader from an introduction to the radioactive waste management field through, it is hoped, a logical progression describing the waste forms and their parameters, regulatory and programmatic aspects, environmental and safety concerns and control measures, and the current technology for managing the various waste forms from generation through disposal. The breadth of the task required that some topics be introduced rather than described in detail. References to additional studies and data are provided at the end of each chapter for those seeking more detailed information. The latter part of the book is devoted to the remediation of radiologically contaminated sites.

We have drawn extensively upon the body of literature documenting the efforts of the responsible governmental agencies in the field and are appreciative of the alacrity with which information has been placed in the public domain. Agency and contractor personnel as well as members of the waste management industry and waste generators have been very helpful in providing data and information on "lessons learned." We appreciate their sharing their experiences with us. Finally, we are indebted to Rose Ann Palestro and Catherine O'Reilly for their help in preparation of the manuscript.

ROBERT E. BERLIN
CATHERINE C. STANTON

New York
April 1988

CONTENTS

RADIOACTIVE
WASTE MANAGEMENT

CHAPTER 1

AN INTRODUCTION TO THE FIELD OF RADIOACTIVE WASTE MANAGEMENT

1.1 GENERAL DEFINITION OF RADIOACTIVE WASTE

Any material that is no longer useful and that contains radioactive isotopes is, by rigorous definition, radioactive waste. A radioactive isotope emits energy in the form of ionizing radiation. The isotopes are characterized by the type of emission—alpha, beta, gamma, neutron—as well as the frequency of decay and energy of the emitted radiation. The half-life of the isotope defines the length of time in which half of the material will have decayed. The health effects that result from exposure to a given isotope are based on all of these factors as well as the chemical and physical form of the material and the method of exposure. That is, impacts will differ depending on whether the source is external to the receptor or it is ingested, inhaled, absorbed through the skin, or introduced by some other method. (For further information, see Appendixes A and B.)

Such a definition was less than useful, as demonstrated when the Oregon legislature passed an ordinance aimed at preventing the development of a radioactive waste disposal facility in the state by banning the burial of radioactive materials. Because we live in a naturally radioactive world the statute, if interpreted literally, would have prevented all human interments as well as radioactive waste disposal facilities. Clearly, some narrower limits are needed to define both the potential risk and practical technologies for management and disposal of radioactive waste.

For purposes of this text *radioactive waste is defined as any material that is no longer useful and that contains radioactive isotopes in amounts recognized by regulatory authorities as posing a potential risk to human health and the environment sufficient to warrant its isolation from the biosphere.*

1

This definition is deceptively simple:

- It condenses into one sentence considerations addressed in dozens of federal and state laws and regulations.
- It allows the consideration of material whether of natural or man-made origin.
- It freezes the frame of reference at the mid to late 1980s.
- It addresses the reality of the involvement of social issues as well as the purely "technical" issues with which the reader may be more used to dealing.

As will be developed in the text, each of these factors plays an important role in defining radioactive waste management as practiced today and as will be practiced in new facilities now being designed and developed. The options available for managing a specific waste stream reflect consideration of the agent responsible for its generation, its current condition, the population at risk, and the cost of isolation (including any necessary intermediate processing) both in dollars and in occupational exposure. The objective of this text is to provide a single source of information on what is in reality many technological specialties. A quick review of the references cited will show that hundreds of volumes have been devoted to the subject of radioactive waste. This text will serve as a guide to much of this more specialized material by providing engineering, regulatory, and health and environmental protection professionals and students with a basic understanding of the characteristics of the materials important to choosing among available management options as well as the reasons for actions already taken or currently underway.

The general definition of radioactive waste provided excludes routine releases of radioactive materials from facilities such as hospitals, power reactors, and industrial installations since regulatory authorities have effectively judged that these releases present potential risks sufficiently low that further isolation (with existing technology) is not required. It includes, however, different sources of waste frequently not considered in a single volume because of different regulatory jurisdictions. For example, the Atomic Energy Act delegates responsibility for waste resulting from nuclear reactor-produced radioisotopes to the Nuclear Regulatory Commission (NRC). Those same isotopes would not be subject to NRC licensing if produced in a linear accelerator.

A key phrase in the definition is "recognized by regulatory authorities." Many of the waste streams described in the text were not initially considered to require subsequent isolation. For example, waste material produced by separating radium from ore and producing luminous products or sealed radiation sources predated the establishment of federal and state regulations and its management was not always well documented. The result has been the identification of several geographic areas requiring evaluation, and large volumes of material that have been subsequently treated and/or relocated. On the other hand, as a result of current efforts to quantify waste streams that contain radioactive materials at "de minimis" levels (levels at which the potential risk from disposal as nonradioactive is below regulatory concern) several of the waste streams discussed in the text may no longer require isolation in the future. One of the main themes of this text is that the field of radioactive waste is dynamic, not static. Management decisions reflect risk–benefit and cost-effectiveness evaluations that will change with time as new lessons are learned from operating experience and technological advances increase the choices available. It is entirely possible, for example, that a new requirement for, or application of, a particular isotope or group of isotopes could make reprocessing of spent fuel and recycling of the isotopes a viable option in the future; waste management needs would therefore change accordingly.

1.2 NATURE AND MAGNITUDE OF THE WASTE MANAGEMENT "PROBLEM"

Radioactive waste has been—and currently is being—produced by both government and private sources. It consists of large volumes of material containing relatively low concentrations of radioisotopes as well as smaller volumes of more highly concentrated materials. The external radiation levels at a container's surface vary from unmeasurable to levels that would provide lethal exposure and therefore require substantial shielding for handling, shipment, and disposal. The length of time for which isolation is required can vary from days to thousands of years, depending on the specific isotopes contained and the amounts in which they are present.

These differences mean that there are, in reality, several radioactive waste management "problems." Detailed characteristics and similarities and differences among the waste types are described in Chapter 2, "Radioactive Waste Forms." This section summarizes several waste forms, making it easier to compare the risks involved, the technical requirements for isolation, and the environmental and social impact of present and future waste management programs.

Compared to other waste management needs with which society is faced, radioactive waste is relatively limited in scope. For example, sanitary landfills used by many municipalities throughout the country are reaching their capacity. Resistance to developing new facilities to replace those filled up, and in some cases the costs for their development, has resulted in waste being transported to other states for disposal. In one extreme example of the problems involved, a town on Long Island, New York, contracted with a hauler to have removed by barge for disposal elsewhere 3,100 tons of garbage that otherwise would have been placed in the municipal landfill. The barge was denied access to the originally planned disposal site in North Carolina at which it was hoped to use the material to demonstrate the feasibility and economics of recovering methane produced from the decomposition of the waste and using the recovered gas for energy. The barge then set out on a months-long odyssey south along the eastern coast of the United States and through the Caribbean Sea and the Gulf of Mexico seeking a state or country that would allow a disposal facility to accept the waste. It eventually returned and anchored in New York Harbor where its further progress was the subject of multiple court rulings and administrative agreements. Eventual disposition was through incineration in a New York City operated incinerator and disposal of the ash in a landfill in the town from which it originated. By comparison, the total low-level waste volume disposed at commercial facilities in 1986 was approximately two million ft^3. Assuming an average density of 30 lb/ft^3, the total weight of low-level waste disposed nationally in 1986 was less than 30,000 tons or less than 10 times the amount on that one barge.

In terms of necessary disposal area, radioactive waste management needs are also relatively small. Current plans for a high-level radioactive waste repository to receive spent fuel rods and defense high-level waste produced through about 2015 are based on use of a 2000-acre site. Similarly, a reference low-level radioactive waste disposal facility capable of accepting about 35 million ft^3 of waste over a 20-yr period would occupy about 200 acres, much of which would be a buffer between the actual disposal area and the site boundary (NRC 1981). In contrast, municipal waste was being generated in the State of New York in the early 1980s at the rate of about 15 million tons annually and required the commitment of approximately 400 acres annually for sanitary landfills (DEC 1982). As will be discussed in detail in Chapter 3, however, it is currently expected that there will be multiple low-level waste sites around the country and the size of each is likely to be at least 100 acres because of buffer zone requirements between the actual disposal area and the site boundary. However, land use considerations are nominal even under such a dispersed disposal system.

With respect to the need to provide a management system to isolate the waste from man and his environment, radioactive waste may be more appropriately compared to hazardous waste than to general municipal waste. Hazardous wastes regulated under the Resource Conservation and Recovery Act are produced at a rate of about 150 million metric tons per year according to the Environmental Protection Agency (EPA 1985).

There is a fundamental difference in the comparison because hazardous waste facilities are designed to retain the material within the disposal unit whereas the underlying philosophy behind radioactive waste disposal (of whatever waste type) is that man-made barriers will eventually fail and the site conditions must be a primary barrier to contact between the contained radioactive material and the biosphere. It should also be noted that the degree of hazard of radioactive waste decreases with time as decay proceeds. However, the time periods for which isolation is planned—on the order of several hundreds to several thousands of years depending on the waste type considered—exceed the lifetime of most structures and even institutions with which we are familiar. The engineering and institutional arrangements required to isolate this material successfully are new and developing rather than being based on a body of knowledge that can be applied from previous experience. Hazardous wastes, on the other hand, effectively have an infinite half-life and so the tasks being addressed in the field of radioactive waste management are far from unique.

A further distinction must be made between the options available to manage waste currently being produced or expected to be produced in the future and those applicable to material either previously disposed and requiring remediation or now in storage and requiring processing prior to final disposition. Both conditions exist for all waste types. High-level waste originally produced with the expectation of long-term care as a liquid in underground tanks is being removed and solidified for subsequent disposal, whereas newly produced material— whether reprocessed defense waste or commercial spent fuel rods—will be encapsulated for disposal in a geological repository. Some previously produced transuranic waste has been processed for disposal along with newly produced waste, whereas other such material, originally disposed of by directly releasing it to the ground, is being evaluated for the usefulness of alternatives such as excavation and removal or fusing it in place by application of electric current to the ground in which it is located. At least one previously operated commercial low-level waste disposal facility will continue to be the subject of efforts to stabilize the contained material and achieve an essentially passive monitoring status. Newly developed facilities will be sited and designed with such conditions built in. Similar considerations will be addressed in new facilities for handling waste material from mining and milling of uranium and thorium ores and phosphate deposits, while methods for retrieving, stabilizing, and/or isolating existing wastes are being applied where necessary to protect public health and safety.

High-level waste currently exists at federal government facilities at the Hanford Reservation in the State of Washington, the Savannah River Reservation in South Carolina, the Idaho National Engineering Laboratory in Idaho, and the former Western New York Nuclear Service Center in West Valley, New York (DOE, 1987a). As of 1986, the Department of Energy estimated that about 13 million ft^3 of such waste existed in a variety of physical forms (liquid, sludge, salt cake, dry calcine) and containers (single shell tanks, double shell tanks, bins). Less than 1% of the volume of this waste was due to the commercial high-level waste in West Valley. Volumes are expected to be about the same in the year 2000 because some production will be offset by solidification of liquids with a related volume reduction. By 2020, although the volumes are projected to be still about 13 million ft^3, the inventory of isotopes in the waste is projected to increase from 1.4 billion Ci in 1986 to 1.8 billion Ci (DOE 1987a).

No commercial spent fuel reprocessing services have been available in the United States since 1971, and spent fuel has been stored in water pools, generally at the reactor site at which the fuel was used, since that time. It is not anticipated that new reprocessing capacity will be commercially developed in the United States in the near future. Therefore, the spent fuel will be disposed as high-level waste in a Federal repository. In response to the unavailability of some offsite storage capacity, utilities have modified the existing spent fuel storage pools and the racks in which the fuel is kept to accommodate extended storage of the fuel onsite. In many cases it has been possible to provide life of plant storage capacity, thus reducing the probability that a plant would have to shut down because of lack of spent fuel storage space. Existing legislative authorization would enable the NRC to license emergency transfer of spent fuel to avoid shutdown of reactors because of this problem. Currently existing inventories of spent fuel are estimated to increase by about a factor of 7 through 2020, by which time it is anticipated that two repositories will be accepting spent fuel for permanent disposal. (See discussion in Chapter 3 on the uncertainties in the program and schedule.)

Table 1-1 summarizes the types and amounts of radioactive waste in existence

TABLE 1-1 Summary of Radioactive Waste Inventories as of December 31, 1986

Waste Type	Volume (ft^3/m^3)	Activity (Ci)	Thermal Power (W)
High-Level Waste (HLW)			
Defense	1.3(7)/3.7(5)[a]	1.4(6)	4.4(6)
Commercial	8.2(4)/2.3(3)	3.1(6)	9.1(3)
Spent fuel	2.1(5)/6.0(3)	1.6(10)	5.9(4)
	1.4(4) metric tons heavy metal		
Transuranic Waste (TRU)			
Retrievably stored			
CH[b]	1.7(6)/4.9(4)	2.9(6)	7.6(4)
		1.8(3) kg	
RH[b]	4.8(4)/1.4(3)	4.7(5)	4.5(3)
		4.0(0) kg	
Low-Level Waste (LLW)			
DOE sites	8.1(7)/2.3(6)	1.2(7)	1.7(4)
Commercial	4.3(7)/1.2(6)	4.6(6)	3.6(4)
Active	3.2(7)/9.0(5)		
Closed	1.2(7)/3.1(5)		
Remediation	3.3(6)/9.5(4)	Not available	
Uranium Mill Tailings			
Active sites	3.6(9)/1.0(8)	Not available	
Remediation	3.7(7)/1.1(6)	Not available	
Phosphogypsum Wastes			
Gypsum piles[c]	1.04(9) metric tons	Not available	

Source: DOE/RW-0006, Rev. 3, "Integrated Data Base for 1987: Spent Fuel and Radioactive Waste Inventories, Projections, and Characteristics," September 1987.

[a] $1.3(7) = 1.3 \times 10^7$.
[b] CH, contact-handled TRU; RH, remotely handled TRU.
[c] Based on NCRP estimates of 1 billion metric tons cumulative through 1984 and an annual production rate of 11 million metric tons.

as of the end of 1986 that meet the definition given in Section 1.1 and that will be subsequently addressed in the rest of this text. As is evident from this table, the many different units by which waste is measured make understanding the overall radioactive waste problem difficult. For some purposes, such as planning and designing new land disposal facilities for low-level waste, volume is a primary unit of interest. In other cases different units are generally used to describe and compare the waste stream. For example, the charge for spent fuel disposal in a federal repository is related to the benefit received from that fuel in terms of kilowatt-hours produced. This is roughly proportional to the amount of uranium used. Therefore, the common method of measuring amounts of spent fuel is in terms of metric tons of uranium (MTU) or heavy metal (MTIHM) initially contained in the fuel. Reprocessed defense high-level waste will also be disposed at the same repository and so a methodology was needed to enable comparison of the different waste types and allocate the costs appropriately. Inventory data in whatever units are used, are the basis for planning, design, and costing of waste management programs. Evaluation of the potential risk to public health and the environment from the radioisotopes contained in the waste requires application of complex and sophisticated computer models that factor in parameters such as initial isotope content, time since emplacement, interaction between the waste and the packaging and surrounding soil or rock, thermal interaction between the host rock and the waste for high-level waste, the probability of inadvertent intrusion several hundreds of years from now for near surface facilities, and the probability and effects of extreme natural phenomena such as earthquakes or tornadoes. Pertinent waste characteristics and expected performance of different disposal sites will be described in detail in Chapters 2, 4, and 6.

1.3 NATIONAL CHARACTER OF RADIOACTIVE WASTE MANAGEMENT

Management of radioactive waste is a national issue primarily for the following reasons:

1. Much of the waste was generated by the production of nuclear materials for weapons uses, and, therefore, the reason for its existence or the "benefit" from its existence is shared by every citizen.

2. The long times necessary for this material to decay to levels that are no longer of concern from a health and safety standpoint exceed those that are generally considered to be realistic in terms of private commercial endeavors and therefore the sites on which such disposal facilities are located are required to be owned by the state or federal government, which are assumed to have a somewhat longer institutional life expectancy. As can be seen from the information in Table 1-2, wastes currently exist throughout the country and there are multiple reasons for these wastes being produced.

3. Perhaps most importantly, public health and safety are enhanced by having the regulation and management of radioactive material as uniform as possible

TABLE 1-2   Location of Existing and Projected Radioactive Waste

Waste Type	Location
High Level Waste	
Defense	Washington, Idaho, South Carolina
Commercial	New York
Spent Fuel	Reactor sites nationwide
Federal Repository	Expected to be Nevada
Transuranic Waste	Washington, Idaho, South Carolina, Tennessee, New Mexico, Colorado, California, Nevada, Illinois, Ohio, Pennsylvania, Kentucky
Low Level Waste	Generated in every state
	Currently disposed in Illinois, New York, Kentucky, South Carolina, Nevada, Washington, Idaho, New Mexico, Tennessee, Ohio, Missouri, Texas, California
	New disposal facilities planned in California, Colorado, Illinois, Michigan, Pennsylvania, Texas, New York, North Carolina, Massachusetts, other states to be determined
Uranium Mill Tailings	Colorado, New Mexico, Texas, Utah, South Dakota, Washington, Wyoming
Phosphogypsum Wastes	Florida, Louisiana, Idaho, Minnesota, North Carolina, Tennessee, Utah, Wyoming

throughout the country based on widely recognized and accepted methods of estimating the potential risks from exposure to radiation and the anticipated performance of a given waste-disposal facility.

The evolution and development of the existing waste management structure in the United States involved all branches of the federal government as well as interaction with states, native Americans, organized citizen groups, and individuals. Table 1-3 illustrates the breadth of involvement in terms of agencies with responsibility under some 18 different federal laws. Dozens of other statutes exist at the state level primarily addressing the management of low-level radioactive waste that was determined to be a state responsibility by the Congress as defined in the Low-Level Radioactive Waste Policy Act of 1980 and the 1985 Amendments Act.

1.4   LEGACY OF PAST DISPOSAL OF RADIOACTIVE WASTES

Management of radioactive waste has ranged from permanently disposing of the material to what is in effect interim storage with a variety of levels of care and attention being given to such stored material. The fundamental criterion for success in waste management is how well the practice isolates the waste from the biosphere as radioactive decay acts to reduce the hazard of the material. By this

TABLE 1-3 Federal Laws and Agencies with Responsibility for the U.S. Radioactive Waste Management Program

Laws	Responsible Agencies
Atomic Energy Act of 1954, as amended	Nuclear Regulatory Commission
	Department of Energy
	Environmental Protection Agency
Energy Reorganization Act	Nuclear Regulatory Commission
	Department of Energy
Department of Energy Organization Act of 1977	Department of Energy
Uranium Mill Tailings Radiation Control Act of 1978	Nuclear Regulatory Commission
	Department of Interior
	Environmental Protection Agency
Hazardous Materials Transportation Act of 1975	Department of Transportation
Federal Water Pollution Control Act, as amended	Environmental Protection Agency
Marine Protection, Research, and Sanctuaries Act of 1972	Environmental Protection Agency
	Department of Transportation
Resource Conservation and Recovery Act of 1976	Environmental Protection Agency
Clean Air Act, as amended	Environmental Protection Agency
Act of 3 March 1849 (9 Stat. 395, 43 u.s.c. 1451)	Department of Interior (Bureau of Indian Affairs)
Federal Land Policy and Management Act of 1976	Department of Interior (Bureau of Land Management)
National Environmental Policy Act of 1969	Council on Environmental Quality
West Valley Demonstration Project Act of 1980	Department of Energy
	Nuclear Regulatory Commission
Low Level Radioactive Waste Policy Act of 1980, as amended	Nuclear Regulatory Commission
	Department of Energy
	States and Interstate Compacts
Nuclear Waste Policy Act of 1982, as amended	Department of Energy
	Nuclear Regulatory Commission
	Environmental Protection Agency
Comprehensive Environmental Reclamation, Compensation and Recovery Act	Environmental Protection Agency
Superfund Amendments and Reauthorization Act of 1974	Environmental Protection Agency

definition, there is at best a mixed record. In particular, it is now recognized that substantially more effort is required to retain wastes for long times after control of the site by the operator or custodial agency has ceased.

For example, management of high-level waste originally included plans for long-term storage of liquids with periodic pumping from one underground tank to another. In effect, this was a decision to postpone choosing a long-term disposal method. As a result, major programs are now underway, particularly at government facilities, to determine how and when such long-term disposal will be accomplished. At least one such facility has recently indicated that it does not expect to choose a disposal method for its tank wastes for several more years (DOE 1987b).

Changing regulatory requirements for transuranic wastes have resulted in a need to review earlier disposal practices to determine which wastes, if any, need to be retrieved, treated and/or repackaged, and disposed under different conditions.

Low-level waste disposal has similarly shifted its emphasis to incorporate as early as possible features that will facilitate minimum active maintenance of the site (such as trench cap repair or replacement) over a 100-yr postclosure institutional control period. Significant advances in the understanding and appreciation of the importance of geotechnical conditions on eventual waste and disposal unit performance have fundamentally changed the process by which new sites are chosen, and facilities designed, constructed, and operated.

Changes in methods for managing uranium mill tailings can be traced to concerns over operational failures of tailings impoundments that distributed liquids on the surface as well as to a better understanding of the long-term impacts of radon evolution from tailings after the site ceases operation. Both currently operating and inactive tailings sites are being managed to stabilize the tailings as soon as possible and to remove any windblown materials from vicinity properties when necessary. Similar considerations have resulted in changes in methods for reclaiming areas in which phosphate deposits were mined and identifying compatible long-term land uses.

Because of earlier radioactive waste management practices
- there is at present a need to perform remedial actions at sites throughout the country, both government and commercial;
- there is a much better understanding of long-term performance of disposal units and the relationship to release, transport, and exposure pathways;
- new facilities are being designed to facilitate waste management, with particular emphasis on long-term stability; and
- waste disposal decisions are no longer considered only on technical grounds. Socioeconomic impacts such as in-migration of construction or operating personnel, changing land use and property values, and involvement of local officials in planning and monitoring disposal facilities are recognized as integral components of a successful program.

1.5   CURRENT STATUS AND PROBLEM AREAS

The high-level radioactive waste management program is based on the use of geologic repositories to achieve isolation of waste from the biosphere for thousands of years. Current plans are for the first repository to be located at Yucca Mountain, Nevada unless there is a technical reason why the site is unable to conform to existing regulatory requirements. Site characterization and waste/package material selection to ensure the long-term integrity of the canister and further refinement of analytical methods to model the performance of the site are major areas that will be developed in the near future. There is also continued investigation of the alternatives available for solidification of liquid waste and decommissioning of structures currently used to store the material. These decisions will be an integral part of site management plans for the several government facilities at which these wastes are now located. Systemwide, transportation requirements will be met by the construction of a government-owned fleet of shielded casks to move spent fuel from the reactor sites to the repository.

Transuranic waste produced by the federal government in defense programs will be disposed at the Waste Isolation Pilot Plant (WIPP), a geologic repository in bedded salt near Carlsbad, New Mexico. First receipt of waste is scheduled for October 1988. It is expected that experience in waste receipt, handling, and emplacement at WIPP will be of direct import to the subsequent high-level waste repository operations. As further described in Chapter 6, some high-level waste will be emplaced at WIPP to demonstrate emplacement and retrieval methods.

The three currently operating low-level waste facilities will restrict access in accordance with Congressionally specified conditions as of December 31, 1992. Currently about 13 states or multistate compacts are in the process of identifying, characterizing, and developing new sites for disposal of radioactive waste subsequent to that time. In addition to the site-selection process, there is also substantial effort being pursued to design new facilities that will resist both natural forces and potential human interaction that might result in waste release for up to 500 years after emplacement. Such efforts include materials selection to maximize the compatibility between waste containment structures and site soil conditions, geohydrologic balances, and local land and water uses. The development efforts are guided, in large part, by lessons learned at the three operating sites and at three other sites that are no longer in operation. Details of this experience, including the status of remedial actions, are provided in Chapter 3.

Tailings resulting from the mining and milling of uranium are the major volume waste stream that must be managed, both from current operations and from inactive facilities. Current production levels are very low and, in fact, the government has issued a finding that the domestic uranium is officially "nonviable." This finding is supposed to result in limits being placed on the amount of uranium that can be imported for use in domestic reactors. However, in the absence of new reactor orders it is not expected that there will be a significant change in uranium tailings production in the immediate future. For that reason, and because of advances in stabilization and closure technology for tailings ponds and more stringent regulations imposed after many of the mines and mills ceased operation, there is a significant level of remediation work in progress and expected to continue. These programs are further discussed in Chapters 3 and 7.

Of concern to both low-level waste generators and regulators is the issue of collateral enforcement by the Nuclear Regulatory Commission and the Environmental Protection Agency. This condition arises because of the presence of both radioactive material and chemically hazardous material in radioactive waste. Responsibility for protecting public health and safety for these materials has been vested in the two agencies in accordance with their enabling legislation and subsequent laws (e.g., the Comprehensive Environmental Reclamation, Compensation and Liability Act, and the Superfund Amendments and Reauthorization Act). Currently generators and disposal facilities handling the "mixed" waste must comply with both agencies' regulations. However, experience is just beginning to be accumulated on actually putting this into practice. To date, no new sites have been selected with this dual guidance as a criterion. It is expected that application of different technology to a portion of the site may be acceptable for disposal of "mixed" waste in a segregated manner.

1.6   RADIOACTIVE WASTE MANAGEMENT IMPLICATIONS OF ACCIDENTS

Events such as the reactor accidents at Three Mile Island Unit 2 in Pennsylvania in 1979 and Chernobyl in the Soviet Ukraine in 1986 present radioactive waste management challenges at different levels. First, there is the required decontamination, cleanup, and removal of highly radioactive equipment and structures in and near the reactor itself. Second, it may be necessary to decontaminate local areas—buildings, land, and road surfaces on which airborne material was deposited either by gravity or due to precipitation. Third, there may be a need to dispose of contaminated foodstuffs that exceed allowable concentrations of dispersed radioisotopes. Finally, there are longer term impacts on the acceptability of the technology and the related waste disposal facilities. These issues are discussed in this section with particular reference to conditions at Three Mile Island and Chernobyl. In addition, the impact of a recent accident with a radioactive isotope source in Brazil and what may have been an accident in a waste storage facility at Khyshtym in the Soviet Union will also be discussed.

The accident at *Three Mile Island Unit 2* occurred on March 29, 1979. The facility is a pressurized water reactor that had recently begun commercial operation. Interruption of flow in a steam generator resulted in temperature and pressure increases in the reactor system. System protective measures—automatic shutdown by rapid insertion of the control rods, pressure relief through a pilot-operated relief valve (PORV), and automatic injection of emergency cooling water—initially performed as designed. The consequences of the accident resulted from a combination of equipment failures and human error. The PORV failed to reseat when pressure was reduced, but this condition was not accurately reflected in the control room instruments. Operators were initially unaware that the PORV was still open and was providing a pathway for coolant to escape from the system. Therefore, they turned off emergency core coolant pumps to prevent what they thought was the possibility of too much water in the system. Once turned off, these pumps failed to restart later in the accident sequence. Fuel cooling continued to deteriorate as the coolant level fell below the top of the elements. Fuel temperatures increased beyond the point at which the cladding and the fuel itself would remain intact. Gaseous and volatile fission products and some fuel material were carried out of the reactor vessel through the open PORV (which was finally reseated 2 hr, 22 min after the accident began). The reactor building sump collected the released water until rising water levels automatically activated a pump that transferred additional liquid to the auxiliary building from which gaseous isotopes, primarily the noble gas ^{133}Xe and small amounts of ^{131}I, were released through the building ventilation system. The President's Commission on Three Mile Island estimated that over 2 million Ci was released. The maximum offsite dose was estimated to be less than 100 mrem. No crop or structural contamination resulted because the isotopes released were primarily the chemically inert xenon isotopes.

The accident at *Chernobyl* occurred on April 26, 1986 in a graphite-moderated reactor called an RBMK. The initiating event was a test of the system's ability to respond to the loss of offsite electrical power. To achieve the desired conditions, operators disabled six safety devices and operated the system below 20% power where the design was inherently unstable. The core essentially self-destructed and dispersed fuel material in the surrounding countryside and, because of atmospheric transport, to Europe. Actually, small amounts of material were distributed throughout the Northern Hemisphere, including the United States. Subsequent releases occurred as a result of burning of the graphite moderator. Estimates of the source term indicate that about 4% of the nonvolatile fission products, 30–50% of the radioactive cesium, and 40–60% of the radioiodine were released during the accident (Warman 1987). Postaccident surveys and reports by the Soviet Union indicate that most of the nonvolatile fission products deposited within the Soviet Union whereas more widespread transport occurred for the radiocesium and radioiodine. Evacuation of 135,000 people was accomplished over a period of about 4 days following the accident. The graphite fire was put out by dropping some 5,000 metric tons of material onto the remaining structure from helicopters. Long-term stabilization of the core involved channeling under the reactor and installing heat exchangers to control the decay heat. A new structure was erected around the core to contain the remaining isotopes. Construction of this barrier was accomplished within about 6 months after the accident.

In 1958 an accident occurred in *Khyshtym* in the Russian Urals that resulted in extensive indication of environmental contamination, relocation of the local population, and extended restrictions on access and use of an area estimated to extend about 25 to 30 miles from the Khyshtym complex at which fuel was reprocessed and waste stored. No formal report of the accident has been prepared and released to the public or the scientific community by the Soviet government. Knowledge is based primarily on reports of Soviet emigres (for example, Medvedev 1977) and details of the exact initiating event are unavailable. Several different scenarios have been postulated to explain the environmental conditions reported. These include a nuclear explosion in a waste disposal area, catastrophic release of stored waste by one of several nonnuclear mechanisms such as an ammonium nitrate explosion, localized fallout from atmospheric nuclear weapons tests, acid rain, and an explosion in a fuel reprocessing plant.

In September 1987 a cesium radiotherapy source was removed from an abandoned clinic in the city of *Goiania* in Brazil. The stainless-steel cylinder containing the source was taken to a local junkyard where scavengers broke it open and extracted the platinum capsule inside that contained the cesium itself. They then sawed open the capsule and found luminescent material described as "carnival glitter." The materials were spread on surfaces, clothing, and skin by children and adults who considered it an adornment and a novelty. The radiation levels incurred were sufficiently high that illness developed almost immediately in the most serious cases. It was several weeks, however, before the reason for the illness was identified and control of the spread of the cesium could be initiated. By early spring 1988, four people had died and it is expected that about the same number will die of radiation effects within the next 5 yr. As of January 1988 249 cases of suspected contamination were identified. About half of these cases involved skin contamination. The other half resulted from clothing or shoes that contained the material. It is estimated that some 40 tons of contaminated material has been shipped to a temporary repository due to cleanup of contaminated homes and other structures from less than 1 oz of radioactive material. By December 1987 over $20 million had been spent in the cleanup and monitoring operations (Petterson 1988).

Facility Recovery. Initial actions in an accident situation must clearly be aimed at limiting the potential for spread of the radioactive material and impact on the population. Such actions must further be performed to protect as much as possible the workers performing them. These two considerations are the foundation for regulation and practice in emergency preparedness as well as facility design and operation. They will also guide subsequent actions required to return the facility to operation or to secure it from the public through external barriers or removal of the contaminated structures. Exactly what recovery operations are feasible will be determined to a large extent by what initial actions were taken.

The sealing process utilized to smother the burning graphite core at Chernobyl effectively committed the operators to retaining the remaining structure in place for a long time period. It also provided a significant radiation shield that permitted recovery team access and onsite cleanup so that other facilities onsite could be returned to operation within about 6 months (compared to the 6-yr delay in restarting the undamaged Unit 1 at Three Mile Island).

Conditions at Three Mile Island resulted in almost all of the released radioactive material being contained within the reactor building and the auxiliary building. Technology development was required to enable the highly radioactive liquid to be processed and stored prior to shipment offsite. Issues such as resin performance in high radiation environments, shipping cask liner designs, and eventual disposal location were among the first to be addressed. Procedures and equipment were developed to enable workers to enter the containment.

Initial entry allowed short-term reconnaissance of damage. High airborne concentrations required full protective clothing and supplied air. Remote cameras, submersible pumps, and robotics were employed to remove collected liquids and "wash down" surfaces to reduce airborne levels.

Technology development includes resins for water cleanup, casks, liners, and storage areas with high radiation levels, evaluation of fuel location, condition, and methods for removing fuel and core hardware, and assessment codes for criticality potential, radiation protection, and environmental impact.

Institutional agreements. The research and development importance of studying the removed spent fuel and core hardware made it practical to transfer the material to the U.S. Department of Energy's (DOE's) Idaho National Engineering Laboratory. Financial and technical needs resulted in a team response including the DOE, reactor vendor, and utility owner. All actions were performed in conformance with NRC license conditions and subject to several public hearings. The pace of recovery operations was sometimes influenced by cash flow, particularly while the Unit 1 reactor at the same site remained shut down (it had been refuelling at the time of the accident), and unavailable to generate power or revenue. Replacement power costs are generally estimated to range between $500,000 and $1 million per day.

Environmental Decontamination. As described in Chapter 4, the extent and location of contamination due to any radioactive materials release depend on the material, the chemical and physical form, the release mechanism, and local environmental conditions. Accidental releases may be to the atmosphere, aquatic environment, or directly to the land surface. The need for remedial action will depend upon the concentration of individual isotopes and probable routes of exposure to people.

At Chernobyl the fire provided a strong heat source that increased the effective release height of the radioactive material. The meteorological conditions present in the area at the time of the accident resulted in the plume's being transported for long distances. Local high ground level concentrations (including crop contamination) resulted from precipitation along the cloud's pathway. Larger diameter particles of the fuel, and those elements that were nonvolatile at the accident temperatures, were deposited nearer to Chernobyl due to gravitational effects. It was this material that was primarily responsible for the need to evacuate residents of nearby villages. By February 1987 14 villages had been resettled and the evacuation zone had been reduced from 35 to 20 km (Wilson 1987). Measures such as washing with chemical solutions, digging up and removing contaminated soil, and/or paving over areas of elevated radioisotope concentration were employed. However, no final decisions on the likely time at which the nearest villages would be suitable for occupation and farming have been reported. Workers continuing the cleanup at the reactor site and operating the other units are bussed in to perform these operations.

The Three Mile Island accident, in contrast to Chernobyl, released only a limited amount of the noble gas ^{133}Xe and ^{131}I. No environmental remedial actions were required as a result of this accident, although a substantial sampling and monitoring program was implemented to verify that no unexpected releases occurred.

Based on the reports of the environmental conditions around the Khyshtym facility, it does not appear that any significant reclamation efforts have been attempted. As previously discussed, the Goiania contamination was widespread and recovery includes decontamination of structures and in some cases razing buildings or rooms that were severely contaminated. Much of the cost incurred in the recovery will result from the government's indicated intention to replace structures that had to be torn down. Final assessment of the extent, cost, and effectiveness of the recovery operations must await completion of continuing efforts.

Contaminated Foodstuffs. The extensive atmospheric dispersion of mixed fission products from Chernobyl and the occurrence of localized precipitation during cloud passage resulted in elevated concentrations of several isotopes in milk and other foodstuffs within the Soviet Union and in eastern and central Europe. Concern over the health effects of consuming this material led the IAEA and individual governments to establish limits on isotopic concentrations in foodstuffs imported from these regions. Beyond the loss of material that did not conform to these limits, there were reported incidents of shipments that transport workers refused to handle because of their point of origin. The available methods for disposing of such material vary with the contained isotope (relatively short half-life material could be stored until the activity decays and the food can then be discarded as if it were nonradioactive) and the material in which it is mixed.

No similar condition existed around Three Mile Island because of the different isotopes released, the smaller amount of material, and differences in local land use and meteorology at the time of each accident. There is no basis for determining what conditions occurred shortly after the Khyshtym accident or exactly when the population was relocated and the land was restricted. In Goiania, contamination was spread by individuals in essentially an urban setting and no contaminated foodstuffs have been reported or would be expected.

Acceptability of Technology and Local Products. There are significant differences in the four accidents considered in terms of the types and amounts of material released, original source of the material, release mechanism, and environmental and public health impacts. Each, however, has had a substantial impact on the perception of the technology, local products, and the area itself. For example, in the United States the name "Three Mile Island" is an object of satire and a rallying cry for antinuclear activists. What is less known, however, is that the local tourist economy has effectively benefitted by the interest in the site and the number of visitors who come to see the facility. On the other hand, the need to dispose of substantial volumes of waste material has complicated Pennsylvania's efforts to secure long-term access to low-level waste disposal capability when the currently operating sites either cease operation or accept only regional waste (see discussion in Chapter 3). Pennsylvania is proceeding to site, develop, and operate a disposal facility that will accept low-level waste from its own generators as well as those in West Virginia, Maryland, and Delaware.

The long-term impact of accident conditions and subsequent remedial actions at Chernobyl has included the very dramatic revision of major portions of the Soviet energy plan in which plans to build several new nuclear facilities were canceled. The residual impact on local grain crops has still not been determined. Detailed plans for management of waste produced during remedial actions have not yet been described in the technical literature.

Significant immediate impact has been reported as a result of the ^{137}Cs contamination in Goiania, Brazil (Petterson, 1988). This included cancellation of reservations in area hotels by individuals and trade shows and conventions. Adverse economic impact was not limited to foodstuffs (which were not contaminated) but extended to textiles and other products manufactured in Goias state. Incidents were reported of residents being denied hotel accommodations in other areas because of fear of contamination. Studies are continuing to determine how long such effects will last and whether they will be affected by the thoroughness and speed with which decontamination and remedial actions will be completed.

1.7 FUTURE DIRECTIONS AND PRACTICES

The currently projected increases in both the volume and radioactivity of several of the different types of radioactive waste described in this text are illustrated in Figures 1-1 and 1-2. These projections are developed and updated annually by the Department of Energy to promote consistency and reflect the impact of changes in waste management practices and programs on the amounts and types of waste to be handled. Table 1-4 lists the major assumptions used to develop these projections. As can be seen from that table, each waste stream is addressed on an individual basis. The impact of varying some of these assumptions can be dramatically illustrated by comparing the results for two major alternative scenarios—the "DOE/EIA Upper Reference Case" illustrated in the figures and the "No New Orders" case that is effectively the lower bound of the range evaluated. As indicated in Table 1-4, a growth rate for nuclear power reactors has

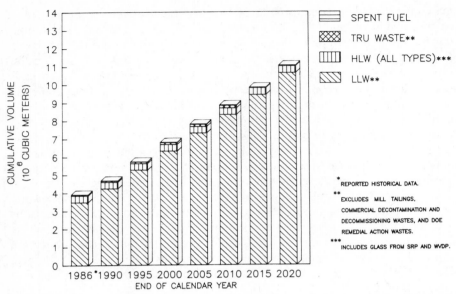

Figure 1-1 Projected cumulative waste volumes for spent fuel, transuranic waste, and defense and commercial high-level waste. Source: (DOE 1987a).

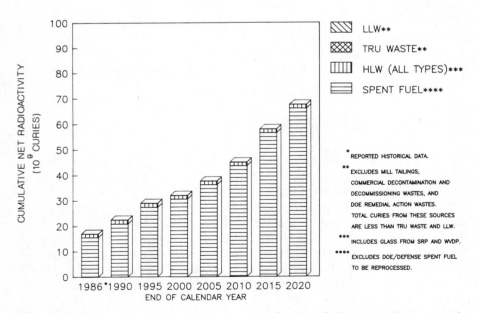

Figure 1-2 Projected cumulative net radioactivity for spent fuel, transuranic waste, and defense and commercial high-level waste. Source: (DOE 1987a).

TABLE 1-4   Assumptions for Projecting Volumes and Activity of Radioactive Waste

Projection Basis

- Projections are made for the years 1987–2020.

Government Activities

- Level of waste-generating activities remains approximately constant.
- Hanford defense reprocessing plant began in 1983 and will conclude operations in 2001.
- WIPP begins operation in October 1988.
- HLW solidification schedules:

> For WVDP, HLW solidification (glass production) starts in March 1989 and is completed by the end of December 1990.
> For SRP, HLW solidification (glass production at the Defense Waste Processing Facility) starts and achieves full production in 1990. Solidification continues through 2020.
> For INEL, HLW solidification (immobilization) starts in 2011, achieves full production by 2014, and continues through 2020.
> For HANF, HLW solidification (borosilicate glass production at the Hanford Waste Vitrification Plant) starts in 1996, and is completed by the end of 2010.

Commercial Activities

- Projections of installed net LWR electrical capacity for the DOE/EIA Upper Reference Case

Year	GW(e)	Year	GW(e)	Year	GW(e)	Year	GW(e)
1987	95.0	1996	107.6	2005	108.6	2014	163.5
1988	97.1	1997	107.6	2006	114.7	2015	169.0
1989	102.5	1998	108.8	2007	120.8	2016	175.2
1990	104.9	1999	108.8	2008	126.9	2017	180.6
1991	104.9	2000	108.6	2009	132.9	2018	187.9
1992	105.9	2001	108.6	2010	139.0	2019	193.7
1993	105.9	2002	108.5	2011	146.7	2020	198.6
1994	105.9	2003	108.6	2012	152.3		
1995	106.3	2004	108.6	2013	157.0		

- By year 2000, spent fuel burnup levels for equilibrium cycles will increase to 30% above the levels for 1984 (27,000 MWd/MTIHM for BWRs and 31,500 MWd/MTIHM for PWRs). From 2000 onward, BWR fuel burnup is 35,100 MWd/MTIHM and PWR fuel burnup is 40,950 MWd/MTIHM.
- For the reference case, spent fuel from commercial reactors is not reprocessed. Thus, a fuel cycle without reprocessing is assumed.

TABLE 1-4   *(Continued)*

- Two quarters of lead time for each fuel cycle activity:

  —Mining and milling   —Conversion   —Enrichment   —Fabrication

- Foreign imports of uranium equal 55% of total demand. This is an average of projected annual imports reported by DOE/EIA for the Upper-Reference-Case scenario.

- Institutional and industrial (I/I) waste generation growth rate (% increase in annual production rate):

Years	Growth rate %
1987–1990	3
1991–2000	2
2001–2020	1

- MRS facility and commercial repository schedules.

The MRS facility is planned for startup in 1998. This facility will have a design spent fuel receipt rate of 2,650 MTIHM/year and a total spent fuel capacity of 14,700 MTIHM. The MRS receipt rate (calendar year) schedule is

   1,200 MTIHM/yr for 1998–2002,
   2,000 MTIHM/yr for 2003, and
   2,650 MTIHM/yr for 2004–2020.

A commercial repository is planned for startup in 2003. The repository will have a design spent fuel receipt rate of 3,000 MTIHM/yr and a total capacity of 70,000 MTIHM. The repository spent fuel receipt rate (calendar year) schedule is:

400 MTIHM/yr for 2003–2005,   1,800 MTIHM/yr for 2007, and
900 MTIHM/yr for 2006,   3,000 MTIHM/yr for 2008–2020.

Source: (DOE 1987a).

been assumed that effectively doubles the installed nuclear capacity between 1986 and 2020. This assumption will affect not only the amounts of spent fuel to be managed but also other waste streams such as tailings from the uranium mined to support those reactors and the low-level waste produced in their operation.

The sensitivity of the projections to this growth assumption is seen in the following comparison, which reflects the most recent Department of Energy projections (DOE 1987a):

Projected Cumulative Spent Fuel Inventory to 2020

Growth Case	Mass (MTIHM)	Activity (10^6 Ci)	Thermal Power (10^6 W)
No new orders	77,800	35,600	126.3
Upper reference	98,300	66,000	246.7

Note that the heat and radioactivity content of the spent fuel increases much more rapidly than the initial amount of uranium from which the power is produced. This reflects design changes to the fuel assemblies that increase the efficiency of power production and reduce the relative amount of waste produced from other parts of the fuel cycle to generate a given amount of power.

Thus, as the amounts of radioactive waste that must be managed are expected to increase with time, that increase reflects increased usage rather than increased waste production by a given user or application. In fact, recent experience has been that individual waste generators are successfully reducing the amounts of waste material produced by their operations. Such reductions have been achieved by combinations of administrative practices (such as strictly controlling the materials that are allowed in radiation areas and that may potentially come in contact with radioactive materials) and process and equipment changes (such as laundering and reusing protective clothing rather than disposing after one use, and compacting trash onsite).

Waste management considerations are receiving more attention than had been the case in the past and at earlier times in the design and operation of facilities and programs. To some extent this has been the result of increased concern over the impact on facility operations of being unable to dispose of waste offsite. For example, having to cease operating a nuclear power reactor due to lack of spent fuel storage space or being unable to use radiopharmaceuticals because the isotope manufacturer is unable to ship low-level waste offsite will result in economic and social consequences that society is unwilling to accept. Further, increased waste disposal costs and the difficulty in siting new facilities make efforts to minimize waste generated, and/or reduce the volumes requiring offsite disposal, a prime goal for waste managers, medical and academic researchers, and facility designers and operators.

Increased social awareness of, and concern over, waste disposal issues in general — municipal landfills, trash to energy incinerators, recycling, and hazardous waste treatment and disposal as well as radioactive waste — means that decisions are increasingly being made in public forums. As a result, technical and economic factors are necessary but not sufficient reasons for any given decision. What might be an acceptable waste management strategy in one community may have no hope of succeeding in another area. Social concern has also raised awareness of the importance of liability for the waste, the facilities at which it is treated or disposed, and the means by which it is transported from the point of generation to the treatment or disposal facility. This has become an issue in the ability of some states to reach agreements to dispose of waste on a regional basis. Further, commercial underwriters are reassessing the basis for their coverage of a site based on the long times involved and the potential for chronic releases rather than catastrophic failure. As part of criteria development for low-level waste disposal facilities, states are presently examining the potential for, and consequences of, waste transportation and disposal site accidents for which coverage will be required. These studies will determine what amounts of liability insurance will be specified in the regulations. The exact instruments by which that protection is provided (e.g., commercial insurance, dedicated funds prepaid by user fees, self-insurance) are likely to vary from state to state.

Waste generators can be expected to continue their efforts to minimize waste generation and facilitate cleanup and recovery of the plant and surroundings should an accident occur. How that will be achieved will vary with the specific waste form(s) generated at a given facility:

Spent fuel is expected to be consolidated to minimize the amount of interim storage volume required and to reduce handling during transport and emplacement at the repository. Maintenance of strict quality control in fuel design, fabrication, and operation minimizes the amount of failed fuel that will have to be disposed. Those facilities currently producing high-level waste (that is, the government run reprocessing plants) are expected to continue to be the primary source of this material in the future. The rate at which waste is produced will be influenced by factors such as arms control agreements and the availability of special nuclear material recycled from older weapons. The rate at which high-level waste is solidified may also vary somewhat from the detailed assumptions provided in Table 1-4. This may be due to budget constraints or revisions of specifications based on conditions at the repository site. Another change may result from the recent NRC proposal that all waste classified as Greater than Class C under 10 CFR Part 61 (see discussion in Chapter 3) should be disposed in a geologic repository although it is not being defined as high-level waste. In general, however, the management of high-level waste is not expected to change substantially in the near future.

The near term milestone in the transuranic waste program is the anticipated startup of the Waste Isolation Pilot Project in October 1988. Continued effort is expected at those facilities producing transuranic waste to minimize the volumes, identify and characterize those wastes previously produced and currently in storage that must be disposed as transuranic waste, and evaluate sites at which transuranic waste was disposed prior to 1970 and, where indicated, plan and perform recovery and disposal of the material.

Low-level waste management practices at existing facilities are in the process of changing to conform to the requirements of 10 CFR Part 61. The major impact of the regulation, however, is expected to be seen in the "next generation" of disposal facilities currently being sited and designed to comply with agreements reached as part of the Low-Level Radioactive Waste Policy Amendments Act of 1985. The trend is definitely away from traditional shallow land burial to disposal facilities that incorporate engineered features such as vaults or modular containers that provide an additional barrier between the waste and the soil. It is expected that detailed designs will vary from site to site reflecting differences in site characteristics and state or regional waste characteristics. Waste generators are expected to continue current efforts to minimize waste volumes and to avoid, whenever possible, production of "mixed" radioactive and hazardous chemical waste.

Regulatory requirements will result in changes to the design and operation of uranium and phosphate tailings impoundments and reclamation of mined areas. The future practices will emphasize stabilization in increments as operation proceeds rather than having such actions performed subsequent to the active life of the facility. The rate at which such changes will be introduced, however, is expected to reflect the current depressed market conditions in the uranium industry and the numbers of facilities that are currently shut down or on standby status.

REFERENCES

(DEC 1982)   Personal communication, Bob Phaneuf, New York State Department of Environmental Conservation to C. C. Stanton, August 1982.

(DOE 1987a)   DOE/RW-0006, Rev. 3, "Integrated Data Base for 1987: Spent Fuel and Radioactive Waste Inventories, Projections, and Characteristics." U.S. Department of Energy, Washington, D.C., September 1987.

(DOE 1987b)   DOE/EIS-0113, "Final Environmental Impact Statement, Disposal of Hanford Defense High-Level, Transuranic and Tank Wastes." U.S. Department of Energy, Washington, D.C., December 1987.

(EPA 1985)   Supplementary Information to 40 CFR Part 191, "Environmental Standards for the Management and Disposal of Spent Nuclear Fuel, High-Level and Transuranic Radioactive Wastes." 50 FR 38065–38089, September 19, 1985.

(Medvedev 1977)   Medvedev, Z. A. "Facts Behind the Soviet Nuclear Disaster." *New Scientist*, 74(1058):761–764 (June 30, 1977).

(NRC 1981)   NUREG-0782, "Draft Environmental Impact Statement on 10 CFR Part 61: Licensing Requirements for Land Disposal of Radioactive Waste." U.S. Nuclear Regulatory Commission, Washington, D.C., September 1981.

(Petterson 1988)   Petterson, J. S. "From Perception to Reality: The Goiania Socioeconomic Impact Model." Presented at Waste Management '88, Tucson, Arizona, March 1, 1988.

(Warman 1987)   "Soviet and Far-Field Radiation Measurements and Inferred Source Term from Chernobyl." Presented by E. A. Warman at the Health Physics Society Symposium on the Effects of the Nuclear Reactor Accident at Chernobyl, Upton, New York, April 1987.

(Wilson 1987)   Wilson, R. "A Visit to Chernobyl." *Science*, 236, June 26, 1987, pp. 1636 ff.

CHAPTER 2

RADIOACTIVE WASTE FORMS

2.1 INTRODUCTION

Radioactive waste is generally described under a variety of classification systems that identify considerations such as the source of the waste (e.g., government, industrial, or academic uses), the relative radiological concerns in handling or disposing of the waste (the high-level and low-level waste distinctions), the actual materials contained therein (such as transuranic wastes and uranium mill tailings), or the procedures that produced the waste (such as decontamination and decommissioning wastes). This classification process is an attempt to group material by characteristics related to methods of production or that may require different methods for packaging, transport and handling, as well as for disposal itself. Characteristics such as external radiation level, half-lives of contained radioisotopes, and, ultimately, the potential health risk from exposures to a specific radioisotope through a specific pathway (such as drinking water, or breathing air containing suspended particulates) are critical inputs to the design, operation, and construction of a successful management and disposal system. In reality, each category includes a range of material types that must often be handled on a case-by-case basis. No single unit such as volume or curies gives enough information by itself on which to base management decisions. However, there are substantial similarities among many waste types and these similarities are used as a point of departure for more detailed studies.

The following sections describe the sources of radioactive waste, the fundamental differences between high-level waste (HLW), transuranic waste (TRU), low-level waste (LLW), and waste produced as a by-product of mining other natural minerals such as uranium and phosphates. This breakdown parallels the organization of agency responsibilities and regulations developed to manage these wastes at the federal and state levels and the discussions of siting and waste management technologies in subsequent chapters.

2.2   SOURCES OF RADIOACTIVE WASTE

Radioactive waste results from a wide range of processes and applications in which radioactive materials are used. Such processes are an integral part of current U.S. society and radioactive waste generators include the federal and local governments, electric utilities, private industry, hospitals and universities, and mining and milling operations in which the waste material (tailings) contains naturally occurring radioactive material, generally uranium and thorium and their decay products. As discussed below, there is considerable overlap in these distinctions and in evaluating data based on them care must be used to avoid double counting of wastes with different descriptions or inaccurate characterization of a given waste stream.

Nuclear fuel cycle wastes can be considered to include any waste material produced incident to generating electric power using nuclear fuel. This definition would consider uranium mine and mill tailings, waste from conversion, enrichment, and fuel fabrication facilities, waste produced during operation of a nuclear power reactor, spent nuclear fuel, and waste from decontamination and decommissioning of nuclear power reactors and other facilities in the nuclear fuel cycle. Figure 2-1 illustrates the nuclear fuel cycle as originally planned, including the reprocessing of spent fuel, recovery of contained uranium and plutonium, and recycle of these fissionable materials to generate additional power in the same or other reactors. As discussed further in Chapter 3, the current fuel cycle in the United States considers spent fuel as a waste rather than an energy source. Provision is made, therefore, for direct disposal of spent fuel. Depending on how data are compiled and reported, some of this material may also be classified as industrial waste and uranium mine and mill tailings. As interest in radioactive waste management has grown in recent years, greater attention has been devoted to identifying and standardizing reports of waste requiring management and disposal. For the portion of the waste identified as "low level" (see discussion in Section 2.4), fuel cycle waste is generally reported as that from nuclear power reactors with the remainder of the fuel cycle being considered as producing industrial waste (EG&G Idaho, Inc. 1985). Uranium mine and mill tailings are considered separately because of the extremely large volumes involved and the history of development and regulation that differed from other materials.

Industrial firms may generate radioactive waste directly as a result of the production processes in which they are engaged, as a result of research into new or improved products, and/or from instrumentation used for quality assurance or process control. Production processes include such diverse operations as manufacture of radiopharmaceuticals or compounds labeled with radioisotopes as well as consumer products such as smoke detectors and luminous watch dials.

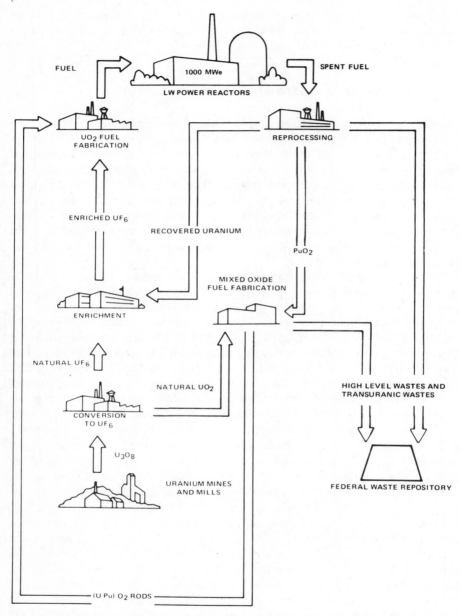

FUEL

1000 MWe

LW POWER REACTORS

SPENT FUEL

UO₂ FUEL
FABRICATION

REPROCESSING

ENRICHED UF₆

RECOVERED URANIUM

PuO₂

MIXED OXIDE
FUEL FABRICATION

ENRICHMENT

NATURAL UF₆

NATURAL UO₂

HIGH LEVEL WASTES AND
TRANSURANIC WASTES

CONVERSION
TO UF₆

U₃O₈

URANIUM MINES
AND MILLS

FEDERAL WASTE REPOSITORY

(U Pu) O₂ RODS

Figure 2-1 Steps in the nuclear fuel cycle with uranium and plutonium recycle. Source: (NRC 1982).

In some cases, particularly for sealed sources used for nondestructive testing (radiography), manufacturers will accept the product back at the end of its useful lifetime and arrange for disposal of the contained, and no longer useful to the customer, radioactive material. The manufacturer may be able to recover and recycle the radioisotopes rather than disposing of them immediately.

Defense wastes are primarily the result of the facilities and processes necessary to maintain the nation's weapons arsenal and fuel nuclear powered naval vessels. They include the high-level waste from extracting the uranium and plutonium used for the weapons themselves, materials containing transuranic wastes in concentrations greater than $100\,nCi/g$ ($1\,nCi = 10^{-9}\,Ci$), and low-level process waste produced in shaping the material and incidental waste such as compacted trash generated from running the production facilities. Defense wastes are frequently discussed as a separate waste type because the activities that produce them, the administrative agencies responsible for their management, the safety rules to which they are subject, and the funds for their management are all separate from those that apply to commercial waste. Preferred waste management strategies and alternatives will frequently be different for defense and commercial wastes because of these institutional differences. In this text, however, defense wastes are discussed in the sections relating to waste with similar characteristics (i.e., HLW, LLW, TRU) rather than as a separate waste type.

Wastes produced at medical and academic facilities are generally sufficiently similar (for purposes of disposal facility design and operation) in chemical and physical form and concentrations of radioisotopes to be considered together as institutional waste. Processes resulting in the production of such waste include medical diagnosis and therapy, usually using radioisotopes that are injected, ingested, or implanted into the patient. Radiation therapy may also involve use of large sources of penetrating radiation (frequently cobalt-60). The useful life of such sources is a function of the isotope's half-life and the exposure times needed to achieve the desired therapeutic results. Medical and academic researchers use radioactive materials and produce radioactive waste in projects that have addressed questions such as crop productivity, nutritive value of foods, sickle cell anemia, and cancer.

Uranium is a relatively abundant mineral in the earth's surface. It is about as common as tin. Unlike many other elements, however, the processes by which uranium deposits were formed resulted in its being fairly widely dispersed among other rocks rather than existing in large concentrated deposits such as occur with copper. Phosphate rock deposits, in particular, frequently contain sufficiently high concentrations of uranium that, when the deposits are processed to extract the nonradioactive ore, the residue or tailings from the process must be managed to achieve isolation of the uranium and daughter products contained therein. There have been times when uranium prices were sufficiently high that it was recovered as a by-product of phosphate production.

Discussion in the following sections will be organized among the functional waste classifications previously described. The sections will address high-level waste, transuranic waste, low-level waste, uranium mill tailings, and phosphogypsum wastes, respectively. This chapter concentrates on the radiological, chemical, and physical characteristics that affect the design of systems for handling, treating, and disposing of the waste. Details of the regulatory requirements for such waste systems are described in Chapter 3.

2.3   HIGH-LEVEL WASTE

Until the 1970s the only specific definition of a radioactive waste type that existed was that for high-level waste (HLW). This definition is contained in Appendix F to the NRC's regulations on power reactors and fuel reprocessing plants (10 CFR Part 50—see discussion in Chapter 3.). Under this definition, HLW is the aqueous waste resulting from the operation of the first cycle solvent extraction system, or equivalent, and the concentrated wastes from subsequent extraction cycles, or equivalent, in a facility for reprocessing irradiated reactor fuels. All other waste was defined by default as "other than high level." This is a functional rather than an analytical definition of the HLW. That is, it is perfectly clear that the product of this portion of the reprocessing operation is "high-level waste." However, the exact isotopic and chemical content of the material depends on the type of fuel reprocessed, the operating history of the fuel (how long it was in the reactor and at what power levels), the length of time between removal from the reactor and reprocessing, and the reprocessing technology and solidification method used.

The definition of "high-level waste" was administratively broadened by the U.S. Nuclear Regulatory Commission (NRC) in 1981 to include (NRC 1981a)

(1) irradiated reactor fuel, (2) liquid wastes resulting from the operation of the first cycle solvent extraction system, or equivalent, and the concentrated wastes from subsequent extraction cycles, or equivalent, in a facility for reprocessing irradiated reactor fuel, and (3) solids into which such liquid wastes have been converted.

This change recognized the fact that in the United States there was no reprocessing capacity commercially available and no prospect for near-term resumption of commercial reprocessing. By this definition, high-level waste includes waste produced from the reprocessing of fuel irradiated in government facilities for weapons production, similar waste produced in a commercial reprocessing facility (operation of which ceased in the United States in 1971), and unreprocessed spent fuel. Reprocessing waste was originally stored in liquid form in underground tanks at the reprocessing facilities. Government facilities [now the responsibility of the Department of Energy (DOE)] are located in Richland, Washington, Idaho Falls, Idaho, and Savannah River, South Carolina. The only commercial reprocessing facility to operate in the United States (from 1966 to 1971) was located at West Valley, New York. Work is underway to solidify all of the liquid waste at each of these sites for eventual disposal in a geologic repository. (The currently planned schedule for waste solidification at the individual DOE sites is given in Table1-4.) The geologic repository being designed to receive spent fuel will also be capable of receiving solidified HLW as further described in Chapter 6. Research is continuing on specific materials to be used in the waste containers for spent fuel and other HLW. Final decisions will need to consider the chemical and physical parameters of the geologic medium chosen for the repository. The following sections describe the chemical, physical, and radiological characteristics as well as current plans for packaging specific types of high-level waste: spent fuel, solidified commercial HLW, and defense HLW.

2.3.1 Spent Fuel

Once fuel has achieved its design power output in a reactor, it is removed from the core and stored onsite underwater in a spent fuel storage pool. The fuel contains highly radioactive fission products and the radiation levels require substantial shielding from workers. The fuel is also physically (thermally) hot as a result of the decay of the fission products. (Residual activity at shutdown produces approximately 6% of the power level of the fully operational core. Because of this residual heat production, fuel rods must be cooled even after the fission reaction ceases. Failure to maintain such cooling was the cause of the fuel damage at the Three Mile Island nuclear power plant.) All onsite operations involving the spent fuel are performed remotely with the fuel remaining underwater to provide both radiation shielding and cooling. Current policy is for the fuel to be transferred eventually to the DOE for disposal as described in Chapter 3. The amount of spent fuel that will require disposal in the first federal repository (originally scheduled to be available in 1998) has been projected by the DOE to be approximately 45,000 MTU (DOE 1985a). Such projections are reviewed annually to evaluate changes resulting from differences in assumptions on power reactor startup dates, reactor performance (power generated), fuel design, and planned residence time in the reactor for a given fuel assembly. The isotopic concentrations in the spent fuel will vary with the type of fuel and the materials in which the fuel is contained (the cladding and fuel assembly hardware such as tie plates and end fittings that contain induced radioactivity resulting from the exposure to neutrons within the reactor core). Radioisotopic concentrations in spent fuel are used to develop the shielding requirements, handling procedures, and package heat loadings for spent fuel during transport to a repository as well as to evaluate postdisposal performance. Computer codes are used to model the initial fuel content and any changes due to irradiation in the reactor and decay subsequent to removal from the reactor. Computer codes such as ORIGEN (the Oak Ridge Isotope Generation and Depletion code) have been benchmarked to actual conditions in operating reactors and in laboratory analysis of waste samples and spent fuel and are good projections of actual radioisotopic distributions that will exist.

Adequate detailed characterization is needed for several purposes including

- developing procedures for packaging and shipping the waste from point of origin to the disposal site in accordance with NRC and Department of Transportion (DOT) regulations;
- designing the disposal package;
- specifying waste acceptance and handling procedures onsite at the repository; and
- projecting long-term (10,000 yr) performance of the waste package in the repository, and of the geologic medium of the repository.

The isotopic analysis of the waste as received is the fundamental input to calculations of the effects of radiation on the waste package, of heat generation and related impact on the host medium (for crystalline rock media, such as granite, the heat load is the limiting condition for the amount of waste that can be emplaced in a given area), and of isotopes present and available for leaching and migration assuming package failure at any given time after emplacement.

Table 2-1 lists the fuel content of major isotopes that contribute significantly to offsite doses from a spent fuel repository and the thermal power produced from one metric ton of uranium (MTU) in pressurized water reactor (PWR) fuel. For perspective, a nominal reactor loading would require about 27 MTU each year (DOE 1980a). As can be seen in Table 2-1, decay of short-lived isotopes reduces the radioactivity and thermal power substantially in the first few years after discharge from the reactor. The activity (measured in curies) of activation products declines by about a factor of 30 in the first 5 yr with most of that decay occurring within the first 2 yr. Fission product activity is reduced by a factor of about 300 and transuranic activity by a factor of over 400 after 5 yr. The thermal output decreases by about a factor of 500 over the same period of time. Short-term (several years) storage of spent fuel reduces the need for shielding and cooling during handling, transport, and disposal. At longer times after discharge, decay is dominated by longer half-life isotopes and therefore the rate of reduction in radioactivity and thermal output is lower. These reductions are illustrated in Figure 2-2, which provides the relationship between radioactivity content and thermal power of spent fuel assemblies as a function of time since removal from the reactor. Data are provided for both boiling water reactor fuel and pressurized water reactor fuel since each has a different design and initial uranium content. Differences due to changes in burnup (extracting different amounts of energy from a given initial uranium fuel loading) are also indicated. The implications of these reductions are further discussed in Chapter 6.

Fuel assemblies for the two major power reactor designs used in the United States, pressurized water reactors (PWRs) and boiling water reactors (BWRs), contain slightly enriched (3–4% ^{235}U) uranium dioxide pellets within zircaloy tubes that are arranged in square arrays and connected and supported by grid structures and end fittings. Figure 2-3 illustrates that BWR and PWR assemblies

TABLE 2-1   Radioactivity and Thermal Power (in watts) in Spent LWR Fuel per One Metric Ton of Uranium in Fresh Fuel[a]

	Years after Discharge		
	0	2	5
Radionuclide Content (curies)			
Important Activation Products			
^{14}C[b]	6.6×10^{-1}	6.6×10^{-1}	6.6×10^{-1}
^{55}Fe	2.0×10^{3}	1.2×10^{3}	5.2×10^{2}
^{60}Co	6.3×10^{3}	4.8×10^{3}	3.3×10^{3}
^{63}Ni	5.5×10^{2}	5.5×10^{2}	5.3×10^{2}
^{95}Zr	2.8×10^{4}	1.2×10^{1}	1.0×10^{-4}
Total Activation Products	1.4×10^{5}	6.7×10^{3}	4.3×10^{3}
Important Fission Products			
^{3}H	5.1×10^{2}	4.6×10^{2}	3.9×10^{2}
^{85}Kr	1.1×10^{4}	1.0×10^{4}	8.3×10^{3}
^{90}Sr	7.8×10^{4}	7.5×10^{4}	6.9×10^{4}
^{106}Ru	5.3×10^{5}	1.3×10^{5}	1.7×10^{4}
^{129}I	3.7×10^{-2}	3.7×10^{-2}	3.7×10^{-2}
^{137}Cs	1.1×10^{5}	1.0×10^{5}	9.6×10^{4}
Total Fission Products	1.4×10^{8}	1.2×10^{6}	4.8×10^{5}
Important Transuranium Products			
^{238}Pu	2.7×10^{3}	2.8×10^{3}	2.8×10^{3}
^{239}Pu	3.2×10^{2}	3.2×10^{2}	3.2×10^{2}
^{240}Pu	4.7×10^{2}	4.7×10^{2}	4.7×10^{2}
^{241}Pu	1.0×10^{5}	9.4×10^{4}	8.1×10^{4}
^{241}Am	8.4×10^{1}	4.0×10^{2}	8.0×10^{2}
^{244}Cm	2.2×10^{3}	2.1×10^{3}	1.8×10^{3}
Total Transuranium Products	3.8×10^{7}	1.0×10^{5}	8.7×10^{4}
Thermal Power (watts)			
	1.0×10^{6}	5.9×10^{3}	2.1×10^{3}

Source: (DOE 1980b).

[a]Calculated with the ORIGEN code for PWR fuel irradiated to 33,000 MWD/MTU at a specific power of 30 MW/MTU.

[b]Based upon 2.5 ppm nitrogen (by weight) in UO_2.

differ somewhat in physical dimensions and hardware components. These physical parameters remain unchanged as a result of irradiation in a reactor. There are also differences in the amount of fuel contained in BWR and PWR fuel elements. At discharge, after about 3–4 yr residence in the reactor, the fission products, uranium, plutonium, and other transuranic elements, are primarily contained within the sealed fuel rods. The activation products are primarily contained in the hardware components.

BOILING-WATER REACTOR SPENT FUEL

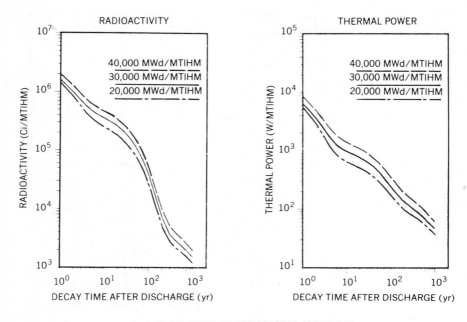

RADIOACTIVITY

THERMAL POWER

40,000 MWd/MTIHM
30,000 MWd/MTIHM
20,000 MWd/MTIHM

DECAY TIME AFTER DISCHARGE (yr)

PRESSURIZED-WATER REACTOR SPENT FUEL

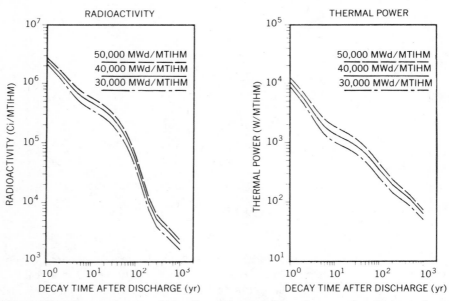

RADIOACTIVITY

THERMAL POWER

50,000 MWd/MTIHM
40,000 MWd/MTIHM
30,000 MWd/MTIHM

DECAY TIME AFTER DISCHARGE (yr)

Figure 2-2 Decrease in spent fuel radioactivity and thermal power after discharge from the reactor. Source: (DOE 1987).

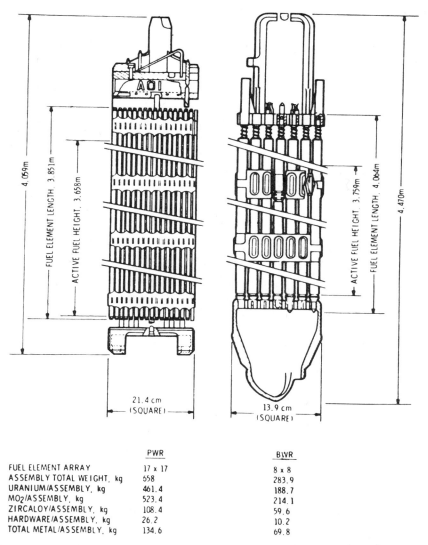

	PWR	BWR
FUEL ELEMENT ARRAY	17 x 17	8 x 8
ASSEMBLY TOTAL WEIGHT, kg	658	283.9
URANIUM/ASSEMBLY, kg	461.4	188.7
MO₂/ASSEMBLY, kg	523.4	214.1
ZIRCALOY/ASSEMBLY, kg	108.4	59.6
HARDWARE/ASSEMBLY, kg	26.2	10.2
TOTAL METAL/ASSEMBLY, kg	134.6	69.8

Figure 2-3 Dimensions and components of typical light water reactor fuel assemblies. Source: (DOE 1980a).

The spent fuel assemblies will be sealed in canisters prior to disposal. Canister materials and packing will be chosen to enhance the performance of the final repository system. That is, container materials will be chosen to minimize the likelihood of chemical reaction between the package and the geologic medium.

The following reference and alternative packing materials are being evaluated (DOE 1985a):

Geologic Medium	Reference Canister	Alternative	Packing
Baselt	Low carbon steel	Iron–chrome– molybdenum steel	Crushed basalt and bentonite
Basalt		Cupronickel copper and alloys	Clay
Salt	Low carbon steel	Carbon steel with a thin titanium alloy veneer	Crushed salt
Tuff	Austenitic stainless steel	Copper and alloys	Possible tuff backfill

The ongoing investigations are addressing corrosion resistance, structural stability, cost, availability, and fabrication requirements of the materials as input into the choice of the actual package. Initially, one assembly will be placed in each canister. Subsequently, it is planned that the hardware will be removed and rods consolidated so that fuel from two rods can be disposed in one waste container. The non-fuel-bearing waste hardware may or may not also be disposed in the geologic repository (DOE 1985a).

2.3.2 Commercial High-Level Waste

Approximately 600,000 gallons of liquid high-level radioactive waste was produced by the operation of the commercial reprocessing facility at West Valley, New York from April 1966 through December 1971. Planned waste management procedures provided for permanent storage in double-shelled tanks underground at the reprocessing site. Leakage and level detection equipment on the tanks would be used to indicate tank failure. Tank contents would then be pumped to another underground tank nearby for subsequent storage. This management program reflected the then-current practice at government facilities, many of which had been designed and constructed under wartime deadlines. Concern with the need for periodic remote handling of the liquid waste and the long time frames required for active care resulted in the (then) Atomic Energy Commission's promulgation in 1970 of a requirement (Appendix F to 10 CFR Part 50) that all future reprocessing facilities be equipped to solidify the liquid high-level waste within 5 yr after production and ship it to a federal repository for permanent disposal. Subsequent development of the high-level waste management program is further discussed in Chapter 3.

Existing facilities were not required to comply with the new rule but would be addressed on a case-by-case basis. For a variety of technical, economic, and political reasons no other commercial reprocessing facilities have operated in the United States and the West Valley facility, shut down for expansion in 1971, was

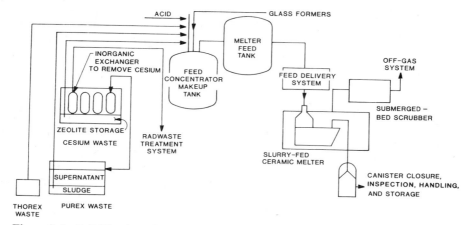

Figure 2-4 Solidification of commercial high-level waste at West Valley. Source: (DOE 1987).

never reopened. This waste is in the process of being solidified in a glass matrix (vitrification), encapsulated and prepared for shipment to the federal repository. It is the only commercial HLW presently expected to exist in the United States.

Both uranium and thorium fuels were processed at the West Valley facility. The resulting wastes are identified as PUREX wastes and THOREX wastes, respectively, and refer to the different chemical interactions used to separate the fission products from the uranium, plutonium, and thorium that would subsequently be recycled. The PUREX waste represents about 95% of the total volume and radioactivity requiring solidification. The waste was neutralized with sodium hydroxide prior to being transferred to the storage tank. Insoluble hydroxides have precipitated out of the neutralized waste resulting in both liquid (about 570,000 gallons of supernatant) and solid (sludge) phases being found in the tank. Table 2-2 lists the radionuclide composition of these wastes as of 1987. The table illustrates that most of the isotopes in the PUREX waste are found in the sludge with the exception of cesium-137, iodine-129, and barium-137 m. The THOREX waste was not neutralized to avoid precipitation of the contained thorium. There are approximately 12,000 gallons of acid THOREX waste stored.

The final solidified waste product will contain three distinct waste feed streams as illustrated in Figure 2-4 (DOE 1987):

- spent zeolite ion-exchange media used to remove strontium and cesium from the PUREX waste supernatant;
- washed PUREX sludge containing the rest of the isotopes listed in Table 2-2; and
- THOREX waste.

This mixture will be combined with glass formers and fed into the melter for vitrification. The product, a borosilicate glass, has been chosen following years of

TABLE 2-2 Radionuclide Composition of Commercial HLW at West Valley (1987)[a]

Radionuclide	Alkaline Waste (PUREX) Liquid(Ci)	Alkaline Waste (PUREX) Sludge(Ci)	Acid Waste (THOREX) Liquid(Ci)	Total(Ci)
^{3}H	9.74E + 01	0.00E + 00	1.74E + 00	9.91E + 01
^{14}C	1.37E + 02	0.00E + 00	1.30E − 01	1.37E + 02
^{55}Fe	0.00E + 00	1.00E + 03	5.63E + 02	1.56E + 03
^{60}Co	0.00E + 00	4.70E + 00	1.14E + 03	1.14E + 03
^{59}Ni	0.00E + 00	8.56E + 01	2.03E + 01	1.06E + 02
^{63}Ni	8.89E + 02	5.34E + 03	2.51E + 03	8.74E + 03
^{79}Se	5.68E + 01	0.00E + 00	3.35E + 00	6.02E + 01
^{90}Sr	2.89E + 03	6.74E + 06	4.54E + 05	7.20E + 06
^{90}Y	2.89E + 03	6.74E + 06	4.54E + 05	7.20E + 06
^{93}Zr	2.56E − 01	2.56E + 02	1.62E + 01	2.72E + 02
^{93m}Nb	1.59E − 01	1.59E + 02	1.02E + 01	1.69E + 02
^{99}Tc	1.60E + 03	0.00E + 00	1.04E + 02	1.70E + 03
^{106}Ru	1.10E − 01	1.10E + 02	6.24E − 01	1.11E + 02
^{106}Rh	1.10E − 01	1.10E + 02	6.24E − 01	1.11E + 02
^{107}Pd	1.09E − 02	1.09E + 01	1.14E − 01	1.10E + 01
^{113m}Cd	2.41E + 00	2.41E + 03	3.75E + 01	2.45E + 03
^{121m}Sn	1.76E − 02	1.76E + 01	5.99E − 01	1.82E + 01
^{126}Sn	1.01E − 01	1.01E + 02	3.11E + 00	1.04E + 02
^{125}Sb	4.90E + 01	1.51E + 04	2.89E + 02	1.54E + 04
^{126}Sb	1.42E − 02	1.42E + 01	4.35E − 01	1.46E + 01
^{126m}Sb	1.01E − 01	1.01E + 02	3.11E + 00	1.04E + 02
^{125m}Te	1.20E + 01	3.70E + 03	7.08E + 01	3.78E + 03
^{129}I	2.10E − 01	0.00E + 00	1.80E − 01	3.90E − 01
^{134}Cs	1.39E + 04	0.00E + 00	3.10E + 02	1.42E + 04
^{135}Cs	1.56E + 02	0.00E + 00	5.47E + 00	1.61E + 02
^{137}Cs	7.26E + 06	0.00E + 00	4.75E + 05	7.74E + 06
^{137m}Ba	6.87E + 06	0.00E + 00	4.49E + 05	7.32E + 06
^{144}Ce	2.09E − 05	9.21E + 00	1.39E − 01	9.35E + 00
^{144}Pr	2.09E − 05	9.21E + 00	1.39E − 01	9.35E + 00
^{146}Pm	4.77E − 02	1.53E + 01	5.07E − 01	1.59E + 01
^{147}Pm	5.71E + 02	1.85E + 05	9.11E + 03	1.95E + 05
^{151}Sm	5.03E − 01	8.15E + 04	4.78E + 03	8.63E + 04
^{152}Eu	4.57E − 02	3.77E + 02	4.82E + 01	4.25E + 02
^{154}Eu	1.44E + 01	1.19E + 05	2.53E + 03	1.22E + 05
^{155}Eu	2.37E + 00	3.54E + 04	8.44E + 02	3.62E + 04
^{207}Tl	0.00E + 00	9.12E − 04	7.50E + 00	7.50E + 00
^{208}Tl	0.00E + 00	4.28E − 02	3.51E + 00	3.55E + 00
^{209}Pb	0.00E + 00	6.61E − 06	2.07E − 01	2.07E − 01
^{211}Pb	0.00E + 00	9.14E − 04	7.52E + 00	7.52E + 00
^{212}Pb	0.00E + 00	1.19E − 01	9.76E + 00	9.88E + 00
^{211}Bi	0.00E + 00	9.14E − 04	7.52E + 00	7.52E + 00
^{212}Bi	0.00E + 00	1.19E − 01	9.76E + 00	9.88E + 00
^{213}Bi	0.00E + 00	6.61E − 06	2.07E − 01	2.07E − 01

TABLE 2-2 *(Continued)*

Radionuclide	Alkaline Waste (PUREX) Liquid(Ci)	Sludge(Ci)	Acid Waste (THOREX) Liquid(Ci)	Total(Ci)
^{212}Po	0.00E + 00	7.62E − 02	6.25E + 00	6.33E + 00
^{213}Po	0.00E + 00	6.47E − 06	2.03E − 01	2.03E − 01
^{215}Po	0.00E + 00	9.14E − 04	7.52E + 00	7.52E + 00
^{216}Po	0.00E + 00	1.19E − 01	9.76E + 00	9.88E + 00
^{217}At	0.00E + 00	6.61E − 06	2.07E − 01	2.07E − 01
^{219}Rn	0.00E + 00	9.14E − 04	7.52E + 00	7.52E + 00
^{220}Rn	0.00E + 00	1.19E − 01	9.76E + 00	9.88E + 00
^{221}Fr	0.00E + 00	6.61E − 06	2.07E − 01	2.07E − 01
^{223}Fr	0.00E + 00	1.26E − 05	1.04E − 01	1.04E − 01
^{223}Ra	0.00E + 00	9.14E − 04	7.52E + 00	7.52E + 00
^{224}Ra	0.00E + 00	1.19E − 01	9.76E + 00	9.88E + 00
^{225}Ra	0.00E + 00	6.61E − 06	2.07E − 01	2.07E − 01
^{228}Ra	0.00E + 00	4.18E − 09	1.48E + 00	1.48E + 00
^{225}Ac	0.00E + 00	6.61E − 06	2.07E − 01	2.07E − 01
^{227}Ac	0.00E + 00	9.14E − 04	7.52E + 00	7.52E + 00
^{228}Ac	0.00E + 00	4.81E − 09	1.48E + 00	1.48E + 00
^{227}Th	0.00E + 00	9.01E − 04	7.42E + 00	7.42E + 00
^{228}Th	0.00E + 00	1.19E − 01	9.76E + 00	9.88E + 00
^{229}Th	0.00E + 00	6.61E − 06	2.07E − 01	2.07E − 01
^{230}Th	0.00E + 00	1.45E − 02	4.38E − 02	5.83E − 02
^{231}Th	6.41E − 03	8.94E − 02	5.17E − 03	1.01E − 01
^{232}Th	0.00E + 00	5.87E − 09	1.64E + 00	1.64E + 00
^{234}Th	5.71E − 02	7.97E − 01	7.11E − 05	8.54E − 01
^{231}Pa	0.00E + 00	2.86E − 04	1.52E + 01	1.52E + 01
^{233}Pa	0.00E + 00	2.30E + 01	3.02E − 01	2.33E + 01
^{234m}Pa	5.71E − 02	7.97E − 01	7.11E − 05	8.54E − 01
^{232}U	3.13E − 01	4.36E + 00	2.74E + 00	7.41E + 00
^{233}U	4.98E − 01	6.94E + 00	2.09E + 00	9.53E + 00
^{234}U	2.80E − 01	3.90E + 00	2.17E − 01	4.40E + 00
^{235}U	6.41E − 03	8.94E − 02	5.17E − 03	1.01E − 01
^{236}U	1.91E − 02	2.67E − 01	9.80E − 03	2.96E − 01
^{238}U	5.71E − 02	7.97E − 01	7.11E − 05	8.54E − 01
^{236}Np	0.00E + 00	9.35E + 00	1.23E − 01	9.47E + 00
^{237}Np	0.00E + 00	2.30E + 01	3.02E − 01	2.33E + 01
^{239}Np	0.00E + 00	3.43E + 02	4.49E + 00	3.47E + 02
^{236}Pu	1.36E − 02	8.24E − 01	1.09E − 02	8.49E − 01
^{238}Pu	1.27E + 02	8.00E + 03	4.80E + 02	8.61E + 03
^{239}Pu	2.54E + 01	1.61E + 03	1.54E + 01	1.65E + 03
^{240}Pu	1.87E + 01	1.18E + 03	8.09E + 00	1.21E + 03
^{241}Pu	1.46E + 03	9.23E + 04	8.50E + 02	9.46E + 04
^{242}Pu	2.54E − 02	1.61E + 00	1.19E − 02	1.65E + 00
^{241}Am	0.00E + 00	5.30E + 04	2.41E + 02	5.32E + 04
^{242}Am	0.00E + 00	2.93E + 02	6.76E + 00	2.99E + 02

TABLE 2-2   *(Continued)*

Radionuclide	Alkaline Waste (PUREX)		Acid Waste (THOREX)	
	Liquid(Ci)	Sludge(Ci)	Liquid(Ci)	Total(Ci)
^{242m}Am	0.00E + 00	2.94E + 02	6.79E + 00	3.01E + 02
^{243}Am	0.00E + 00	3.39E + 02	7.83E + 00	3.47E + 02
^{242}Cm	0.00E + 00	2.43E + 02	5.59E + 00	2.49E + 02
^{243}Cm	0.00E + 00	1.44E + 02	2.34E − 01	1.44E + 02
^{244}Cm	0.00E + 00	8.56E + 03	1.37E + 01	8.57E + 03
^{245}Cm	0.00E + 00	8.62E − 01	2.00E − 02	8.82E − 01
^{246}Cm	0.00E + 00	9.87E − 02	2.29E − 03	1.01E − 01
Total	1.42E + 07	1.41E + 07	1.86E + 06	3.01E + 07

Source: (DOE 1987)

[a]Includes all radionuclides >0.1 Ci prior to the year 3090.

research at government facilities. It provides structural integrity even under the heat and self-irradiation conditions to which it will be subjected in the repository, and high-leach resistance to water that might be present in the geologic conditions of the repository. The glass will be sealed in a stainless-steel canister that will be enclosed in a disposal container of whatever material is appropriate to the host geologic medium as described above. It is projected that some 300 canisters, each 2 ft in diameter by 10 ft tall and containing 18.7 ft^3 of waste-bearing glass, will be produced by the solidification of the West Valley waste (DOE 1985a).

2.3.3   Defense High-Level Waste

Reprocessing of fuel and recovery of plutonium produced in a weapons program production reactor are fundamental to the weapons production process. HLW is produced incident to this reprocessing. Another source of defense HLW is the reprocessing of cores used for naval propulsion and recovery and recycle of the highly enriched uranium fuel contained therein.

There are several differences between defense HLW and the commercial HLW described in Section 2.3.2. The major differences are

- Reactors operated primarily to produce plutonium rather than electric power have different operating cycle lengths, power densities, and fission product inventories from those found in commercial power reactors.
- Propulsion reactors use highly enriched uranium fuel rather than the 3–4% enriched fuel in commercial reactors to minimize the amount of fuel needed—and consequently weight and space requirements—to produce a given amount of power.

TABLE 2-3 Comparison of Defense and Commercial HLW

	Canisters Required[a]	Heat Output (kW/canister)[a]	Radionuclide Content (Ci/canister)[a]					
			^{90}SR	^{137}CS	^{238}Pu	^{239}Pu	^{241}Am	^{244}Cm
Defense HLW[b]								
Savannah River	8.0×10^3	0.2	1.5×10^4	1.5×10^4	1.4×10^2	2.9	8.2	8.2
Idaho Falls	1.2×10^4	0.09	7.3×10^3	7.4×10^3	4.0×10^1	4.0×10^{-1}	6.0×10^{-1}	3.1×10^{-1}
Hanford	2.6×10^4	0.06	5.2×10^3	4.8×10^3	2×10^{-2}	9.2×10^{-1}	6.5	5.4×10^{-1}
Total	4.6×10^4							
Commercial HLW[c]								
	1.0×10^5 to 2.8×10^5	3.2 to 1.2	1.4×10^5 to 5.0×10^4	2.0×10^5 to 7.1×10^4	1.8×10^2 to 6.5×10^1	4.3 to 1.5	1.7×10^3 to 6.1×10^2	1.4×10^4 to 5.1×10^3

Source: (DOE 1980c).

[a]Nominal values, assuming uniform distribution of waste radionuclides among the canisters.

[b]Estimated data for 1990. Treated waste volumes (assuming a waste form having a 25% loading of waste oxides) and radionuclide contents supplied by J. L. Crandall and W. R. Cornman of the High-Level Waste Lead Office at Savannah River. Canister requirements based on 0.6-m-diameter × 3-m-long canisters, 80% full of treated waste. Heat outputs based on the contained radionuclides.

[c]Data for the reprocessing of spent fuel containing 2.4×10^5 MTHM (Case 3) and radioactivity at 6.5 yr after reactor discharge. Canister requirement dictated by the heat output allowed by the disposal systems.

These differences result in defense HLW having a lower radioisotope concentration than commercial HLW and a correspondingly lower heat output per canister (see Table 2-3). This means that the size of the defense HLW canister, particularly the diameter, can be larger than a commercial HLW canister for which the heat transfer capability between the package and the host rock is more restrictive. The similarities between defense HLW, commercial HLW, and spent fuel are more significant than the differences when considering the requirements for handling and disposal of the waste. Therefore, as further discussed in Chapters 3 and 6, the DOE plans to dispose of all these wastes in a single repository (DOE 1985b).

Volumetrically, there are over 10 times more defense HLW than commercial HLW. As indicated in Table 2-3, these wastes are produced (or have been produced and are in storage at) government installations in Hanford, Washington, Idaho Falls, Idaho, and Savannah River, South Carolina. The combined defense and commercial HLW volumes are expected to be on the order of 12% of the spent fuel waste volumes accepted for disposal at the repository (DOE 1985a). In contrast, DOE estimates for the amounts of radioactivity present in spent fuel and HLW accumulated through 2000 indicate that about 95% of the total is due to spent fuel (DOE 1987).

2.4 TRANSURANIC WASTE

Radioisotopes heavier (having more protons) than uranium are called transuranic isotopes (TRU). Very low naturally occurring concentrations of TRU have been measured in uranium bearing ores (Rankama 1954). Most TRU in existence today, however, has been artificially produced by irradiation of nuclear fuel. It results from the interaction of uranium and thorium with neutrons and subsequent β decay. For example, plutonium, the best known TRU element, is produced by the following interaction:

$$^{238}_{92}U + ^{1}_{0}n \rightarrow ^{239}_{93}Np + ^{0}_{-1}e + ^{239}_{94}Pu + ^{0}_{-1}e$$

Such materials include isotopes that have long half-lives and are highly radiotoxic becuase they decay by emitting high energy alpha particles. Table 2-4 lists the primary elements of interest in TRU waste. Such wastes are primarily produced as a result of reprocessing spent fuel and subsequent recycling of plutonium and uranium. TRU may be present in solidified liquid waste as well as incorporated into fuel cladding hulls and other components filters, sludges, and trash. Decontamination and decommissioning of facilities for reprocessing spent fuel and fabricating plutonium will also result in TRU waste production. Materials that contain TRU in concentrations below $100\,nCi/g\,(1\,nCi = 10^{-9}\,Ci)$ are eligible to be disposed of as low-level radioactive waste in accordance with 10 CFR Part 61. This level was chosen because it is similar to the level of naturally occurring TRU in ore. At concentrations above $100\,nCi/g$ the longevity and health impacts concerns are such that the wastes are subject to the requirements

TABLE 2-4 Radionuclides of Interest in TRU Waste

Isotope	Half-Life (yr)	Principal Means of Production	Decay Products
^{237}Np	2.14×10^6	^{238}U(n, 2n), ^{237}U(β^-)	^{233}Pa + 4.9 MeV α
^{238}Pu	86.4	^{237}Np(n, γ), ^{238}Np(β^-), ^{242}Cm(α)	^{234}U + 5.6 MeV α
^{239}Pu	24,400	^{238}U(n, γ), ^{238}U(β^-), ^{239}Np(β^-)	^{235}U + 5.2 MeV α
^{240}Pu	6,580	Multiple n capture	^{236}U + 5.2 MeV α
^{241}Pu	13.2	Multiple n capture	^{241}Am + 0.021 MeV α
^{242}Pu	2.79×10^5	Multiple n capture, ^{242}Am(α)	^{238}U + 4.9 MeV α
^{241}Am	458	^{241}Pu(β^-)	^{237}Np + 5.6 MeV α
^{243}Am	7,950	Multiple n capture	^{239}Np + 5.4 MeV α
^{243}Cm	32	Multiple n capture	^{239}Pu + 5.8 MeV α
^{244}Cm	17.6	Multiple n capture	^{240}Pu + 5.9 MeV α

Source: (DOE 1983).

of the Environmental Protection Agency's (EPA) rule 40 CFR Part 191, "Environmental Standards for Management of Spent Nuclear Fuel, High level and Transuranic Wastes." TRU waste is produced primarily by the defense program, in the absence of commercial reprocessing of spent fuel.

Some TRU waste has physical and chemical characteristics in common with the HLW described in Section 2.2. Other TRU waste sources are similar to the low-level waste that is discussed in Section 2.4. The extent to which isotope separation is performed on plant waste streams will determine what waste must be classified as TRU. That is, if extensive separation is performed, as was once considered for possible commercial recovery of isotopes such as ^{237}Np, it is possible that a waste stream may be disposed as low-level waste rather than as TRU waste. TRU waste may contain sufficiently high concentrations of γ-emitting nuclides that remote handling is required to maintain occupational exposures as low as reasonably achievable (ALARA). Other TRU waste may contain primarily α-emitting nuclides and the dose rate at the package surface does not make remote handling necessary. DOE estimates that approximately 97% of TRU in retrievable storage at the end of 1986 was capable of being contact handled. As indicated below, there is also a substantial volume of TRU waste that was buried under rules in effect through the early 1970s. (See discussion in Chapter 3.)

Inventories and Characteristics of DOE Defense TRU Waste in Retrievable Storage as of December 31, 1986[a]

	Volume (10^6 ft^3)	Mass TRU (kg)	Radioactivity (10^3Ci)	Thermal Power (10^3W)
Contact handled	1.75	1826	2989	76.3
Remotely handled	0.05	40	472	4.5
Buried	6.77	770	232	5.7

[a]Adapted from (DOE 1987). Radioactivity includes β and γ activity from non-TRU isotopes.

2.5   LOW-LEVEL WASTE

Low-level waste (LLW) is frequently defined and described on the basis of what it is *not* rather than what it is. The Low-Level Radioactive Waste Policy Act of 1980 (Public Law 96–573) defines LLW as

> radioactive waste not classified as high-level radioactive waste, transuranic waste, spent nuclear fuel, or by-product material as defined in Section 11(e)(2) of the Atomic Energy Act of 1954.

This last exclusion refers to the waste (or tailings) produced by the mining and milling of uranium and thorium. In establishing its regulations for land disposal of radioactive waste (10 CFR Part 61), the NRC essentially repeated the definition of the act but specified that such waste must be "acceptable for land disposal." This qualification referred to a series of concentration limits for selected isotopes contained in the waste. (See Chapter 3 for a detailed discussion of the 10 CFR Part 61 waste classification system.)

LLW is produced—or potentially produced—as a result of any action in which radioactive material is used. Information on sources of LLW, the chemical and physical form, and the radioisotope concentrations is needed to develop adequate handling procedures at the site of generation, during shipment, and at the disposal facility. LLW characterization is a moving target requiring periodic review and update. This is very different from the case of HLW the characteristics of which remain relatively stable even though the amount of waste produced in a given time may vary. The variation in LLW characteristics reflects the very large number of applications to which radioactive materials may be put, the many individual and independent users, and the development of new applications and packaging and treatment procedures. Regulatory requirements and economic viability further influence the amount and characteristics of LLW produced and requiring disposal at any given time. For clarification, it is important to note that there is a substantial difference, in volume and radioisotopes contained, between LLW "produced" and that "requiring offiste disposal." There are several management options available to an LLW generator and the mix of actions used depends on the isotopic distribution and the chemical and physical form of the waste. As discussed further in Chapter 3, there are allowable limits on releases to air and water from a licensed facility and some material is handled in this manner. Storage onsite for decay and eventual disposal of the material as nonradioactive waste is a practice that is particularly applicable to short half-life isotopes such as those often found in medical waste. Several surveys of radioactive materials licensees have been conducted to identify the types and amounts of wastes so handled (CRCPD 1982, 1984). Response to the survey questionnaires varied among different states and the results were not used to develop a nationwide assessment of onsite radioactive waste management practices. These practices will be further discussed in Chapter 5 but the volumes and waste characteristics described in this section will address the waste that has historically required

TABLE 2-5 Typical LLW Streams by Generator Category

	Power Reactors	Institutional[a]	Industrial	Government
Compacted trash or solids	×	×	×	×
Dry active waste	×			
Dewatered ion-exchange resins	×			
Contaminated bulk	×		×	×
Contaminated plant hardware	×		×	×
Liquid scintillation wastes		×	×	×
Biological waste		×		
Absorbed liquids		×	×	×
Animal carcasses		×		
Depleted uranium MgF_2			×	

Source: (EG&G Idaho, Inc. 1985).
[a] Medical and academic generators.

offsite disposal and not the wastes that are treated and/or released at the site of generation.

LLW characteristics are described in the following sections for the several major categories of waste generators. This discussion is substantially based on material compiled in support of the NRC rulemaking on land disposal of radioactive waste (Wild 1981). The waste types produced by generators in different categories are summarized in Table 2-5 (EG&G Idaho, Inc. 1985).

2.5.1 Nuclear Power Reactors

Control of radioactive materials in a power reactor is generally achieved by removing material from process streams, concentrating it in a relatively small volume and disposing of that volume as LLW. Projected volumes for new power plants were estimated by NRC for licensing evaluation purposes to be 17,000 ft^3 for PWRs and 29,000 ft^3 for BWRs (NRC 1982). Actual waste generation rates have decreased substantially over the past 5 yr at many power reactors. This reduction was a response to both the rising costs of disposal services and the uncertainty of their availability as discussed in Chapter 1.

Small amounts of radioactive material are present in the coolant of a power reactor from the fuel as well as from corrosion of the system's metallic components and impurities in the coolant that have been activated by neutron bombardment. The radioactive waste treatment systems are designed to remove these materials on an ongoing basis through filtration and ion-exchange resins in both the primary system (in contact with the fuel) and secondary or auxiliary systems that treat liquids with which primary coolant may have come in contact (e.g., through steam generator or valve leakage).

Ion-exchange resins use small (about 1-mm-diameter) organic beads or granules to remove radioisotopes from liquids. The resins may be specifically designed to remove anions or cations or may contain both cation- and anion-removing resins (this is called a deep bed or mixed bed resin or demineralizer). The resins are generally packed in cylindrical containers through which the liquid streams flow. As the waste flows through the resin bed, ions present in the resin selectively exchange with those in the waste at rates and in amounts that are dependent on differences in charge on the ions and concentrations in the waste and resin. Once the ion-exchange capacity of a resin bed has been exhausted, the resins may be either replaced or regenerated. Resin regeneration is accomplished by washing the resin with a concentrated solution of the ion originally present (generally H_2SO_4 for cation resins and NaOH for anion resins). Regenerant solutions may be further concentrated by evaporation and solidification prior to shipment. Spent resins are transferred as a slurry to shipping containers where they are dewatered (to 42–55% water absorbed in the resin) prior to shipment. Resin densities have been reported to range from 0.67 to 0.91 g/cm^3 (Phillips 1979). Some facilities may solidify the resin in cement or a polymer prior to shipment. Radionuclide concentrations in spent resins are generally sufficiently high that shielded shipping containers are required. Gas generation (CO_2, NO_x, SO_x) due to chemical, radiolytic, and biological decomposition may occur in the resin (Clark 1978). It must be considered in designing the waste package and disposal unit because it may provide a transport mechanism for radionuclides from the waste to the biosphere.

Filters used in nuclear power reactors are generally either cartridge filters or precoat filters. The primary difference is that cartridge filters contain disposable filter elements made of woven or wound fabric, or pleated or matted paper supported by a stainless-steel mesh. Precoat filters have filter aids such as diatomaceous earth, powdered mixtures of cation- and anion-exchange resins (POWDEX resins), and high-purity cellulose fibers deposited as a thin cake on the initial, reusable filter medium. Once exhausted, these filter aids are backflushed from the filter and disposed as a dewatered but unsolidified sludge with an average density of 0.86 g/cm^3. For comparison, the average cartridge filter density is 0.6 g/cm^3 (Phillips 1979). Cartridge filters are more commonly used in PWRs and precoat filters in BWRs.

Concentrated liquids are produced at some power reactors using evaporators to reduce the volume of liquid waste to be disposed. The concentrated liquids are also known as evaporator bottoms. They have a high solids content and an average density of 1.0–1.2 g/cm^3 prior to solidification.

Dry active waste is the term generally applied to a wide variety of waste products such as cleaning materials, glass, filters, concrete, miscellaneous wood, and metal. It may be compactible (such as wood, glass, fiber) or noncompactible and combustible (such as paper) or noncombustible (pipe or hardware). The extent to which such material is segregated so that distinctions can be made between the waste types identified in Table 2-5 as compacted trash or solids, dry active waste, contaminated bulk, and contaminated hardware will vary with operations at a given facility and the volumes of waste being produced. For example, during a refuelling outage or facility modification there will probably be greater than normal numbers of workers using protective clothing and/or cleanup solutions and rags. They will produce sufficient amounts of waste of a given type (e.g., concrete block) that waste packages may contain only one waste type. During normal operations a single package (drum, crate, or liner) may contain a mix of several of these waste types. Similarly, compaction may be performed routinely on all waste generated or may be contracted on an as-needed basis for periods of high-volume production. Accurate characterization of this waste stream, therefore, must be based on facility and shipment-specific data rather than a generic model.

Nonfuel reactor components such as fuel channels, control rods, and in-core instrumentation are relatively low volumes of waste that require special handling (remote operation and shielding) onsite, in-transit, and at the disposal facility because of the high radiation levels (primarily due to ^{60}Co) and long half-lives of contained radionuclides (such as ^{63}Ni).

Decontamination of plant vessels and equipment may be performed periodically during the operating life of a reactor facility to reduce the occupational exposure that would otherwise be incurred during major equipment repair and/or replacement operations. Chemical decontamination has been performed on the primary cooling system of a BWR (NRC 1980) and repair by sleeving or capping of steam generator tubes in PWRs has included prior removal of buildup on metal surfaces in contact with primary coolant water. LLW is produced by decontamination operations primarily as spent ion-exchange resins through which the decontamination fluid is processed. The resins are expected to be similar to resins produced during plant operations with higher concentrations of activation products (e.g., iron, nickel, cobalt, chromium) found in reactor steel and fuel components. The resins may also contain large quantities of chelating agents. As discussed in Chapter 3, there are strict limits on the amounts of chelates permitted in LLW because of their tendency to concentrate and mobilize radioisotopes. Special packaging and disposal requirements may be imposed on wastes exceeding the regulatory limits for chelate concentration.

Decommissioning nuclear power reactors at the end of their useful life will produce large volumes of LLW, much of it as contaminated concrete and metal vessels and piping. Decommissioning waste will contain the same isotopes as plant hardware and noncompactible dry active waste produced during plant operation. Size, weight, and rate of transfer to a disposal facility will be the main differences from earlier practice. Planning for new disposal facilities has been based on the assumption that nuclear power facilities will be available for decommissioning at the termination of the initial full power operating license, which is valid for 40 yr from issuance of a construction permit. (For a fuller discussion of the licensing procedures for a nuclear power reactor, see Lamarsh 1983, Chapter 11.) As is true of most other projections, there is considerable uncertainty in this timetable. Several of the small, early power reactors have already ceased operation because changes in regulatory requirements on issues such as seismic protection or emergency core cooling would require modifications so costly that producing power is no longer economic. Larger units more representative of current technology, however, are being studied to determine if life extension (beyond 40 yr) is feasible with some equipment repair and/or replacement. Initial estimates of the volumes and activities of decommissioning waste were based on studies undertaken for the NRC (Smith 1978; Oak 1980). The studies considered three alternative decommissioning scenarios:

DECON: immediate dismantlement (extending over several years)
            sufficient to enable the site to be released for unrestricted use;
SAFSTOR: delay of up to 100 yr after shutdown to permit decay of
            shorter lived nuclides prior to dismantlement; and
ENTOMB: encasement of radioactive materials in place in concrete or
            some other material and release of the site for unrestricted use
            once the contained radioactivity has decayed to acceptable
            levels.

The volumes and activities estimated for immediate dismantlement of the reference facilities are listed in Table 2-6. Deferring dismantlement for 100 yr after shutdown results in approximately a 10-fold reduction in the volumes of decommissioning waste requiring disposal (NRC 1983). At the time these analyses were performed, the volume of decommissioning waste was estimated to be approximately equal to the total volume produced over the entire operating life of the facility. Since that time, the average annual waste volumes have been reduced significantly. Continuing research into techniques for decontamination, salvage, and recycle may also reduce the volumes of decommissioning waste

TABLE 2-6   Summary of LLW from Decomissioning[a]
Reference Nuclear Power Reactors[b]

	PWR		BWR	
Waste Stream	Volume (ft^3)	Activity (Ci)	Volume (ft^3)	Activity (Ci)
Activated metal	97,100[c]	4,841,300	4,900	6,552,300
Activated concrete	25,000	2,000	3,200	200
Contaminated metal	192,000	900	549,200	8,600
Contaminated concrete	374,700	100	59,200	100
Dry solid waste (trash)	50,600	—	119,500	—
Spent resins	1,100	42,000	1,500	200
Filter cartridges	300	5,000	—	—
Evaporator bottoms	4,700	—	18,300	43,800

Source: PWR (Smith 1978); BWR (Oak 1980).
[a] Assumes immediate dismantlement (DECON).
[b] Reference PWR, 1175 MWe; reference BWR 1155 MWe.
[c] Note: All figures are rounded to the nearest hundred.

actually requiring disposal as LLW. There is, at present, no more certain estimate than that decommissioning waste volumes are likely to be about equal to the total volumes produced over the plant lifetime.

2.5.2 Institutions

Medical and academic institutions use radioactive materials in research, diagnosis and therapy and employ a variety of radioisotopes and material forms to achieve their purposes. Surveys of institutional waste generators (Anderson 1978; Beck 1979; CRCPD 1982, 1984) have identified four specific waste streams that were contributing significant waste volumes in the late 1970s and early 1980s. These are liquid scintillation vials, other organic and inorganic liquids, biological wastes, and trash.

Smaller volumes of accelerator targets and sealed sources were also produced. The isotopes generally contained in these wates were ^{51}Cr, ^{192}Ir, ^{35}S, ^{125}I, ^{32}P, ^{14}C, ^{90}Sr, ^{3}H, ^{57}Co, ^{99m}Tc, and ^{60}Co. Isotopes with shorter half-lives, such as 8-day ^{131}I, are generally stored onsite until sufficient decay has occurred that the wastes may be discarded as nonradioactive (typically after 10 half-lives).

Volumes of *liquid scintillation waste* requiring disposal have decreased significantly in the last few years. The scintillation fluids are generally used to detect β-emitting nuclides and many are primarily composed of flammable organic solvents such as toluene. Such fluids are therefore a mixed (radioactive and hazardous chemical) waste. Liquid scintillation vials containing specified small amounts of ^{14}C and ^{3}H have been exempted by the NRC from having to be disposed as LLW and may be disposed as a chemical waste instead, frequently by incineration. Use of aqueous-based scintillation fluids rather than organic materials enables the fluid to be discharged to the sanitary sewer in accordance with 10 CFR Part 20 rather than as LLW. Lack of coordinated regulations (between NRC and EPA) results in mixed wastes being ineligible for disposal by land burial solely in accordance with 10 CFR Part 61. Such wastes are currently being stored at the point of generation pending resolution of regulatory requirements (see additional discussion on mixed waste regulations in Chapter 3). Similar storage is being used for radium sources that are not subject to 10 CFR Part 61 regulation. Such wastes are currently being accepted at one disposal site if special preapproved packaging is used.

Other liquids produced at institutions from preparation and analysis of samples such as elution of Tc generators, radioimmunoassay procedures, and radioactive tracer studies are generally shipped in packages containing absorbent material equal to twice the volume of liquid to comply with transport regulations (10 CFR Part 71) designed to limit the release of radioactive material in transit.

Biological wastes are produced by research programs at hospitals and universities and consist of animal carcasses, tissues, animal bedding and excreta, vegetation, and culture media (Wild 1981). Small amounts of pathogenic and carcinogenic substances may also be included in these wastes depending on the initial research problem. The materials are generally of very low specific activity. Care must be taken in packaging, storage, and disposal of these materials because the potential gas buildup due to biodegradation of the waste can overpressurize the containers and cause them to fail. If this happened in storage or in transit, time-consuming cleanup with additional occupational exposure would be required. After disposal it is possible that the gas may provide a transport mechanism for waste from the disposal cell (Matuszek 1982). General practice is to ship carcasses packed with absorbent material and lime in a 30-gallon drum within a 55-gallon drum and place absorbent material between the two drums. As discussed in Chapter 5, incineration rather than direct burial would reduce the volume of biological waste requiring disposal and preclude the problem of gas generation within the waste. Questions of offgas control for radioactive and other materials as well as the economics of operation at different volume levels are major determinants of when incineration is used for this waste stream as well as other combustible materials such as trash.

Trash produced at institutional facilities differs from that produced at power reactors in that it is primarily composed of materials such as paper, rags, glassware, packaging, and mops that are compactible and/or combustible. When treated onsite, it is generally compacted although some larger volume generators have installed LLW incinerators.

Volumes of LLW produced by medical and academic institutions have decreased substantially since 1979. Much of this reduction has resulted from the regulatory changes exempting some liquid scintillation fluids from the need for disposal as LLW. Increased onsite storage for decay, more careful waste segregation to remove nonradioactive waste previously disposed as LLW for convenience and conservatism, and compaction of trash are the other major reasons for this reduction. A recently completed study of medical waste generators (Weir 1986) anticipates small continued reductions in the overall volumes of medically related waste requiring disposal. With the exception of a few specific instances (e.g., use of ^{111}In rather than ^{99m}Tc for antibody labeling), this study does not anticipate that the characteristics of the waste will be significantly different from current experience.

Accelerator targets are used to produce radionuclides through direct interaction with charged particle beams or indirectly through the interaction of induced radionuclides and other materials. The targets are generally titanium foils containing absorbed tritium. *Sealed sources* have radioactive materials in the form of foils or beads encapsulated to prevent leakage during use. They may be relatively low in activity and used as calibration or reference sources for radiation detectors and analytical instrumentation used in research or clinical laboratories. High-activity sources may be used for medical radiotherapy or for research such as investigation of radiation effects on materials. These targets and sources may be disposed directly by the institution or they may be returned to the manufacturer at the end of useful life. Depending on the levels of activity and isotopes in a given source, the manufacturer may recycle the contained material or simply dispose of the used sources.

2.5.3 Industry

Industrial processes that result in the production of LLW include the production and distribution of radioisotopes for medical, academic, or industrial use, manufacture of materials containing radioisotopes, and the use of radioisotopes for research or testing and in gauges or instrumentation. Wild *et al.* identified several distinct industrial waste streams (Wild 1981) that varied in the concentration of activity, contained isotopes, and volumes produced. These waste streams are summarized in Table 2-7 and the following paragraphs.

Medical isotope production is achieved through irradiation of highly enriched uranium fuel and the separation and purification of the resulting fission products. LLW generated from medical isotope production includes solidified aqueous liquids and trash produced in the separation, cleanup, and shipping of the radioisotopes. The solidified aqueous liquids contain a radioisotope distribution similar to spent fuel with several isotopes (particularly shorter lived isotopes such as ^{90}Mo, ^{131}I, ^{133}Xe, and ^{125}I) being selectively removed. Isotopes present in the solidified aqueous liquids include uranium, transuranics at concentrations less than 100 nCi/g, and fission products such as ^{90}Sr, ^{137}Cs, and ^{3}H. Compaction is used to reduce volumes generated and storage is used to reduce activity shipped. The solidified solids are stored onsite for approximately 9 months after production. The final waste package shipped for disposal consists of a small metal container of the solidified aqueous salts packed within a drum containing low specific activity trash.

Although technically an industrial source, manufacture of clinical and research *radiopharmaceuticals* produces waste that is very similar in isotopic distribution and concentrations to that present in institutional waste streams. Testing of the biological uptake and transport of new pharmaceuticals frequently involves use

TABLE 2-7 Estimated Volumes and Activities of Industrial LLW Streams

Waste Stream	Circa 1980 Volume (ft^3)	Gross Specific Activity (Ci/ft^3)
Medical isotope production	6,800[a]	16
Industrial tritium	3,500	66
High-activity waste (> 1 Ci/ft^3)		
Sealed sources	200	160
Other	2,600	6
Low-activity waste (< 1 Ci/ft^3)		
Source and special nuclear material[b]	1,103,500	8×10^{-5}
Other	162,700	8×10^{-5}
Total	1,279,300	

Source: Based on Table 3-5 of (Wild 1981)

[a]All Figures rounded to nearest hundred cubic feet.

[b]Nonreactor fuel cycle facilities are added to the industrial facilities described in the original report.

of radioisotope tracers, often ^{3}H because it follows general bodily fluids. The tritium wastes described below reflect such applications and the wastes produced therefrom.

High specific activity industrial waste (defined in Wild 1981 as greater than 0.1 Ci/ft^3) is the name used by NRC to describe activated metal and equipment produced by accelerators, research reactors, and neutron irradiation capsules. The isotopes expected in this waste stream are primarily the activation products ^{14}C, ^{55}Fe, ^{59}Ni, ^{60}Co, ^{63}Ni, and ^{94}Nb produced by the interaction of neutrons with trace elements in the metal. The distribution of these isotopes is expected to be similar to that observed in decommissioned non-fuel-reactor components (NRC 1981).

Tritium is widely used in biological research and medicine as well as in commercial products such as paints and dials because of its luminescent properties. Because of its many applications there are relatively large volumes of tritium wastes, from the production of tritium (which results in tritium, fluoride, and trash), from incorporation of the tritium into biological compounds (labeling), and fabrication of luminous products such as watch dials. This is the only one of the 25 generic waste streams characterized and evaluated by the NRC in the LLW rulemaking (10 CFR Part 61) that contains only a single isotope. Based on surveys of previously disposed tritium waste, the NRC estimated that the average concentration of tritium in the waste was 66 Ci/ft^3 (NRC 1981b).

Source and special nuclear material wastes contain the isotopes of uranium, ^{235}U and ^{238}U, and small amounts of their daughter products. They are produced in the early steps of the nuclear fuel cycle, at the conversion facilities where milled uranium dioxide is processed into gaseous uranium hexafluoride, at the enrichment facilities where the ^{235}U concentration is increased from the naturally occurring 0.711% to approximately 3–4%, and at the fuel fabrication facilities. Non-fuel-cycle producers of these wastes are primarily industrial facilities that process depleted uranium. Both economics and health and environmental impact considerations have resulted in fuel cycle facilities being designed to recycle and recover as much uranium as possible from the process streams. It is estimated (NRC 1981b) that about 50,000 ft^3 of process waste is shipped for disposal as solid waste annually from existing conversion facilities. This is less than 1% of the plants throughput. The concentration of uranium isotopes in the process waste is estimated to be 1.1×10^{-5} Ci/ft^3, making this a very low-activity waste stream.

The gaseous diffusion uranium enrichment complex in the United States is a government run process. Small amounts of uranium are contained in liquids from equipment cleanup that is routed to settling ponds onsite where it precipitates as sludge. This material is retained onsite rather than shipped for commercial disposal. Should site operations and uses change in the future, these areas would be analyzed to determine what recovery actions would be appropriate, if any.

Fabrication of fuel produces LLW in the form of dry solids of CaF$_2$ containing low concentrations of enriched uranium and other low-activity waste that is shipped offsite for disposal. Other uranium-bearing waste in the form of liquids and sludges are being stored onsite pending decisions on the timing and extent of waste treatment and uranium recycle operations or decontamination of the storage area and removal of the material for offsite disposal. (See Chapter 7 for detailed discussion of remediation activities at fuel cycle facilities.)

2.5.4 Government

Government operations produced 2% of the LLW disposed at commercial facilities in 1984 (EG&G Idaho, Inc. 1985). These wastes are similar to wastes being shipped by other generator categories. For example, Veterans Administration Hospital waste is substantially the same as other medical waste, and government-produced used luminous equipment dials are the same as those from industrial facilities. LLW produced by "atomic energy defense activities" and "federal research and development activities" are not subject to the provisions of the Low Level Radioactive Waste Policy Act requiring the states (or compacts) to provide disposal capability. They are disposed of at government facilities as described in Chapter 1. These wastes are produced by (Act 1980)

   (i)   naval reactors development and propulsion,
  (ii)   weapons activities, verification and control technology,
 (iii)   defense materials production,
 (iv)   inertial confinement fusion,
  (v)   defense waste management, and
 (vi)   defense nuclear materials security and safeguards.

2.6 URANIUM MINE AND MILL TAILINGS

The NRC estimated (NRC 1979) that premature deaths due to radon releases from 1978 to 2000 (some 20 million Ci) from uranium mining and milling operations would total 181 fatalities. Of this total, 60% was estimated to be due to mining operations. This value is considered an upper bound on potential effects because domestic uranium production has not increased as quickly as projected in this analysis.

Because uranium is a very concentrated energy source, it is economic to process even very low concentration ore. Average ore grades of approximately 0.113% U_3O_8 are currently processed (DOE 1983). The corollary to that statement is that substantial amounts of waste material are produced at the mine and mill for every pound of uranium recovered. Most regulatory and analytical attention to date has focused on the milling operation rather than the mine because the tailings material is more geographically concentrated and in a physical form (crushed stone and sand-like fines) that is more readily dispersible by wind and water.

Management of uranium mine and milling wastes is needed because approximately 6% of the original uranium content is not recovered in the milling process and is retained in the waste (DOE 1983). Figure 2-5 shows the uranium decay chain. Two members of this chain, radium-226 and radon-222, have received particular attention in terms of potential health impact. Both are α emitters with slow biological turnover once they enter the body. Radium settles preferentially in bone and the mobile gas radon produces daughter products that lodge in the lung. Mill tailings are a very large volume source of radioactive material existing in a very accessible and dispersible condition. They are, therefore, a potentially large radiation exposure source.

The milling process physically consists of crushing and grinding the ore, leaching the uranium with either an acid or alkaline solution (depending on the lime concentration of the original ore), and processing the uranium-bearing liquid through a series of tanks that permit settlement of suspended solids (source of the tailings) followed by solvent extraction and precipitation of the concentrated U_3O_8. The U_3O_8 is dried and packaged for shipment as yellow-cake (see Figure 2-1).

The actual amount of tailings generated per pound of yellow-cake produced depends upon the initial uranium content (grade) of the ore, and the recovery rate achieved in the milling process. Facility throughput then determines the total amount of tailings produced per year. Table 2-8 lists the chemical and radiological constituents of tailings from a model mill analyzed by the U.S. NRC in the *Generic Environmental Impact Statement on Uranium Milling* (NRC 1979). This facility was modeled as representative of industrial practice at existing facilities and any anticipated changes at new facilities. The tailings (produced at a rate of 1800 MT/day) are considered to exist as sand (solids larger than 75 μm), slimes (solids smaller than 75 μm), and liquid solutions of chemicals from the ore and process reagents. The slimes are estimated to contain 35% (by weight) of the tailings. They also contain 85% of the radioactivity of the tailings.

Physically, the tailings are slurried to an onsite impoundment or tailings pond. During mill operation, the tailings in the pond are initially kept wet, reducing the potential for wind transport. The earthen embankments used to form the tailings pond are increased in height and thickness as operation continues to permit additional tailings to be stored as produced. The final tailings pond for the model mill analyzed by NRC is 3100 ft long at the centerline. The embankments are 33 ft high, 43 ft broad at the crest, and 174 ft broad at the base. The pond is estimated to contain 1,630,000 yd^3 of tailings to a depth of about 26 ft within approximately a 200 acre area (75 acres are wet during operation) (NRC 1979). Once operations cease, the tailings pond will dry out. Stabilization and closure procedures are intended to minimize the dispersion of particulate material primarily through wind erosion and the release of radon gas. Release limits are further discussed in Chapter 3. Site closure plans and stabilization and remedial actions are discussed in Chapters 6 and 7, respectively.

In estimating the amount of tailings waste to be managed and methods, responsibilities, and schedules for such management, distinction is made among uranium mill tailings sites as being either active or inactive. Active sites are those at which milling operations are still being conducted. This means that the tailings

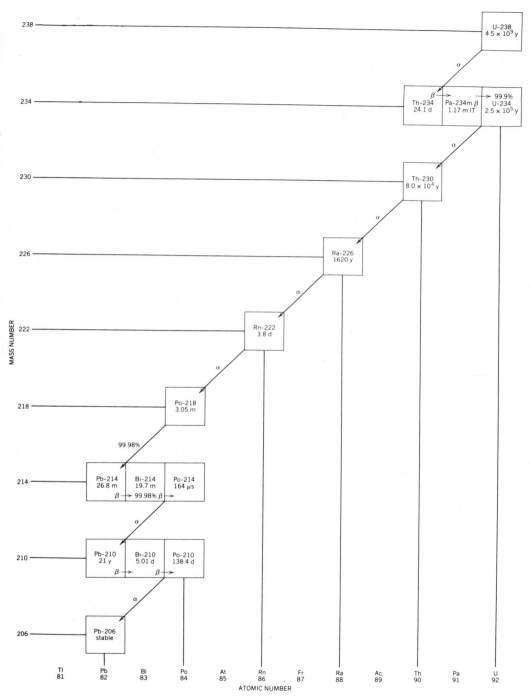

Figure 2-5 Uranium-238 decay series. Source: (NRC 1979).

TABLE 2-8 Chemical and Radiological Properties of Reference Uranium Mill Tailings[a]

Parameter	Value
Dry Solids	
U_3O_8, wt %	0.011
U nat, pCi/g[b]	63
^{226}Ra, pCi/g	450
^{230}Th, pCi/g	430
Tailings Liquid	
pH	2
Aluminum, g/L	0.0
Ammonia, g/L	0.5
Arsenic, g/L	2×10^{-4}
Calcium, g/L	0.5
Carbonate, g/L	—
Cadmium, g/L	2×10^{-4}
Chloride, g/L	0.3
Copper, g/L	0.05
Fluoride, g/L	5×10^{-3}
Iron, g/L	1.0
Lead, g/L	7×10^{-3}
Manganese, g/L	0.5
Mercury, g/L	7×10^{-5}
Molybdenum, g/L	0.1
Selenium, g/L	0.02
Sodium, g/L	0.2
Sulfate, g/L	30.0
Vanadium, g/L	1×10^{-4}
Zinc, g/L	0.08
Total dissolved solids, g/L	35.0
U nat, pCi/L	5.4×10^3
^{266}Ra, pCi/L	4×10^2
^{230}Th, pCi/L	1.5×10^5
^{210}Pb, pCi/L	4×10^2
^{210}po, pCi/L	4×10^2
^{210}Bi, pCi/L	4×10^2

Source: (NRC (1979)).

[a]Based on the following:
M. B. Sears *et al.*, "Correlation of Radioactive Waste Treatment Costs and the Environmental Impact of Waste Effluents in the Nuclear Fuel Cycle for Use in Establishing as Low as Practicable Guides—Milling of Uranium Ores." ORNL-TM-4903, 1975.
"WIN Reports on Uranium Ore Analysis," U.S. AEC Contract, 49-6-924, various reports 7 January 1957 through 10 July 1958.
"United States Mineral Resources." Geological Survey Professional Paper 820, 1973.
"Mineral Facts and Problems." U.S. Bureau of Mines Bulletin 667, 1975.
[b]A picocurie of natural uranium (U nat) weighs 1.5 μg and contains 0.49 pCi each of ^{238}U and ^{234}U and 0.023 pCi of ^{235}U.

are generally wet to a large extent and the the retention walls are being monitored for stability. There is also an ongoing income source that can be used for necessary waste management actions and personnel to accomplish them. Inactive facilities, on the other hand, must first be analyzed to determine the amount of material present, and the condition of the impoundment and surrounding area that may have elevated uranium levels because of dusting from the tailings impoundment. Many of these sites are being managed by the U.S. DOE in accordance with federal law (Uranium Mill Tailings Radiation Control Act) as further discussed in Chapter 3.

Active mill tailings sites exist in seven western states in which most of the U.S. uranium production has occurred to date. DOE estimates that some 3.48 billion ft^3 of commercial tailings had been accumulated through 1982 (DOE 1983). This material contains over 408 thousand Ci of radioactivity and produces about 9000 W of thermal power as a result of radioactive decay. New Mexico accounts for about half of this total amount and Wyoming almost 30%. The five other states in which commercial tailings exist are Colorado (6.1%), Utah (5.8%), Texas (4.4%), Washington (2.2%), and South Dakota (1.2%). Data on the operational status (as of the end of 1986), of commercial milling facilities, the amount of tailings present at the facilities, and the condition of the impoundment are presented in Table 2-9 for conventional uranium milling sites. Production from unconventional sources in 1984 and 1985 is listed in Table 2-10. Future tailings production depends on the rate of growth of nuclear power capacity, the extent of uranium impacts, and the efficiency of recovery processes. In the 1987 report, DOE estimates that over 5 billion ft^3 of tailings will exist by 2000.

In addition to wastes existing at active mill sites, tailings exist and must be managed at DOE sites and at inactive sites, many of which were originally operated in support of government weapons production programs. Several other sites were operated to recover radium from uranium ores. These facilities predated any of the currently existing regulatory agencies and requirements and the details on waste amounts, characterizations, and sometimes locations have had to be carefully recreated from a variety of sources and with mixed success. DOE's 1987 waste inventory report identified over 608 million ft^3 of mill tailings subject to remedial action programs. Another 340 million ft^3 of soil and stabilization material subject to contamination by windblown tailings may also need to be handled. Most (91% by volume) of the material was produced at some 24 sites in 10 states, mostly in the west. It is being managed by DOE under the Uranium Mill Tailings Remedial Action Program. The one non-western site is in Canonsburg, Pennsylvania where radium was extracted from carnotite ore from 1911 to 1922. Operation from 1930 to 1957 produced radium and/or uranium from ores and scrap. The balance of the material consists of about 40 million ft^3 of tailings generated at a government-owned mill in Utah and about 2 million ft^3 of material removed from structures in the Grand Junction, Colorado area where mill tailings were used as construction material from 1952 to 1966. (See discussions in Chapter 1 and Section 7.5).

TABLE 2-9 Status of Conventional Uranium Mill Sites (as of the End of 1986)

Location	Operator	Rated Capacity (Tons ore/day)	Status: Operations	Status: Tailings[a]	Tailings Storage Area (ha)	^{226}Ra Activity (pCi/g)	Total Tailings: Volumes[b] (10^3 m^3)	Total Tailings: Mass (10^3 tons)	Total Tailings: Government Portion[c] (10^3 tons)
Colorado									
Canon City	Cotter	1,090	Shutdown 1985	Unstabilized	81	780	562	900	290
Uravan	Umetco Minerals	1,180	Shutdown March–September 1981; resumed May 1983; shutdown 1984	Partially stabilized	44	476	5,560	8,900	5,170
Subtotal		2,270			125		6,122	9,800	5,460
New Mexico									
Cebolleta	Sohio Western Mining	1,450	Shutdown May 1981	d	73	504	938	1,500	0
Church Rock	United Nuclear	2,720	Shutdown May 1982	d	83	290	1,750	2,800	0
Grants	Anaconda	5,440	Shutdown March 1982	Partially stabilized	199	620	15,000	24,000	8,020
Grants	Quivira Mining	6,350	Shutdown 1985	Fenced	142	615	15,625	25,000	9,100
Grants	Homestake Mining	3,080	Shutdown 1984	Unstabilized	105	385	12,000	19,200	10,350
Marquez	Bokum Resources	1,815	New (on standby)	0	0	0	0	0	0
Subtotal		20,885			602		45,313	72,500	27,470
South Dakota									
Edgemont	TVA	680[e]	Shutdown August 1974; decommissioning since September 1983	Partially stabilized	50	705	1,313	2,100	1,470
Subtotal		0			50		1,313	2,100	1,470
Texas									
Falls City	Continental Oil/ Pioneer Nuclear	3,080[e]	Shutdown July 1981 (dismantled)	d	89	d	3,500	5,600	0
Panna Maria	Chevron Resources	2,270	Active	d	101	d	875	1,400	0
Subtotal		2,270			190		4,375	7,000	0
Utah									
Blanding	Umetco/Energy Fuels Nuclear	1,810	Shutdown February 1983; resumed 1985	Partially stabilized	135	d	338	540	0
La Sal	Rio Algom	680	Active	d	14	560	1,000	1,600	0
Moab	Atlas	1,270	Shutdown 1984	Unstabilized	>80	540	4,375	7,000	5,390
Hanksville	Plateau Resources	680	New (on standby)	0	28	0	0	0	0
Subtotal		4,400			>257		5,713	9,140	5,390

Location	Operator		Status						
Washington									
Ford	Dawn Mining	410	Shutdown 1982	Wood chip covering	43	d	1,250	2,000	1,060
Wellpinit	Western Nuclear	1,810	Shutdown August 1982; resumed March 1983; shutdown 1984	d	17	d	938	1,500	0
Subtotal		2,220			60		2,188	3,500	1,060
Wyoming									
Gas Hills	American Nuclear	860	Shutdown September 1981	Unstabilized	52	420	2,608	4,173	1,890
Gas Hills	Pathfinder	2,270	Shutdown 1985	Unstabilized	55	420	5,375	8,600	2,580
Jeffrey City	Western Nuclear	1,540	Shutdown June 1981	Interim stabilization	34	429	6,804	10,886	3,040
Natrona	Umetco	1,270	Shutdown 1984	Unstabilized	70	309	4,762	7,620	1,910
Powder River	Exxon	2,900[e]	Shutdown; decommissioned in 1984	Partially stabilized	81	450	4,531	7,250	0
Powder River	Rocky Mountain Energy	1,810	Active	Unstabilized	61	420	1,688	2,700	0
Shirley Basin	Pathfinder	1,630	Active	d	94	540	3,313	5,300	0
Shirley Basin	Petrotomics	1,360[e]	Shutdown 1985	Unstabilized	65	570	1,750	2,800	660
Red Desert	Minerals Exploration/ Union Energy Mining	2,720	Shutdown May 1983	Partially stabilized	121	d	562	900	0
Subtotal		12,100			633		31,393	50,229	10,080
1983 total for all sites[f]		48,445			>1,917		96,417	154,269	50,930[g]
1984 total for all sites[f]		45,545			d		98,864	158,184	50,930[g]
1985 total for all sites[f]		44,185			d		99,958	159,934	50,930[g]
1986 total for all sites		~37,691			d		100,697	161,117	50,930[g]

Source: (DOE 1987).

a On August 15, 1986, EPA issued its final rules on ^{222}Rn emissions from tailings piles. Mill owners have 6 Yr (subject to certain extension) to phase out the use of large existing tailings piles. New tailings piles may be contained in small impoundments (< 16 ha) or disposed of continuously by dewatering and burial (i.e., no more than 4 ha are uncovered at any one time).

b Calculated from reported mass using density = 1.6 tons/m^3.

c These tailings are from government contracts only.

d Not available.

e This capacity no longer available (see column labeled "Operations" under "Status" for reason).

f These values are cumulative totals.

g Total at the end of government-contracted deliveries in 1970.

TABLE 2-10 Production of Uranium from Unconventional Facilities in 1984 and 1985

Uranium Recovery Plant	Plant Location	Estimated Annual U_3O_8 Production (tons) 1984	1985
Solution Mining Operations[a]			
Caithness Mining Company (McBryde)	Hebbronville, TX	53	0[b]
Conoco (Trevino)	Hebbronville, TX	<180	0[b]
Everest Mineral Corporation (Las Palmas)	Hebbronville, TX	<70	0[b]
Everest Mineral Corporation (Mt. Lucas)	Dinero, TX	230	230
IEC Corporation[c] (Lamprecht)	Oakville, TX	0	>10
IEC Corporation (Zamzow)	Three River, TX	90	0[b]
Mobil Oil Corporation (El Mesquite)	Bruni, TX	>295	680
Tenneco Uranium, Inc. (West Cole)	Bruni, TX	>10	90
U.S. Steel Corporation (Clay West/Burns Ranch)	George West, TX	1010	<1010[d]
Uranium Resources, Inc. (Benavides)	Bruni, TX	20	90
Subtotal		1960	>1645[e]
Phosphoric Acid, Copper, or Rare Earth Industry By-product and Heap Leaching (Low-Grade Ore, Uranium, and Copper Tailings or Recovery from Mine Water)			
AGRICO Chem-Williams/Freeport Uranium	Donaldson, LA	0[f]	0[f]
Freeport Uranium Recovery Corporation	Uncle Sam, LA	<450	<450
IMC Corporation (New Wales)	Plant City, FL	<225	<225
Uranium Recovery Corporation	Mulberry, FL	<900	g
Wyoming Mineral Corporation	Lakeland, FL	<200	g
Gardinier Big River Corporation	Tampa, FL	90	g
Rhone-Poulenc Company	Freeport, TX	45	45
Anamax Mining Company	Sahuarita, AZ	<450	0[h]
Wyoming Mineral Corporation	Bingham Canyon, UT	g	g
Union Carbide Corporation (Natrona)	Riverton, WY	g	g
Cotter Corporation (Schwartzwalder Mine)	Golden, CO	<5	<5
Subtotal		430[e]	<725
Estimated actual production[i]		2390	2370

Source: (DOE 1987).

[a]Currently, there are 20 licensed solution mining facilities. Of these, only 10 have operated at least part-time during 1984–1985. A total of eight are now under restoration, four (two new) are on standby, six are operating, one (for research and development) is being built, and one has never been built.
[b]Now under restoration.
[c]Formerly owned by Wyoming Minerals Corporation.
[d]Limited production.
[e]By subtracting the best estimate from the reported total.
[f]Does not produce yellow-cake; crude product is shipped to Freeport Uranium Recovery Corporation for refinement.
[g]Not available.
[h]An application for decommissioning has been filed.
[i]DOE/EIA. Total U_3O_8 production by unconventional uranium facilities in 1986 is estimated to be 2030 MT. Site-by-site data for 1986 were not available.

2.7 PHOSPHOGYPSUM WASTES

Phosphate rock contains relatively high concentrations of uranium. When the phosphate ore is mined, the uranium-bearing material is brought to the earth's surface, crushed, and processed by screening and flotation. The waste products include slimes and sands similar to uranium mill tailings.

Phosphate mines have operated in the United States in Florida, Tennessee, Idaho, Montana, Utah, North Carolina, and Wyoming. The Florida deposits occur near the surface and strip mining and return of the tailings as fill in the mined-out areas are common practices. Treatment of the phosphate rock to produce phosphoric acid for use in fertilizers produces insoluble gypsum (calcium sulfate), which is also generally deposited with the mine tailings. The National Council on Radiation Protection and Measurements reports that the slimes contain about 50% of the radioactivity present in the original ore. An additional 12% is retained in the sand (NCRP 1984). Radium concentrations of 30–40 pCi/g have been measured in the gypsum by-product.

Several facilities have been operated to process the phosphoric acid stream to remove the contained uranium for commercial sale. Use of this by-product production for uranium reduces the amount of activity present in the tailings piles and mitigates the waste management demands (NRC 1979). It also reduces the uranium concentration of the ammonium phosphate fertilizer product.

As discussed with respect to uranium mill tailings, the phosphogypsum wastes represent a large volume of material with relatively high concentrations of radionuclides easily accessible to humans. Homes have been built over reclaimed phosphate mine areas in Florida and the gypsum piles are estimated to be accumulating at a rate of 11 million MT per year. Cumulative gypsum piles contained about one billion MT through 1983 (NCRP 1984). NRC estimates that 3.6×10^4 Ci of radon are released annually from land reclaimed after phosphate mining (NRC 1979).

REFERENCES

(Act 1980)   94 STAT.3347, Public Law 96-573 "Low-Level Radioactive Waste Policy Act." December 22, 1980.

(Anderson 1978) Anderson, R. L., et al. "Institutional Radioactive Waste." NUREG/CR-0028, Prepared for U.S. NRC by University of Maryland, March 1978.

(Beck 1979)   Beck, T. J., et al. "Institutional Radioactive Wastes, 1977." NUREG/CR-1137, Prepared for USNRC by University of Maryland, October 1979.

M. J. Bell *ORIGEN-The ORNL Isotope Generation and Depletion Code.* U.S. AEC report ORNL-4628, Oak Ridge National Laboratory, Oak Ridge, TN (May 1973).

(Clark 1978)   Clark, W. E., "The Use of Ion Exchange to Treat Radioactive Liquids in Light-Water-Cooled Nuclear Reactor Power Plants." NUREG/CR-0143, prepared by Oak Ridge National Laboratory for U.S. NRC, August 1978.

(CRCPD 1982)   "1982 Low-Level Radioactive Waste Management Survey Summary." Prepared for U.S. DOE by Conference of Radiation Control Program Directors, 1979.

(CRCPD 1984)   "The 1983 State-by-State Assessment of Low-Level Radioactive Wastes Shipped to Commercial Disposal Sites," DOE/LLW-39T, December 1984.

(DOE 1980a)   Nuclear Proliferation and Civilian Nuclear Power. U.S. Department of Energy, DOE/NE-0001/9, Volume 9, 1980.

(DOE, 1980b)   Final Environmental Impact Statement on U.S. Spent Fuel Policy, Volume 2, Storage of U.S. Spent Power Reactor Fuel. DOE/EIS-0015, May 1980.

(DOE 1980c)   Final Environmental Impact Statement, "Management of Commercially Generated Radioactive Waste." DOE/EIS-0046 F, U.S. Department of Energy, Washington, D.C., October 1980.

(DOE 1983)   "Spent Fuel and Radioactive Waste Inventories, Projections, and Characteristics," DOE/NE-0017/2, U.S. Department of Energy, September 1983.

(DOE 1985a)   Mission Plan for the Civilian Radioactive Waste Management Program. DOE/RW-0005, Volume 1, June 1985 (Figure 1-1).

(DOE 1985b)   *An Evaluation of Commercial Repository Capacity for the Disposal of Defense High-Level Waste.* DOE/DP-0020, Office of Defense Programs, January 1985.

(DOE 1987)   "Integrated Data Base for 1987: Spent Fuel and Radioactive Waste Inventories, Projections, and Characteristics." DOE/RW-0006, Rev 3, U.S. Department of Energy, September 1987.

(EG&G Idaho, Inc. 1985)   "The State-by-State Assessment of Low-Level Radioactive Wastes Shipped to Commercial Disposal Sites." DOE/LLW-50T, December 1985.

(Lamarsh 1983)   Lamarsh, J. R. "Reactor Licensing, Safety, and the Environment," Chapter 11. *Introduction to Nuclear Engineering*, 2nd ed. Addison Wesley, 1983.

(Matuszek 1982)   Matuszek, J. M. "Radiochemical Measurements for Evaluating Air Quality in the Vicinity of Low-Level Waste Burial Sites—The West Valley Experience." Presented at the NRC Symposium on Low-Level Waste Disposal, June 1982, NUREG/CP-0028 CONE-820674 Volume 2.

(NCRP 1984)   "Exposures from the Uranium Series with Emphasis on Radon and Its Daughters." National Council on Radiation Protection and Measurements, Report No. 77, March 1984.

(NRC 1979)   "Generic Environmental Impact Statement on Uranium Milling." NUREG-0511 U.S. NRC, April 1979.

(NRC 1980)   USNRC, "Final Environmental Statement Related to Primary Cooling System Chemical Decontamination at Dresden Nuclear Power Station Unit 1," Docket No. 50-10, NUREG-0686, October 1980.

(NRC 1981a)   U.S. Nuclear Regulatory Commission 1981. Notice of Proposed Rulemaking, 10 CFR Part 60: "Disposal of High-Level Radioactive Wastes in Geologic Repositories," *Federal Register*, 46 (130), pp. 35280 ff.

(NRC 1981b)   U.S. NRC. "Draft Environmental Impact Statement on 10 CFR Part 61 Licensing Requirements for Land Disposal of Radioactive Waste." NUREG-0782, September 1981.

(NRC 1982)   U.S. NRC. "Final Environmental Impact Statement on 10 CFR Part 61, Licensing Requirements for Land Disposal of Radioactive Waste." NUREG-0945, November 1982.

(NRC 1983)   U.S. NRC Office of State Programs Memorandum to State Officials. "The Impact on Low-Level Radioactive Waste Burial Sites of Waste from Decommissioning Commercial Light Water Power Reactors." March 1983.

(Oak 1980) Oak, H. D. et al. "Technology, Safety and Costs of Decommissioning a Reference Boiling Water Reactor Power Station." NUREG/CR-0672, Prepared by Battelle Pacific Northwest Laboratory for U.S. NRC, June 1980.

(Phillips 1979) Phillips, S., et al. "A Waste Inventory Report for Reactor and Fuel Fabrication Facility Wastes." ONWI-20, Prepared by NUS Corporation for Battelle Office of Nuclear Waste Isolation, March 1979.

(Rankama 1954) Rankama, K. *Isotope Geology.* McGraw-Hill, New York, 1979.

(Smith 1978) Smith, R., et al. "Technology, Safety and Costs of Decommissioning a Reference Pressurized Water Reactor Power Station." NUREG/CR-0130, Prepared By Battelle Pacific Northwest Laboratory for U.S. NRC, June 1978.

(Weir 1986) Weir, G. J., Jr. "Characteristics of Medically Related Low-Level Radioactive Waste." DE-FGO7-5IDO12605, Prepared for U.S. DOE by the American College of Nuclear Physicians, July 1986.

(Wild 1981) Wild, R. E., Oztunali, O. I., Clancy, J. J., Pitt, C. J., and Picazo, E. D. "Data Base for Radioactive Waste Management." NUREG/CR-1759, Waste Source Options Report, Prepared by Dames & Moore for USNRC Division of Waste Management, November 1981.

CHAPTER 3

REGULATION AND SITING OF WASTE MANAGEMENT FACILITIES

3.1 HISTORICAL PERSPECTIVE

As with many other technological advances, initial enthusiasm and effort among the scientists, government officials, industrialists, and technicians involved with radioactive materials were narrowly channeled into the primary challenge of separating materials, determining their properties, and developing devices that would allow useful applications of the materials. Handling and disposal of waste materials were often considered ancillary procedures that could be accomplished using methods in existence for nonradioactive materials. Even in those cases in which special attention was given to the safe containment of highly radioactive materials, plans were made for storage in liquid form in underground steel tanks that provided short-term isolation and radiation shielding. Provision was made for spare tanks into which liquid waste could be pumped when a primary storage tank reached the end of its useful life. More detailed management plans did not evolve until the 1960s.

The timing of changes in the waste management programs was the result of the interaction of several different forces:

- Continued experience provided the data necessary to characterize waste streams and identify physical, chemical, and radiological factors important to the design, construction, and operation of waste management facilities.
- Waste was accumulating to the point at which initial storage would have to be expanded and therefore it was appropriate to reassess then available treatment and disposal technologies to provide long-term isolation in the second-generation facilities. Experience at some first-generation facilities

indicated a need for additional controls in subsequent operations or facilities.

- The transition to a commercial nuclear industry, made possible by the Atomic Energy Act of 1954 (the Act) was beginning to occur. Cooperative government/industry research, development, and demonstration programs were being followed by completely commercial ventures.
- Commercial operations, in turn, required the promulgation and enforcement of regulations. Several states negotiated Agreements for Licensing with the (then) Atomic Energy Commission as authorized in Section 274 of the Act. Some waste management activities were therefore regulated by state agencies under rules compatible with federal requirements. In other states there was direct regulation by the Atomic Energy Commission. Naturally occurring and accelerator-produced material (NARM) was not subject to the Atomic Energy Commission's control but was regulated by the states in accordance with their responsibility to protect the health and safety of their citizens.
- Improvements in computing capabilities and continued refinement of health impacts models and environmental transport models made detailed assessment of long-term performance and impact of waste management facilities a more exact science.

Interest in and support for the development of improved waste management technologies increased as the decade progressed. This was consistent with the increasing awareness of general waste management issues and environmental concerns nationwide. Projections indicated a rapid and accelerating growth in the private nuclear industry. Regulators wanted to consolidate the lessons learned from existing governmental and industrial facilities and apply them to new facilities as well as to upgrade existing practice at government and early industrial sites. One major change was to require solidification of high-level liquid waste for eventual disposal in a federal repository. Without both the schedule and secrecy constraints of wartime operations, the siting and design of new facilities would address waste management as an integral part of the overall process or installation. Passage of the National Environmental Policy Act of 1969 (NEPA) required additional analysis of the planned interaction of the facility and its environment if a major federal action—such as issuing a license to construct, modify, or operate—was involved. These analyses were documented in Environmental Impact Statements (EISs) that became a part of the licensing record.

Establishment of the Environmental Protection Agency (EPA) by Reorganization Plan No. 3 in 1970 broadened the number of agencies with responsibility for waste management. Interagency agreements, known as Memoranda of Understanding (MOU), are one method by which different agencies negotiate and clarify their respective responsibilities and provide guidance on how each will fulfill its legislative mandate. Other changes were also made in the governmental structure and personnel responsible for waste management, sometimes for reasons totally unrelated to waste management, per se. For example, the separation of the Atomic Energy Commission into the Nuclear Regulatory Commission (NRC) and the Energy Research and Development Administration (ERDA) by the Energy Reorganization Act of 1974 addressed concerns about the agency's being able to fulfill its charge both to promote the peaceful uses of atomic energy and to regulate those uses to protect the national defense and security and public health and safety. It also broadened the scope of the agency's research and development activities to other energy sources, an important consideration in the aftermath of the 1973 Arab oil embargo and the uncertainty about future availability and cost of conventional energy sources. For waste management, as with other programs, the separation further isolated decisions made regarding waste produced and managed by the government, now through ERDA, and

TABLE 3-1   Chronology of Major Events Affecting Radioactive Waste Management in the United States

Year	Event
1954	Passage of the Atomic Energy Act
1963	First commercial disposal of LLW
1966	Colorado Department of Health identifies high radon levels in structures in Grand Junction
1967	DOE sites begin to store TRU retrievably
1970	National Environmental Policy Act becomes effective
	Environmental Protection Agency formed
	Surgeon General issues Guidelines for Corrective Action at Grand Junction
	AEC issues regulation on solidification and disposal of HLW (10 CFR Part 50 Appendix F)
	AEC requires waste containing TRU at concentrations greater then 10 nCi/g to be stored retrievably rather than disposed as LLW
1974	Atomic Energy Commission split into NRC and ERDA
1975	WIPP proposed as unlicensed defense TRU disposal facility
	West Valley LLW disposal facility closed
1977	President Carter deferred reprocessing pending review of the proliferation implications of alternative fuel cycles
1979	Report to the President of the Interagency Review Group on Radioactive Waste Management
	Three Mile Island 2 reactor accident
1980	Low Level Radioactive Waste Policy Act
	Commercial disposal of TRU waste ends
1982	Nuclear Waste Policy Act
	10 CFR Part 61 issued as final regulation for LLW
1985	Low Level Radioactive Waste Policy Amendments ACT
1986	Chernobyl reactor accident
1987	Nuclear Waste Policy Act Amendments provide for sequential rather than simultaneous HLW site characterization; Yucca Mountain, Nevada is the first site to be characterized

those applied to licensees of the NRC. Further, the former Atomic Energy Commission was responsible to a single congressional body, the Joint Committee on Atomic Energy (JCAE), which simplified program review and appropriations. The newly created agencies were subject to oversight by several committees in both the House of Representatives and the Senate.

Table 3-1 lists major milestones affecting radioactive waste management in the United States. Prior to 1954, activities producing radioactive waste included the concentration and application of radium for medical and industrial uses (e.g., the production of luminous timepieces and dials) and processing of uranium and thorium ores for pigment and luminescence in glassware and ceramics. Regulation of such activities was the responsibility of local or state agencies, generally the Health Department. The bulk of the mining, milling, and processing of uranium during this period was primarily performed on behalf of the U.S. Army's Manhattan Engineering District (or the AEC after 1946) for weapons production. Some foreign uranium was also processed for the Manhattan Engineering District.

The Atomic Energy Act of 1954 provided a framework for commercial development of radioactive materials applications and regulation of those applications by the AEC or Agreement States. The regulations provided that transfer of licensed material could be made only to a licensed recipient. Other possible methods for disposing of radioactive materials included release to air, water, or the sanitary sewer system in stringently controlled amounts and concentrations. Provision was also made to allow onsite disposal of material if it could be shown to the regulator's satisfaction that limits on exposure of individuals would not be exceeded due to the material so disposed.

In 1966 the Colorado Department of Health determined that uranium mill tailings had been routinely removed from a mill in Grand Junction, Colorado and used as landfill and construction material. The tailings were attractive because they were clean, accessible, relatively homogeneous, and cheap. As a result, several hundred thousand tons of tailings had been removed during the period 1952–1966. The problem posed by such action was twofold:

- direct exposure due to γ radiation from the uranium and daughter isotopes present in the tailings; and
- buildup within enclosed spaces (such as basements and upper levels of homes and schools built on top of, or with, tailings materials) of gaseous radon-222, a decay product of uranium.

The second factor was the major concern in terms of public health impact. Radon gas is easily respirable and decays to daughter products that emit high-energy α particles that can impart a relatively high lung dose. (This is also the basis for the current attention being given to indoor radon concentrations due to natural sources.) Field investigations by the State of Colorado and the U.S. Public Health Service were conducted to identify the extent and magnitude of the local contamination and the Surgeon General issued guidelines for taking corrective actions as a function of external γ radiation levels and radon concentration in structures. These guidelines provided that corrective action was indicated when the external γ radiation levels in dwellings exceeded 0.1 mR/hr or the radon daughter concentration exceeded 0.05 Working Level (a Working Level is defined as the concentration of radon daughter products per liter of air resulting in the eventual emission of 1.3×10^5 MeV of α energy). Action may be suggested at external γ radiation levels between 0.01 and 0.1 mR/hr or radon daughter concentrations between 0.01 and 0.05 Working Level. It is estimated that over 700 structures qualified for corrective action based on these Guidelines (DOE 1980). Cleanup began in 1973 as a state-run program with significant financial support from the federal government as authorized by Congress in 1972 (P.L. 92–314). This provision was agreed to since much of the material had been produced incident to the government's uranium procurement program. The Grand Junction experience provided insight into the public health concerns with unregulated use of uranium mill tailings, development of corrective action guides, and cooperative federal/state remediation on a large scale. It also provided experience with the length of time and amounts of money involved in such programs. As noted in Table 3-1, it was 4 yr after the problem was identified by the Colorado Department of Health that the Surgeon General issued guidelines for identifying the areas and structures that should be subject to corrective action. Legislation passed in 1972 (and subsequently amended in 1978) provided that 75% of the cost of the program would be borne by the U.S. DOE and 25% by the State of Colorado. The state was responsible for actually performing the corrective actions. The total program was expected to last some 20 yr since the problem was identified and to cost over $20 million (DOE 1979).

Production and processing of uranium during the 1950s and early 1960s was primarily performed in support of the government's weapons program. Uranium deposits are generally small and dispersed at relatively low concentrations compared to other minerals such as coal or copper. Such deposits are economically minable because uranium is a concentrated energy source. Mills were developed to support several nearby mines. Because of the small scale of operation, however, many deposits and related milling capacity were opened, mined, and closed at a time when the public health implications of the resulting tailings were not fully understood or appreciated. Local development subsequent to the closure of some of the facilities resulted in potential exposure to large numbers of people. However, because the facilities were closed and the owners had complied with then-existing rules (when such owners could still be located), it was unclear how corrective action could be accomplished and financed. The benefit to national defense and security of producing the material was further cited in support of assigning responsibility for the sites to the federal government. The issue was finally addressed by Congress in 1978 with the passage of the Uranium Mill Tailings Radiation Control Act (P.L. 95–604) that made the cleanup and recovery of these sites the responsibility of the DOE. The lessons learned with the inactive sites and the stricter standards for stabilization of the tailings were extended to those sites still in operation to prevent the same conditions from being repeated at different locations.

Similar circumstances surrounded sites at which material had been processed to support the operations of the Manhattan Engineering District (MED) and subsequent government procurement and processing programs. Waste was generally disposed onsite. This practice produced several benefits to the individual program including maintaining government control over the material, minimizing handling and related occupational exposure, minimizing transport-ation, and minimizing cost. Many of these facilities were decontaminated to then-applicable limits and released for other use. Other facilities continued to be owned by the government but were no longer used. Studies performed in the mid-1970s by the then-AEC identified 31 sites that had once been used by the MED or AEC and required some kind of remedial action. This is being accomplished under the Formerly Utilized Sites Remedial Action Program (FUSRAP). An additional 20 sites were the ongoing responsibility of DOE and decontamination and decommissioning of these facilities are being accomplished under the Surplus Facilities Management Program (SFMP).

Part of the problem with radioactive waste management results from the long lead times needed for evaluation of technologies, site characterization, and development. During that time period, other events may occur that, while peripheral to waste management per se, have a fundamental effect on how the program proceeds. This was the case with high-level waste that was originally stored as a liquid in underground tanks at the site of the reprocessing facilities (both government run and commercial). In 1970 the then-AEC promulgated a regulation (10 CFR Part 50 Appendix F) that required HLW be solidified within 5 yr after production and shipped to a federal repository for disposal within 10 yr after production. The agency also announced plans to develop the first such repository in a salt deposit in Lyons, Kansas. The Lyons site had been the location of studies in the mid-1960s of the effects of emplacing spent fuel rods in a geologic setting (Project Salt Vault was conducted from 1965 to 1967). The announcement of the site, however, preceded detailed site investigation. The site was rejected in 1972 because of the presence in the vicinity of wells previously used for solution mining of the salt deposit and uncertainty over the effects of those operations on the long-term integrity of the repository. The Lyons experience has resulted in increased skepticism and resistance on the part of state and local officials and individuals to siting any waste repository.

Subsequent actions to select a repository site and waste form and package have been subject to several revisions reflecting changes in agency responsibility and authorization, independent reviews, and congressional directives. Re-organization of the AEC into the Nuclear Regulatory Commission (NRC) and the Energy Research and Development Administration (ERDA) was a logical point for internal review of the program pace and direction. Several reviews of available technology, the need for additional data, and institutional requirements for successfully implementing the program were performed in the late 1970s by committees of the National Academy of Sciences (NAS 1979), the American Physical Society (APS 1978), and several federal government agencies (IRG 1979). Together with the Nuclear Waste Policy Act of 1982 (P.L. 97–425), these studies have been instrumental in structuring the high-level waste program that exists today and that is discussed in Section 3.3. It must be remembered, however, that the program is continually subject to change for technical, social, economic, and political reasons.

Without agreement on the required waste form and a schedule for availability of a federal repository, it was impossible for the only commercial reprocessing venture (the Nuclear Fuel Services facility at West Valley, New York) to comply with the requirements for solidification and transfer within the times stipulated in Appendix F to 10 CFR Part 50 as described above. Further, after shutting the facility down to expand capacity and incorporate building and equipment changes required since the original operating license was issued, Nuclear Fuel Services decided not to continue in the reprocessing business. Decisions on facility shutdown requirements and the long-term responsibility for the high-level waste existing in the underground tanks involved not only the operator and the NRC as the licensing agency but also the State of New York as the owner of the site and the Department of Energy as the sponsor of the demonstration thorium fuel that was the source of the THOREX waste, 5% of the waste volume in storage. Complex negotiations among all the parties resulted in the passage of the West Valley Demonstration Project Act (P.L. 96–368) in 1980. This law provides that solidification of the high-level waste and closure of the site will be accomplished by the Department of Energy as a demonstration of the feasibility of the technology. Other planned commercial reprocessing facilities were not brought into operation for a variety of reasons. General Electric's facility in Morris, Illinois was not opened as a result of a company internal decision regarding the feasibility of the technology they intended to use and the fact that the facility had been planned for contact rather than remote maintenance. As the Allied Gulf Nuclear Services (AGNS) facility in Columbia, South Carolina was nearing completion, the government policy on reprocessing was changed to accommodate review of the proliferation (potential for spread of material used in nuclear weapons) implications of reprocessing and recycling of plutonium fuel. By the time that review [the Nonproliferation Alternative Systems Assessment Program (NASAP) in the United States and the International Nuclear Fuel Cycle Evaluation (INFCE) worldwide] was completed and the official government position (as stated by President Reagan in October 1981) once again supported reprocessing, the economics were such that the company decided not to proceed on its own and no viable joint (government/industry or international) program was ever developed.

The lack of commercial reprocessing capacity, in turn, required extended onsite storage of spent fuel underwater in pools at the power reactor facilities. Such storage capacity was originally planned to accommodate about one full core plus one reload batch of fuel. This would allow for some 6 months of decay between removal from the core and shipment for reprocessing. The plant would also have the flexibility to manage emergency discharge of all the fuel in the core if necessary to protect health and safety. Such spent fuel pool storage volumes were inadequate in light of long-term absence of reprocessing capability. Many utilities responded by installing new, specially designed spent fuel storage racks incorporating neutron poisons. With the new racks, closer fuel spacing is possible without the potential for unplanned criticality. Onsite storage space was significantly expanded in this way and near term shutdowns of power reactors averted. Storage requirements are now defined by the schedule for availability of the federal high-level waste repository. Contracts between the utilities and the DOE specify that the government will begin to accept spent fuel—which has become high-level waste by default—in 1998. The pertinent program plans and regulatory requirements for disposal of spent fuel are discussed in Section 3.3.

Low-level radioactive waste (LLW) produced at federal facilities was initially disposed onsite. This practice provided benefits in terms of continuing government control over the site and material, not distracting from the primary projects of the facilities, and being the most cost-effective means of managing the waste. Shallow land burial was the disposal method of choice because it was an extension of conventional practice with other waste materials, it provided shielding for those waste materials with substantive external radiation levels, soil properties would generally retard the migration of contained radionuclides, and the need for transportation of waste was reduced.

There are five primary sites managed by the DOE at which LLW produced in federal government operations is disposed: Oak Ridge National Laboratory (Tennessee), Los Alamos Scientific Laboratory (New Mexico), Hanford Reservation (Washington State), Savannah River Plant (South Carolina), and Idaho National Engineering Laboratory (Idaho). Operations at these facilities began during or immediately following World War II and the sense of urgency and security surrounding early operations results in some uncertainty regarding total amounts and characteristics of waste disposed at these sites. In addition, smaller volumes of government waste were disposed of at other facilities around the country including the Nevada Test Site (Nevada), the Pantex Plant (Texas), Sandia Laboratory (New Mexico), Paducah Gaseous Diffusion Plant (Kentucky), Feed Materials Production Center (Ohio), Portsmouth Gaseous Diffusion Plant (Ohio), Weldon Springs Site (Missouri), Lawrence Livermore Laboratories (California), National Lead Company (New York), Brookhaven National Laboratory (New York), and the gaseous diffusion facilities in Oak Ridge (Tennessee). Not all government facilities dispose of LLW onsite, however. The Mound Laboratory (Ohio), Argonne National Laboratory (Illinois), Bettis Atomic Power Laboratory (Pennsylvania), and Rocky Flats facility (Colorado) are among DOE facilities whose LLW is now transported to other government sites for disposal.

Initially, the small volumes of LLW produced commercially were also disposed at government sites or licensed ocean disposal sites. In the early 1960s several commercial sites were licensed for operation (Beatty, Nevada and Maxey Flats, Kentucky in 1962, West Valley, New York in 1963, and Richland,

Washington in 1964. The Sheffield, Illinois site began operation in 1967. The Barnwell, South Carolina site opened in 1971.) Commercial sites accepted both commercial and government LLW during the 1960s and 1970s. This practice was discontinued in 1979 when commercial disposal capacity was recognized as being in potentially short supply (see subsequent discussion). Since that time, government-produced LLW has primarily been disposed at one of the five major facilities identified above. It should be noted that some waste produced at "government" facilities continues to be disposed at commercial sites. As described in Chapter 2, such waste inludes material from Veterans Administration hospitals and used luminous gauges from military equipment. It is similar in activity and physical and chemical characteristics to waste produced at civilian hospitals or industrial generators. Waste disposed at the government sites, on the other hand, may differ from the general commercial LLW streams because of the specialized processes producing it and the relatively long time over which the sites have operated. During the approximately 40 yr of operation for many of these sites, administrative changes and operating experience (documented in NUREG/CR-1759) have resulted in revisions to waste management practice as discussed in the following paragraphs.

LLW management at the Oak Ridge National Laboratory (ORNL) includes storage and disposal of material produced both onsite in government operations (beginning in 1943) and offsite in commercial applications. Six local Solid Waste Disposal Areas (SWDA) have been identified within the ORNL site. General characteristics of the site that have influenced the performance of the SWDAs include relatively shallow groundwater, high annual precipitation (about 55 in.), and a relatively humid environment. Significantly less moisture (31–35 in.) evaporates (measured as potential evapotranspiration) from the area than is incident upon it and therefore there is an important component of infiltrating water available as a possible leaching and mobilization mechanism for the contained radioisotopes in the disposal areas. Because there is no significant regional aquifer below the disposal areas and the site soils have relatively high ion-exchange capability, isotope migration rates are low and impact on the biosphere is limited. A variety of conditions experienced at the several SWDAs on the ORNL site have resulted in changes in current and planned practice and, in some cases, entailed remediation of existing disposal units.

Intrusion of groundwater into disposal trenches has been observed in several of the areas. In particular, placing of additional fill on top of closed disposal trenches in SWDA 4 altered the preexisting groundwater contours and resulted in saturating the emplaced water. In some sections of the same SWDA, differences in permeability between the undisturbed soil around the trenches and the backfilled cover material has resulted in infiltrating water being in contact with waste for prolonged periods of time, leaching contained radioisotopes and eventually being discharged at the surface. This condition resulted in the release of an estimated 1–2 Ci/yr of ^{90}Sr in the mid-1970s. Similar surface discharges (seeps) have been observed at other sites within SWDA 5. The mechanism in this case is the "bathtub" effect in which water builds up in a trench over long periods of time due to the higher rates of infiltration into the trench than percolation out of the sides and bottom. For some of these trenches there was a difference in elevation of either end following the site's natural topography. This produced the driving force for surface seepage. In addition to ^{90}Sr (at the rate of about 1 Ci/yr), the transuranic elements ^{238}Pu and ^{244}Cm have also been observed in one of these discharges.

Water infiltration into the trenches increased as emplaced compressible waste degraded and created voids into which trench covers settled producing pathways for surface and groundwater intrusion. Initial mitigative actions have included combinations of altering surface drainage (by constructing ditches to divert surface water away from the disposal trenches), upgrading trench covers by installing impermeable barriers to water infiltration [both polyvinyl chloride (PVC) and a bentonite-shale layer have been used], and installing concrete dams in two trenches to reduce the flow from one end of the trench to the other. Postmitigation monitoring has indicated that the PVC liner and concrete dam combination has reduced the ^{90}Sr discharge rate from SWDA 5. Ongoing evaluation will enable later conclusions to be reached concerning the effectiveness of the bentonite-shale seals. It is important to recognize that DOE site limitations have always been complied with in spite of the localized radioisotope migration.

Several different disposal technologies have been used and are still in use at the ORNL facility. The largest portion of the waste volume has been disposed in unlined trenches in what is termed "traditional" shallow land burial. Particular wastes, notably α-contaminated materials, have been emplaced in concrete lined trenches and augured holes capped with concrete as well as unlined trenches with a concrete cap. The site changed its α-contaminated waste disposal practices in 1971 in keeping with the AEC directive on disposal of transuranic waste (TRU). Since that time TRU in concentrations greater than 10 nCi/g have been stored in structures for subsequent removal and permanent disposal. Liquid wastes have been placed in onsite storage ponds as well as injected as a slurry into shales underlying the site.

Waste disposal operations at Los Alamos Scientific Laboratory (LASL) benefit from the site's location and characteristics because the combination of low precipitation rates, generally deep groundwater levels, and relatively impermeable soil with good adsorptive properties means there is not a strong transport mechanism for migration via groundwater pathways. With the exception of possible long-term erosion near canyon edges, erosion is not a substantive force that might result in dispersion of the emplaced wastes. A variety of disposal technologies, including traditional shallow land burial, liquid waste ponds, and deep augured shafts, have been used at the site since disposal started in 1944. Problems encountered at the site included inadequate record keeping and identification of the earliest disposal areas. This inadequacy was common to all sites operating at the time. (It is also a characteristic that will be subject to continued refinement and upgrade as needs indicate or technological developments allow.) Changes in site operating procedures have included installation and application of volume reduction programs including sorting and segregating waste and compacting trash. Overall volume reductions of 20–25% have been achieved. Liquid waste is no longer disposed into seepage pits. Such waste is now mixed with cement and disposed in deep augured holes. The previous practice of disposing of hazardous chemical waste with radioactive waste has been identified as a probable contributor to fires in disposal trenches, which resulted in temporary airborne contamination. Current practice provides for segregated disposal of hazardous and radioactive waste and covering waste with soil shortly after emplacement. One no longer used disposal area has been surfaced with asphalt and reused as a parking lot by the County of Los Alamos which leases the area from LASL.

The large size (365,000 acres) of the Hanford Reservation (HR) is a major influence on waste disposal operations there. Without the space constraints of later sites in more densely populated areas it was possible to use disposal methods that might otherwise not be available. For example, two tunnels onsite house railway spurs containing flat cars loaded with large, heavy, or highly contaminated (primarily with activation products although TRU and non-TRU materials are present) equipment. Over 65 individual disposal sites have been used in several main areas since disposal began in 1944. Wastes disposed include equipment and machinery from the production reactors, construction debris, and liquids as well as waste more typical of industrial operations utilizing radioactive materials.

Like the LASL site, HR has a relatively low annual average precipitation rate (6.3 in.) and is in a dry climate so potential evapotranspiration exceeds precipitation. Site soils are relatively homogeneous and have high adsorptive capacities. The depth to groundwater varies but is large in the areas currently used for disposal. (Previously used areas may be inundated under probable maximum flood conditions.) The unsaturated soils at the site, however, have relatively high permeabilities and are subject to substantial wind erosion that is of concern in ensuring long-term site stability.

A major difference between disposal at this site and at other government sites is the continuing (although at reduced rates from past practice) disposal of liquid waste in settling ponds and cribs. (A crib is a long ditch about 20 ft deep that is backfilled with rock and covered with an impermeable membrane and soil. Liquid waste is uniformly dispersed into the crib through a perforated pipe.) The combination of distance to groundwater and soil adsorptive properties provides significant reduction in radioisotope concentrations as the liquids percolate through the soil to the groundwater.

In addition to the traditional shallow land disposal trenches and the liquid disposal in cribs or ponds, underground caissons of reinforced concrete are used for disposal of higher activity waste. Other variations are utilized as necessary because of the size or activity of the waste or for security reasons. TRU wastes in retrievable storage are stacked (in drums and boxes) on asphalt pads separated by fire-retardant plywood. The stack is covered with a polymer membrane and over 4 ft of overburden. High-activity waste is stored retrievably in underground caissons.

During the course of waste management at HR over the past 40 plus yr problems have arisen from spills and leaks of liquid waste resulting in surface contamination as well as in fires in disposal trenches and a waste storage area. Trench cover designs now incorporate a layer of stone to minimize wind erosion and reduce intrusion by both burrowing animals and plants. These two phenomena had been identified as the transport mechanism of waste from the disposal areas to the local (onsite) environs. There is an extensive environmental monitoring program onsite and research programs on the behavior of the waste released into the biosphere.

The Idaho National Engineering Laboratory (INEL) occupies over 500,000 acres of land on the Snake River Plain in Idaho. Most disposal and storage of solid LLW and TRU waste occur within the 143-acre radioactive waste management complex (RWMC). About 88 acres have been designated as a subsurface disposal area (SDA) for LLW. About 20 acres of this area is still available for use. The transuranic storage area (TSA) occupies about 55 acres of which 44 are still available. The specific portion of the INEL site selected for waste disposal was chosen for a combination of ease of access (that is, it did not require major new road construction), substantial thickness (about 20 ft) of unconsolidated sediment with good adsorptive capacity, a single area encompassing tens of acres, and soils that would be easily excavated. Waste disposal at the RWMC began in 1952. The area has received both waste from operations at INEL (primarily mixed fission products and activation from research reactor operations onsite) and waste from other government sites. TRU-containing waste from the Rocky Flats, Colorado site began to be shipped to INEL in 1954. The LLW portion of the site was expanded from an original 12 acres to the present 88 acres in 1957. Another small disposal area was opened and operated from 1961 to 1962 to receive material produced during the recovery and decontamination following the SL-1 reactor excursion. The primary purpose of using the additional site was to reduce occupational exposure to the highly radioactive material during handling and transportation.

Procedures for record keeping, handling, and marking of disposal trenches were formalized in 1957 with the issuance of AEC-ID Manual Chapter 0500–7. These procedures include definition of routine waste as anything that emitted less than 500 mR/hr at a 1-m distance from the surface of the package. Any waste emitting more than 500 mR/hr at 1 m or requiring special equipment, hauling, or handling was defined as nonroutine waste. Source material, liquids, and slurries were all considered nonroutine waste.

The transuranic storage area (TSA) provides the ability to store TRU waste as required by the AEC in 1971. Storage is accomplished by stacking waste on aboveground asphalt pads that are placed over a layer of crushed gravel. Some TRU waste is also stored at the intermediate level transuranic storage facility (ILTSF) in concrete-filled carbon steel vaults. Some TRU buried prior to the rule change has been exhumed and placed into storage. Exhumation was conducted within double containment. That is, a structure was erected around the area of exhumation and the whole operation took place within an air-supported weather shield.

Experience at the RWMC included several fires in open disposal trenches and surface spills resulting in local contamination. The disposal areas have been flooded several times. During one flood in 1962 uncovered waste at the SL-1 disposal area was transported outside the area. The waste was successfully recovered. Changes in waste management practices include increasing the minimum trench depth, increasing the trench cover thickness, more frequent covering of emplaced waste to prevent fires, and employing compaction to improve the waste stability. Changes have also been made in surface drainage systems to enhance flood protection.

The Savannah River Plant (SRP) in South Carolina, like ORNL, is located in a humid environment. Precipitation is responsible for 20–22 in./yr of recharge to the groundwater. The soil in the area is relatively permeable and consists mostly of clayey sand with good adsorptive capacity. The normal depth to the water table ranges from 20 to 60 ft in the LLW disposal area which occupies about 195 of the 192,000 acres at SRP. Waste disposal and storage began at SRP in 1953. In addition to waste produced by the onsite facilities (production reactors, fuel fabrication, and reprocessing facilities) the site has received ^{238}Pu process waste from Mound Laboratory and LASL and radioactive debris from two United States airplane accidents involving nuclear weapons.

Waste disposal is primarily accomplished using traditional shallow land burial with waste sorted and segregated according to surface radiation levels. A minimum soil cover of 4 ft is placed over the waste and surface radiation levels are reduced to less than 6 mR/hr. Liquid wastes are released into seepage basins from which the liquid percolates into the ground. It reaches groundwater after several years of travel during which radioactive decay and ion exchange with the soil reduce the concentration of all isotopes but tritium to insignificant levels. Tritium levels in the closest streams (Upper Three Runs Creek and Four Mile Creek) are well within allowable limits.

TRU was first disposed at SRP in 1965. At that time waste containing <0.1 Ci/package continued to be disposed nonretrievably in designated "α" trenches. Packages containing >0.1 Ci were placed in concrete containers (or encapsulated in concrete) prior to burial. Subsequent to 1971, TRU waste containing >10 nCi/g has been stored in concrete containers, steel boxes, and galvanized steel drums stacked on reinforced concrete pads. Filled pads are covered with a multilayer barrier that includes sand, soil, plastic sheeting, and 4 ft of overburden. This covering is then topped with a layer of asphalt and additional soil and a vegetative cover is sown. Other TRU exists in spent solvent from the reprocessing facilities (about 150,000 gallons containing about 45 Ci as of 1977) stored in underground tanks.

Experience at SRP has been similar to that at other government LLW disposal facilities in that several fires have occurred in disposal trenches and a few spills and overflow of rainwater collected in open trenches have occurred. The resulting surface contamination was removed and disposed in all cases. Cover thickness has been increased over disposal trenches to reduce uptake of radioactivity into plant roots. Deep-rooted vegetation has been replaced with short-rooted species to reduce the possibility of plant uptake further.

The NRC licensed a portion of the West Valley site for disposal by burial of radioactive waste produced in the reprocessing facility. Monitoring of this area identified migration of radionuclides from the trenches. The major contributor to the movement was the presence of organic liquids from the reprocessing operation. Nuclear Fuel Services also operated a commercial LLW disposal facility at another location on the West Valley site. This operation, however, was directly regulated by the State of New York under its licensing agreement with the Nuclear Regulatory Commission. One of the key site characteristics affecting LLW disposal at West Valley was the highly impervious clay till in which the wastes were disposed. The soil characteristics meant that water would not move away from the disposal trenches quickly and the clay provided strong ion-exchange capability to remove most radionuclides that might be present in any leachate. It became evident as operation proceeded, however, that the rate of water infiltration into the filled trenches was greater than that leaving the trench (through the soil at the sides and bottom). This was due in part to slumping of the trench covers as biodegradable waste decomposed and the covers filled in gaps left among the waste. As a result, water was collecting in the trench (the "bathtub" effect) and leaching radioisotopes from the emplaced waste. While the reprocessing facility was in operation, water accumulating in trenches was periodically pumped for treatment in the liquid waste system prior to release. After shutdown of the reprocessing plant, this was no longer performed on a routine basis. In 1975 water overflowed a filled trench and was discharged to the surface at the rate of about 1 gallon/day. Nuclear Fuel Services ceased accepting LLW at that time. Measurements have also indicated that gaseous products of biological decomposition of the waste containing tritium and ^{14}C have evolved through the trench covers and may represent a significant release pathway for this site (Matuszek 1982). Further studies are planned to determine the extent of such releases. Subsequent care involved pumping and treating the leachate prior to release of water and changes to trench cover construction to reduce the rate of water infiltration. Efforts are continuing to stabilize the trenches and provide continuing monitoring and care to document that the site continues to isolate the contained radioisotopes effectively.

At the Maxey Flats, Kentucky LLW disposal facility, a combination of site characteristics (high annual rainfall, low soil permeability, complex geology), waste form (easily degradable containers, loosely packed waste, chemically mobile species), and operating practices (burial of waste containers with significant void spaces, allowing rainwater to accumulate in trenches during burial, imprecise marking of trench locations after closure) resulted in water infiltrating the covers of the closed trenches and accumulating in contact with the disposed waste. The water management program instituted at the site included pumping of leachate from the trenches and processing through an evaporator to concentrate the radioisotopes. The concentrates were subsequently solidified and disposed as LLW. This site has been the subject of studies by a number of agencies including the State of Kentucky, the Environmental Protection Agency, and the NRC. Field measurements have identified ^{3}H, ^{14}C, ^{89}Sr, ^{90}Sr, ^{134}Cs, ^{137}Cs, ^{238}Pu, and ^{239}Pu in the unrestricted environment. The agencies concluded, however, that the levels involved did not represent a significant risk to public health and safety (NRC 1975; NUREG-0217; Blanchard 1978; KDHR 1976). Acceptance of waste for disposal at Maxey Flats ceased in December 1977 and the State of Kentucky assumed responsibility for its long-term care in May 1978. Since that time, custodial contractors operating on behalf of the state have continued to remove and evaporate leachate from the disposal units as well as performing large-scale remediation in the form of trench cover improvements. Current efforts to define, implement, and pay for the actions needed to decontaminate the site and environs and stabilize the disposal units are being carried out under the provisions of the Comprehensive Environmental Reclamation, Compensation and Liability Act (CERCLA) and the auspices of the U.S. EPA because it was determined that hazardous chemicals were present as well as radioactive materials.

The Sheffield, Illinois LLW disposal site was also closed in 1978 when the existing licensed area was filled. Field investigations at this site have identified ^{3}H migration into the unrestricted environment (ITF 1979). Site problems include a geology that was significantly more complex than originally perceived. For example, sand lenses that acted as conduits for rapid groundwater movement with little ion exchange and radionuclide retardation were identified subsequent to the start of site operation. Significant subsidence and erosion also require ongoing active maintenance. Responsibility for long-term care of the site rests with the State of Illinois. Acceptable conditions for transfer of responsibility and control from the site operator to the State were the subject of detailed consideration as part of the proceeding to terminate the NRC license.

Experience at the three operating LLW disposal facilities (at Barnwell, South Carolina, Beatty, Nevada, and Richland, Washington) has not identified problems related to site characteristics. There have been concerns over management control of material onsite as well as of the condition of packages and trucks arriving at the sites. Such concerns were dramatized by the simultaneous closing of all three sites by the respective host state governors in 1979. Increased enforcement action at the federal level addressed existing requirements of both NRC and DOT and improved the performance of the packages in-transit and upon receipt at the disposal site. The host states also expanded their programs to certify waste shippers and to enforce packaging and shipping standards. The State of South Carolina also imposed a cap on the volume of waste that could be received annually at the Barnwell facility. The level was approximately half that received in 1979. To accommodate this limit, the site operator prorated this capacity among then-existing customers. Therefore no new customers could be accommodated nor could increased volumes from a particular generator. The

State of Nevada instituted an extensive third-party inspection system whereby the waste characteristics and packaging were independently verified. The costs of this inspection were borne by site users and resulted in significantly higher disposal costs at the Beatty facility than at either of the other two operating facilities. As would be expected, the volumes received at the Beatty facility are the lowest of the three operating sites. Statements by the Governors of the three states and referenda proposals to close the operating facilities clearly indicated an unwillingness on the part of these three states to continue to accept all of the nation's LLW for disposal.

The possibility of repeated unilateral shutdown of the existing LLW disposal sites by the host states raised concern over the impact on health and safety of material stored at the site of generation. Recognizing the benefits of the processes that result in producing the waste (for example, medical and academic research, consumer products such as smoke detectors and luminous safety signs, generation of electric power, and radiopharmaceuticals), there was a need to develop a system to provide for continued safe disposal of the waste. With the passage of the Low-Level Radioactive Waste Policy Act of 1980 (P.L. 96–573) Congress assigned to the individual states the responsibility for ensuring the availability of adequate disposal capacity for the LLW generated within their borders. This included government waste except for that owned or generated by the Department of Energy or the U.S. Navy (from decommissioning nuclear vessels) or produced from research, development, testing, or production of atomic weapons. States were encouraged to adopt interstate compacts to form multistate regions to manage all of the waste at a single site in the region rather than having a site in each state. There are several advantages to a regional approach:

1. It better matches practical site size to the rate of waste production than having many small sites.
2. It provides cost advantages because much of the disposal costs are essentially unaffected by the amount of waste the site receives.
3. It enhances public protection by limiting the demand for trained operating personnel nationwide.
4. Several regional facilities will result in lower transportation requirements than would fewer facilities nationwide.

Compact formation was encouraged by a provision that congressionally approved compacts could exclude out of region waste as of January 1, 1986. This gave particular incentive to negotiate with states in which a site was already operating since the time needed to develop new facilities would make it difficult to meet the January 1, 1986 deadline. As it became evident that new disposal capacity would not be available by that time, there was strong action in Congress not to ratify those compacts that had been submitted. Without a ratified compact, the host states could not exclude out of region waste because that would violate the Interstate Commerce Act. Subsequent amendments to P.L. 96–573 were

TABLE 3-2 Milestones in the Low-Level Radioactive Waste Policy Amendments Act of 1985 (P.L. 99–240)

Date	Action Required
July 1, 1986	Either ratify compact legislation or pass legislation or provide Governor's certification that the state will develop system to manage its own waste
January 1, 1988	Identify the host state for the region's disposal facility or select a developer for the facility and the site to be developed
	Regardless of which approach is used for host state selection, develop a siting plan for the facility including schedules and procedures for having the facility available when needed and delegate authority to implement the plan
January 1, 1990	File a complete license application to operate the facility or provide Governor's written certification that the state will be capable of providing for, and will provide for, the storage, disposal, or management of waste after 12/31/92. A description of actions to be taken to ensure that such capacity exists is also needed
January 1, 1992	File a complete license application to operate the disposal facility
January 1, 1993	Begin to accept waste from the region or state

passed in late 1985. The Low-Level Radioactive Waste Policy Amendments Act of 1985 (P.L. 99–240) revised the time at which Compacts could deny access to waste from outside the region to January 1, 1993. The Amendments Act further specified detailed interim milestones measuring a state's or compact region's progress toward meeting the 1993 deadline. A system of economic incentives was provided for meeting the milestones and having new disposal capacity available as soon as possible. Provision was also included for the imposition of penalty surchages for missing an interim milestone. Table 3-2 summarizes the milestone requirements of the Amendments Act.

A graduated system of surcharges was imposed on out of region waste to encourage development of new capacity as quickly as possible and help compensate the states of South Carolina, Nevada, and Washington for the additional access to existing facilities by out of region generators. Out of region disposal in 1986–1987 is subject to a $10/ft^3 surcharge. This rises to $20/ft^3 for 1988–1989 and $40/ft^3 for 1990–1992. As an incentive to meet the interim milestones, 25% of the surcharge is held in an escrow account by the Department of Energy to be refunded upon a finding that the compact region or individual state has met the milestone. An independent determination of compliance is made by the host states of the operating facilities (known as the "sited states"). This determination affects whether penalty surcharges are imposed or access is denied.

If a compact or unsited state fails to meet the January 1, 1988 milestone, generators will be charged double surcharges (i. e., $40/ft^3) for waste shipped for disposal between January 1, 1988 and June 30, 1988. From July 1 to December 31,

1988 the surcharge is quadrupled (i.e., $80/ft^3). A triple surcharge is applied for failure to meet the January 1, 1992 milestone (i.e., $120/ft^3). Immediate denial of access will be imposed if a host state is not chosen and a siting plan developed and submitted by January 1, 1989 and if the January 1, 1990 milestone is not met, either by submitting a complete license application or having the Governor certify that the state will assume responsibility as of January 1, 1993 for waste generated in the state.

There has been a wide range of responses to the requirements of the Low-Level Radioactive Waste Policy Amendments Act. Compact agreements with a state in which an operating facility already exists were relatively more desirable than those in which a new site was required. The State of Washington plans to allow long-term operation of the Richland site for Northwest compact region waste. It is expected that new sites will be developed in the Southeast and Rocky Mountain Compacts and that the Barnwell and Beatty sites will close. Several states (notably Texas, New York, and Massachusetts) have chosen to develop facilities to accept waste only from their own state. Others have joined interstate compacts that vary in numbers of member states and amount of waste requiring management and disposal. Appendix D identifies the current LLW compacts and outlines significant characteristics of the regions and the siting and development programs underway.

The choice among different treatment and/or disposal technologies—and the rate at which the waste management program for a given waste type proceeds—reflects decisions about the relative risks and costs of the alternatives, the environmental impact, the need for action, and the general availability of funds. As will be discussed in detail in the following sections, there have been changes in both the rate and the direction of all of the programs and it is expected that such changes will continue to occur in the future. The fundamental principles and technical considerations underlying waste management decisions as described in succeeding chapters are expected to remain relatively constant. The particular combination of actions applied, however, will reflect then-current concerns and opportunities.

3.2 REGULATORY FRAMEWORK FOR GOVERNMENT CONTROL

The United States Environmental Protection Agency (EPA) has, since it was formed in 1970, exercised authority and responsibility to develop and promulgate generally applicable environmental standards for protection of public health and the environment from radioactive materials. Such standards may include limits on radiation exposures to workers and members of the general public, and concentrations or quantities of radioactive materials in an uncontrolled area. The radiation control requirements within a licensed facility on the other hand—or the area controlled by the licensee around the facility—are the responsibility of the agency issuing the facility license in accordance with regulations that implement EPA's standards. At the federal level this is primarily the Nuclear

Regulatory Commission and the Department of Energy. DOE does not issue licenses but does regulate the actions of its contractors in accordance with formal requirements provided in the DOE Manual. These requirements are similar to those in the NRC regulations. There may be differences because of the specific work being accomplished or site conditions at a particular facility. State regulatory programs may reside in a variety of agencies such as the Health Department, Department of Environmental Protection, the Labor Department, or some combination thereof.

EPA's basic responsibility is derived from the Reorganization Plan No. 3 of 1970 that transferred authority to "advise the President with respect to radiation matters, directly or indirectly affecting health, including guidance to Federal agencies in the formulation of radiation standards" from the former Radiation Protection Council to the EPA. It was this authority that was exercised in January 1987 to establish revised criteria for occupational radiation exposure control. In addition, the agency is charged with exercising the authority previously delegated to the AEC to set generally applicable environmental standards. The agency's task is therefore to evaluate the data on the behavior of radioactive materials in the environment and the biological effects of exposure to radiation and identify levels that achieve a desired level of protection.

There are a variety of sources of information on which the agency bases its standards. These include the International Commission on Radiological Protection (ICRP), the United Nations International Atomic Energy Agency (IAEA), the United Nations Scientific Committee on the Effects of Atomic Radiation (UNSCEAR), the National Council on Radiation Protection and Measurements (NCRP), the National Academy of Sciences (NAS), in particular its Committees on the Biological Effects of Ionizing Radiation (BEIR Committee) and Radioactive Waste Management (in conjunction with the National Academy of Engineering), radiation protection agencies in other countries, agency-sponsored research, and the general scientific literature including the results of research performed by or for other agencies. As new studies are completed, the results are analyzed to determine the impact, if any, on standards in place or under development. Such results may also indicate a need for a new standard.

EPA standards for limiting exposure due to radioactive materials in the environment are published as Title 40 of the Code of Federal Regulations (40 CFR). The parts of these regulations of particular importance for radioactive waste management are 40 CFR 190, *Environmental Radiation Protection Requirements for Normal Operations of Activities in the Uranium Fuel Cycle*; 40 CFR 191, *Environmental Standards for Management and Disposal of Spent Nuclear Fuel, High Level and Transuranic Radioactive Wastes*; 40 CFR 192, *Environmental Standards for Management of Thorium and Uranium Mill Tailings*; and the forthcoming 40 CFR 193, *Environmental Radiation Protection Standards for Low Level Radioactive Waste Disposal*.

The fundamental standards to which any management and storage system for high-level waste, spent fuel, and transuranic waste must conform according to 40 CFR Part 191 is that "the combined annual dose equivalent to any member of the public in the general environment resulting from: (1) Discharges of radioactive material and direct and indirect radiation from such management and storage and (2) all operations covered by Part 190 [the uranium fuel cycle] shall not exceed 25 millirems to the whole body, 75 millirems to the thyroid, and 25 millirems to any other critical organ" (Section 191.03). As defined in this standard, transuranic waste is any waste containing more than 100 nCi of α-emitting transuranic isotopes, with half-lives greater than 20 yr, per gram of waste.

Essentially the same criteria (25 mrem whole body and 75 mrem to any critical organ) are to be met with "reasonable expectation" for undisturbed performance of a disposal system for 1000 yr after emplacement of the high-level waste, spent fuel, or transuranic waste. Further, the facility design shall provide a "reasonable expectation" that releases due to all significant processes and events will be limited for 10,000 yr after disposal. The 10,000 yr reference time was chosen because, as shown in Figure 3-1, after that length of time the spent fuel or

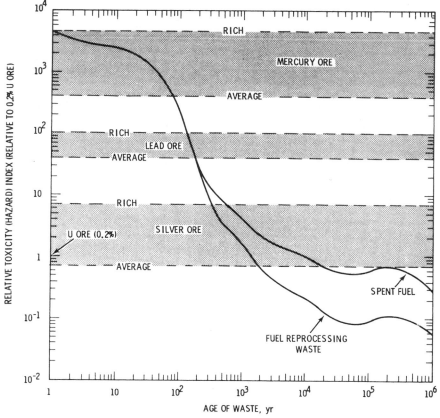

Figure 3-1 Toxicity of high-level waste and spent fuel as a function of waste age. Source: (DOE 1980a).

associated high-level waste will be similar in toxicity to the amount of uranium initially needed to produce the fuel. Therefore, there is effectively no long-term change in risk from the conditions that would exist if the ore had not been mined and the power not produced.

Recognizing the inherent uncertainty in projecting system performance for such long periods, the standard provides that a judgment must be reached that there is less than 10% probability that cumulative releases will exceed the quantities in Table 3-3 and less than 0.1% probability that the releases will exceed 10 times that amount. The term "unit of waste" is defined to enable the release limits to be appropriately adjusted for differences in characteristics of the several waste types addressed in the standard (i.e., spent fuel, high-level waste, and transuranic waste).

Groundwater contact with the waste emplaced in the repository may provide a mechanism for leaching of the contained radionuclides and transport through the geologic medium to the accessible environment. The following limits (as found in Section 191.16) are placed on the allowable concentrations of radionuclides in a "special source of groundwater" (defined as a Class I groundwater

TABLE 3-3 EPA HLW Standard for Cumulative Releases for 10,000 Years

Radionuclide	Release Limit per 1000 MTHM or Other Unit of Waste (Ci)
Americium-241 or -243	100
Carbon-14	100
Cesium-135 or -137	1,000
Iodine-129	100
Neptunium-237	100
Plutonium-238, -239, -240, or -242	100
Radium-226	100
Strontium-90	1,000
Technetium-99	10,000
Thorium-230 or -232	10
Tin-126	1,000
Uranium-233, -234, -235, -236, or -238	100
Any α-emitting radionuclide with a half-life greater than 20 yr	100
Any other radionuclide with a half-life greater than 20 yr that does not emit α particles	1,000

Source: (EPA 1985).

identified in accordance with EPA procedures that are within the controlled area of the disposal system or an area up to 5 km beyond the controlled areas, that supply drinking water for thousands of persons at the time a site is chosen, and are irreplaceable as a source of drinking water for that population).

For undisturbed performance of the disposal system for 1,000 years after disposal there must be a reasonable expectation that the annual average concentrations in water from a special source of ground water will not exceed:

(1) 5 picocuries per liter of radium-226 and radium-228;
(2) 15 picocuries per liter of alpha-emitting radionuclides (including radium-226 and radium-228 but excluding radon); or
(3) the combined concentrations of radionuclides that emit either beta or gamma radiation that would produce an annual dose equivalent to the total body or any internal organ greater than 4 millirems per year if such an individual consumed 2 liters per day of drinking water from such a source of ground water.

Development of the LLW standards in 40 CFR Part 193 is continuing at EPA and it is anticipated that a proposed regulation may be promulgated sometime in 1988. The regulation may substantially reduce the volume of material that is technically classified as LLW by identifying an amount or concentration level below which the risks of disposal because of the radioactive characteristics of the waste are of no regulatory concern. This determination, described in the literature as below regulatory concern (BRC) or "de minimis," is an attempt to channel finite economic and regulatory resources to achieve the most public protection possible. It is based on the legal principle originally expressed in Latin that "de minimis non curat lex" or "the law does not concern itself with trifles." In concept it is similar to the radiation principle of "As Low As Reasonably Achievable" (ALARA) because it recognizes that social and economic effects must be considered in decisions on how strictly to limit radiation exposure. It would effectively establish a floor for exposures below which reductions would not be necessary. Also to be included in the Part 193 standard are requirements for worker protection during predisposal management procedures, protection of the public after the closure of the disposal facility, and protection of groundwater from possible releases from a disposal facility. The standard will also provide requirements for disposal of certain types of naturally occurring and accelerator produced (NARM) waste in a manner similar to LLW regulated under the authority of the Atomic Energy Act of 1954, as amended. Authority for NARM regulation is based on provisions of the Toxic Substances Control Act.

The Nuclear Regulatory Commission (NRC) is responsible under the provisions of the Atomic Energy Act of 1954, as amended, for regulating the receipt, possession, use, and transfer of radioactive materials to protect public health and safety and the common defense and security of the United States. NRC regulations are contained in Title 10 of the Code of Federal Regulations (10 CFR). Part 20 of the NRC regulations (10 CFR Part 20) provides the basic standards for protection from radiation that underlie specific provisions in other parts that deal with particular types of material or processes for use (such as nuclear power reactors that are licensed in accordance with Part 50 of Title 10). Any waste management procedures must be conducted in conformance with the 10 CFR Part 20 requirements to limit public and occupational exposure. This is accomplished by a combination of limiting the amounts and concentrations of individual isotopes (or mixtures of isotopes) released to the environment, controlling the concentrations of isotopes within process equipment (and thereby the direct radiation exposure rate in the vicinity of the equipment), installing permanent or temporary shielding around equipment, and limiting the length of time a person is in an area in which it is possible to accumulate significant radiation exposure. Pertinent sections of 10 CFR Part 20 are summarized in the following paragraphs. [*Note*: The provisions cited here are those contained in a proposed revision to 10 CFR Part 20 that incorporates updated scientific information on the biological risk of radiation exposure and adopts more detailed methods for computing the effective dose equivalent resulting from both internal and external exposures. Application of these limits is intended to result in maintaining risk (based on estimated radiation-induced fatal cancers and serious hereditary disorders) to 10^{-4} per year for workers and 10^{-6} to 10^{-5} per year for members of the public.]

Occupational Limits
5 rem/yr (0.05 Sv/yr). This includes both external and internal exposure to the whole body and individual organs.

3 rem/quarter (0.03 Sv/quarter).

Exposures should be as far below the limits as reasonably achievable (ALARA).

Public Limits
0.5 rem/yr (5 mSv/yr) for individuals from all sources (internal and external and food pathways).

An individual licensee may demonstrate compliance with the previous limit by demonstrating that any radiation source "under the licensee's control will not result in an individual member of the public receiving a dose in excess of a 0.01 rem (1 milliSv) annual reference level" (Section 20.301).

Collective Dose
0.001 rem/yr (0.01 mSv/yr) per person cutoff level for evaluating collective doses to general population.

Monitoring
Required at 10% annual limit for deep dose equivalent.

Required at 30% of annual limit of intake (ALI).

Records
Occupational exposure history.

Current dose experience.

Reports

Immediate notification of NRC if a dose of five times annual dose limit is incurred.

Notification to NRC if an individual in an unrestricted area exceeds 0.5 rem (5 mSv) in 1 yr.

Notification to NRC for exceeding 0.1 rem (1 mSv) level to members of the public without prior approval.

Some licensees must submit annual statistical summaries of individual monitoring reports and termination reports.

Any reports to NRC are available to individuals involved upon request. Overexposure and termination reports are transmitted to individuals involved routinely.

Disposal of licensed material is regulated by 10 CFR Part 20, Subpart K, Sections 20.1001–20.1006 and 20.1008 (Records of Waste Disposal), and Appendix F (requirements for low-level waste transfer for disposal at land disposal facilities and manifests). Acceptable methods for disposal of radioactive material (whether or not it is waste) are

1. Transfer to another person licensed to receive it.
2. Storage for decay and disposal as nonradioactive.
3. Release in effluents below concentrations specified to ensure that public exposures are within the limits specified in Section 20.301 and 20.303 (discussed above).
4. Release into sanitary sewer of a readily soluble liquid containing isotopes below concentrations (averaged over a month) specified in Table B of Appendix B of 10 CFR Part 20 up to a limit of 5 Ci/yr (185 GBq/yr) of hydrogen-3 (tritium) and 1 Ci/yr (37 GBq/yr) of all other radioactive materials.
5. Incineration if authorized under Section 20.1002.
6. For material containing $0.05\,\mu$Ci (1.85 kBq), or less, of hydrogen-3 or carbon-14/g of medium used for scintillation counting or in animal tissue (averaged over the weight of the entire animal), no restrictions are placed on disposal based on the radioactivity contained therein. Such animal tissue may not be used as food for humans or as animal feed and restrictions based on other toxic or hazardous properties of the material still apply. Records of the disposal of such material must be retained until the license is terminated.

Two other parts of the NRC regulations deal specifically with the disposal of radioactive material: 10 CFR Part 60, "Disposal of High-Level Radioactive Wastes in Geologic Repositories," and 10 CFR Part 61, "Licensing Requirements for Land Disposal of Radioactive Waste".

Details of these regulations are discussed in succeding sections of this chapter on the programs for managing particular waste types.

The NRC provides supplemental guidance on acceptable methods for interpreting and implementing its regulations in the form of Regulatory Guides, Branch Technical Positions, and generic notices to licensees. Regulatory Guides are developed by the NRC to illustrate at least one method of complying with a given regulation. License applicants may choose other methods but must provide sufficient documentation to enable the NRC to reach the conclusion that the regulatory requirements are being met by that alternative method.

Other agencies also have responsibility to develop regulations for controlling the receipt, possession, use, and transfer or disposal of radioactive materials by agency personnel, contractors, or licensees in conformance with the standards promulgated by the EPA. The DOE provides rules in the DOE Manual (10 CFR Parts 900 and following) for use in facilities owned and operated either directly by the Department or by any of its contractors. These rules govern, for example, disposal of low-level waste at DOE sites such as Oak Ridge National Laboratory in Tennessee. Specific DOE waste program requirements are discussed in the succeeding sections on individual waste types.

Transportation of radioactive materials is primarily regulated by the U.S. Department of Transportation (DOT) as specified in 49 CFR Parts 173–178. These regulations specify required performance criteria for packaging, package activity limits, driver training requirements, and routing restrictions for some large quantities of radioactive materials. (Transportation regulations are discussed in detail in Chapter 5.)

State responsibility for radioactive waste management and disposal varies among waste types as well as from state to state. Repositories for HLW and TRU waste must be developed on federally owned land and will be the responsibility of the U.S. DOE in compliance with NRC regulations in 10 CFR Part 60. The states' role is defined as one of "consultation and concurrence," particularly during the siting portion of the program. The situation is almost completely reversed for LLW. The states have the responsibility to provide for the safe management of LLW generated within their respective borders. The states that eventually host LLW disposal facilities (see discussion of compacts in Appendix D) may license the facility if they have entered into an Agreement for Licensing with the NRC as provided for in Section 274 of the Atomic Energy Act of 1954, as amended. Such states will regulate the facility in accordance with state-promulgated regulations that must be compatible with, but not necessarily the same as, the NRC's requirements in 10 CFR Part 61. Without such an agreement, licensing of the facility will be the responsibility of the NRC. It would be accomplished in compliance with the provisions of 10 CFR Part 61. In addition to the requirements for radiological protection, there are other portions of the siting, development, and operation of an LLW disposal facility that are the responsibility of the host state. These include general state policies, procedures, and/or regulations for siting major facilities. They may be exercised by an individual agency (such as the Department of Environmental Protection) or by an independent siting board. Such reviews generally consider the environmental impact of the proposed facility, particularly groundwater impact, and compliance with overall development plans for major facilities in the state. General environmental effects of construction—such as noise, dust, and erosion control—and worker protection during construction and operation will also be performed in compliance with state requirements. The state may also be involved in developing a program of assistance to offset the impact on localities in which the facility is sited. Such impact may include population changes due to the presence of construction or operating personnel, the related need for services such as schools, needed road upgrades to accommodate shipments to the facility, and training and equipment for service personnel such as police, fire, and ambulance attendants.

Responsibility for regulating disposal of naturally occurring and accelerator produced (NARM) radioactive waste is not addressed in the Atomic Energy Act. It currently rests solely with the states as an exercise of their role in protecting public health and safety. Such wastes include radium-contaminated soils, radium needles, accelerator targets, and accelerator maintenance or decommissioning wastes. It is expected that the forthcoming EPA standard on management of LLW (40 CFR Part 193) will identify levels of NARM waste appropriate for disposal as LLW. Whether other requirements are imposed is a decision for the individual states in which the waste is produced and in which it will be disposed.

It is important to note that there are a number of other agencies at the federal level that will be involved in decisions on radioactive waste management and disposal although not primarily in a regulatory role. As indicated in Table 1-3, several different bureaus within the Department of the Interior as well as the Coast Guard (Department of Transportation) and Council on Environmental Quality have responsibilities that will involve them in radioactive waste management programs. There are also several offices within EPA in addition to the office of Radiation Programs that will influence decisions both for new disposal capability and for remediation of existing sites.

3.3 HIGH-LEVEL WASTE MANAGEMENT PROGRAM

The Nuclear Waste Policy Act of 1982 (NWPA) establishes a framework for developing and maintaining a system for high-level waste (HLW) management and disposal in the United States. Passage of the Act was a recognition that the nation must move beyond the interim storage of such wastes at dispersed sites and provide for long-term isolation of the waste from the biosphere in facilities that "will provide a reasonable assurance that the public and the environment will be adequately protected."

A fundamental principle of the NWPA is that all costs of waste disposal will be borne by those receiving the benefit from its production. That is, the government will recover all costs incurred in the program from fees levied on the users of the disposal facilities and ancillary operations. A Nuclear Waste Fund was established in the U.S. Treasury to receive payments of these fees. Utilities are currently making payments into this fund at the rate of one mill per kilowatt-hour

(1 mill/kWh) produced with nuclear fuel. This unit was chosen as a measure of the benefit derived from the fuel (DOE 1983).

The major element of the system established under the NWPA is the development of geologic repositories to receive (by January 31, 1998) and isolate spent fuel and HLW. The DOE was given responsibility to accomplish this task. This includes siting, obtaining a license (from the NRC in compliance with the provisions of 10 CFR Part 60), and constructing and operating at least one and maybe more repositories. Safe transportation of waste to the repository is integral to such a system since the DOE will take title to the spent fuel at the reactor site. Further study of monitored retrievable storage of HLW was also mandated by the NWPA. Authorization was also included for the DOE to provide for a limited amount (up to 1900 MTU) of interim spent fuel storage at a federally owned and operated facility if it were needed to prevent shutdown of nuclear power reactors with no ability to provide additional onsite storage. Costs of such interim storage would be borne by those using the service.

Recognizing that "State and Public participation in the planning and development of repositories is essential in order to promote public confidence in the safety of disposal of such waste and spent fuel," the NWPA established a detailed system of public involvement. The system includes negotiation of written consultation and concurrence agreements between the federal government and individual states and Indian tribes affected by the HLW program.

It is important to remember that while passage of the NWPA represents a significant milestone in establishing a system for the long-term isolation of HLW and spent fuel, it was made possible by extensive technical investigation and interaction among various federal agencies, elected representatives in Congress, states, Indian tribes, and members of the public. For example, the interagency Review Group (IRG 1979) report to the President issued in 1979 recommended use of mined geologic respositories for long-term disposal of HLW. This supported earlier recommendations made by committees of the National Academy of Sciences and agency reports. Alternative technologies were reviewed and an Environmental Impact Statement prepared on Management of Commercially Generated Radioactive Waste (DOE 1980b) concluded that such a choice was reasonable on environmental grounds. The formal decision to use geologic repositories as the preferred disposal method was published in 1981 (DOE 1981a) EPA's standards development program and NRC's licensing requirements were also proceeding in parallel with the actions to reach agreement and enact legislation giving additional support and impetus to the HLW disposal program. NRC published proposed procedures for licensing a geologic repository in December 1979 (NRC 1979) and draft technical criteria in May 1980 (NRC 1980). EPA issued proposed standards in December 1982 (EPA 1982). Detailed comments on the proposed standard, as well as input from public hearings, were reflected in final standards issued in September 1985 (EPA 1985). DOE's General Guidelines for the Recommendation of Sites for the Nuclear Waste Repositories (10 CFR Part 960) were published on December 6, 1984 (DOE 1984) following review and concurrence by the NRC.

The DOE HLW Management Program is conducted by the Office of Civilian Radioactive Waste Management (OCRWM). DOE was required by the NWPA to develop a Mission Plan describing the elements of the program and providing an assessment of information still to be developed. The description of the program provided in the following paragraphs draws heavily from that Mission Plan (DOE 1985) with supplementary material as necessary because of changes occurring since its publication.

The requirement to begin accepting waste at a geologic repository by January 31, 1988 may be met in a number of different ways as illustrated in Figure 3-2. This flow diagram identifies the basic program as originally authorized by Congress in the NWPA. That is, development of a geologic repository to accept spent fuel and commercial and defense HLW in limited amounts and at limited

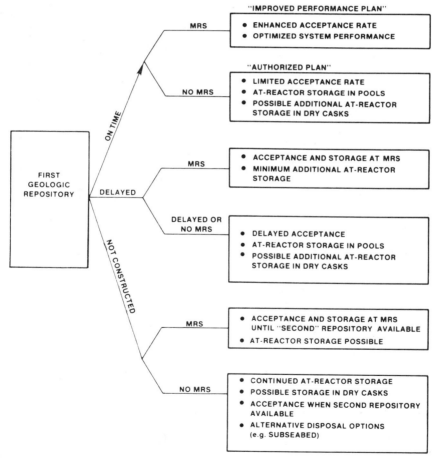

Figure 3-2 Components of possible alternative high-level waste programs. Source: (DOE 1985).

rates, providing interim storage if required, developing a transportation system and developing a second repository if authorized by Congress. DOE was also charged with evaluating the need for and feasibility of including a monitored, retrievable storage system (MRS) in the overall management system. Based upon its investigations DOE has recommended inclusion of an MRS to achieve the following improvements compared to the basic program:

1. Improved transportation efficiency by consolidating fuel at the MRS and reducing the number of shipments to the repository and related environmental impact and risks.
2. Increased flexibility to regulate the flow of waste to the repository.
3. Increased confidence that waste would be accepted by January 31, 1988 since the MRS would be scheduled for earlier operation.
4. Reduced need for at-reactor storage capacities required by limited receiving rates during the initial years of geologic repository operation.
5. Concentration of repository licensing efforts on demonstrating long-term isolation capability of the site by performing functions such as waste preparation at the MRS.

DOE is further providing flexibility to meet the needs for accepting waste on schedule by evaluating a range of contingency scenarios that may be implemented if the need arises. Such needs could result from delay in resolving technical or environmental issues involved with siting and constructing the geologic repository or in achieving consensus between the federal government and the host state or Indian tribe. Such states and Indian tribes have the right under the NWPA to file a notice of disapproval of a site recommendation submitted by DOE to Congress. A joint resolution of Congress would be required to overturn a notice of disapproval. If the disapproval is not overturned, the President must recommend a new site within 1 yr subject to the same review and approval procedures. Depending on the reasons for delay of the first repository and the extent of the delay, waste management needs for spent fuel may be accommodated by acceptance at the MRS, by additional at-reactor storage in pools, additional at-reactor storage in dry casks (see discussion in Chapter 5 on spent fuel storage options), initial acceptance at what was originally planned to be the second repository, or, in the extreme, development and use of an alternative disposal technology.

As described in Chapter 2, a determination has been made (as required by the NWPA) that both defense HLW and civilian waste (spent fuel and the separated HLW currently at the West Valley site) will be disposed in a single repository. Should the NRC determine that other wastes currently not defined as HLW require the same length and degree of isolation, they may also be accepted at the same facility. A second repository is being considered because of capacity limitations placed on the first repository by the NWPA. The first repository is authorized to receive only 70,000 MTU of spent fuel (or HLW resulting from reprocessing a similar quantity of spent fuel) before operation of a second repository. As indicated in Section 2.3 of Chapter 2, this level is projected to be reached in 2006 and some 130,000 MTU is projected to require disposal by 2020. Therefore, a second repository will be necessary within that time period. As would be expected, however, identification of potential repository sites is a focus for objection from the states and areas so identified. In late 1987, Congress passed amendments to the Nuclear Waste Policy Act that identified Yucca Mountain, Nevada as the preferred site for the first repository unless the site is found to be technically unacceptable during site characterization studies. Exactly what gets built where and when may differ from current plans because of the continuing need for congressional authorization and appropriation, which is exactly why alternative or contingency plans are also being continuously evaluated. For planning purposes, the Mission Plan identified the elements in Table 3-4 as basic assumptions regarding the amount and characteristics of the waste, the necessary transportation system, the MRS, and the first and second geologic repositories. The date and rate at which spent fuel will be accepted will vary depending on whether or not an MRS facility is developed, and when it is available, as well as when the first and second repositories begin operation. Two possible variations are illustrated in Tables 3-5 (without MRS) and 3-6 (with MRS). Each of these schedules assumes operation of the first repository will occur in 1998 and the second in 2006. (More recent schedules assume initial repository operation begins in 2003.) In both cases a finite amount of spent fuel and HLW is planned for but it is noted that additional material can be accepted by extending operation of the first two repositories or developing additional repositories. Such flow sheets would be the basis for planning the operations schedule and equipment needs at the several facilities and the related transportation requirements. It is important, too, for utility managers responsible for planning the amount of onsite spent fuel storage capacity required to understand the variation that may exist in the time at which fuel may be shipped from their facility in accordance with the terms of the contract for disposal services between the utility and DOE. These contract provisions include publication of projected receiving capacities and power reactor ranking. In 1991 DOE will begin to publish firm waste acceptance schedules for individual reactors, including shipment allocations.

The process of developing a geologic repository is one of successive screening and additional detailed site information as a function of time. The first step is to identify states in which potentially available sites exist so that interaction with officials and members of the public potentially affected could begin as soon as possible. At least five sites must be nominated for characterization for both the first and second repositories. Each nomination requires an environmental assessment that evaluates the sites with respect to previously published siting guidelines. At least three sites in two different host media (e.g., salt, tuff, or basalt) must be subject to detailed site characterization (see Chapter 6). The results of the characterization studies, together with public input from hearings and comments on intermediate documents, and a final environmental impact statement will be the basis for the recommendation of a specific site for development as the first

TABLE 3-4 Program Assumptions about HLW Characteristics and Management System Components

I. Assumptions about the Waste

A. Waste Types
1. Spent fuel from commercial nuclear reactors is the dominant waste form. The spent fuel assemblies are assumed to be consolidated and packaged before emplacement in a geologic repository.
2. Existing commercial high-level waste (from the West Valley Demonstration Project) is assumed to be accepted for disposal. No new commercial high-level waste is assumed to require disposal.
3. Defense high-level waste requiring disposal in a geologic repository will be accepted by the civilian waste management system.
4. Other wastes determined by the Nuclear Regulatory Commission to require permanent geological disposal will be accepted.

B. Waste Quantities
1. The Energy Information Administration (EIA) "middle-case" forecast of 130,000 metric tons of uranium (MTU) for the cumulative spent fuel generation through 2020 is assumed. The EIA forecast assumes a constant level of fuel burnup in the reactor for the entire projection period; the design implications of extended fuel burnup are being examined. Other projections are being monitored.
2. Approximately 650 MTU equivalent of commercial high-level waste from West Valley will be accepted at the repository. No additional commercial reprocessing of spent fuel is assumed, although system designs wil include the capability to handle solidified high-level waste from commercial reprocessing.
3. Defense waste will be accepted at approximately the rate of its generation from the atomic energy defense facilities. It is assumed that approximately 20,000 canisters of solidified high-level defense waste will be available for geological disposal by 2020. The actual quantities and the acceptance rate have not yet been determined; they are the subject of current analyses by the DOE.

C. Waste Age
The spent fuel received for disposal is assumed to have a minimum cooling age of 5 yr, although certain system-component designs are based on receiving fuel that has a maximum decay-heat rate of 1 kW per metric ton of uranium, which is equivalent to current pressurized water reactor spent fuel that has been discharged from the reactor for approximately 10 yr. This assumption will be reviewed to determine the effects of age on system performance and cost effectiveness.

II. Assumptions about Transportation

A. Shipment Mode
From reactors to the repository or to the MRS facility (if authorized by Congress): 70% by rail and 30% by truck. From the MRS facility to the repository: 100% by rail.

TABLE 3-4 *(Continued)*

B. Distance Traveled

Average distances for potential routes from reactors to a repository in the indicated host rock or to the MRS facility are presented below in the first two columns. For waste shipped first to the MRS facility, the last column shows distances from the MRS facility to the repository.

	Distance (miles)		
	Reactors to Facility		MRS to Repository
Facility	Truck	Rail	Rail
Basalt repository	2190	2360	2590
Salt repository	1320	1450	1250
Tuff repository	2040	2270	2380
Crystalline rock repository	760	900	610
MRS facility	790	920	—

III. Assumptions about an MRS Facility

A. Proposed Capacity

The initial authorized capacity will be proposed to be 15,000 MTU. If necessary, the DOE will request congressional approval to modify this capacity.

B. Waste Receipt Rate

The waste-acceptance rate will be approximately 3000 MTU/yr of spent fuel or equivalent high-level waste.

C. MRS Startup Dates

The MRS facility will begin operations as early as 1996. In 1998, after a 2-yr ramp-up period, the MRS facility will achieve the design acceptance rate.

IV. Assumptions about Geologic Repositories

A. Number of Repositories

Two.

B. Design Capacity for Each Repository

The design capacity for each of the two repositories is assumed to be 70,000 MTU. The capacity for both repositories can be expanded if the geologic conditions permit, but the capacity of the first repository cannot exceed 70,000 MTU until the second repository begins operation.

C. Repository Startup Dates for the Authorized System

The operation of the first repository will be conducted in two phases, with phase 1 starting operation in 1998 and phase 2 starting operation in 2001. The second repository will start receiving spent fuel in 2006.

D. Host Rock Options

Basalt, salt, and tuff for the first repository; two of the first-repository host rocks and crystalline rock (granite) for the second repository.

E. Other Design Assumptions

Designs include provision to maintain capability to begin waste retrieval for up to 50 yr from the date on which the first waste package is emplaced.

Source: (DOE 1985).

TABLE 3-5 Waste-Acceptance Schedule—Without MRS [Metric Tons of Uranium (MTU) per Year]

| Year | Spent Fuel Generation[a] | | First Repository | | | | Second Repository | | Cumulative Spent Fuel Acceptance | Spent Fuel Backlog |
	Annual	Cumulative	Spent Fuel	High-Level Waste[b,c]	Total	Cumulative	Spent Fuel	Cumulative		
Pre-1998		40,100								40,100
1998	2900	43,000	400		400	400			400	42,600
1999	3000	46,000	400		400	800			800	45,200
2000	3000	49,000	400		400	1,200			1,200	47,800
2001	3000	52,000	900		900	2,100			2,100	49,900
2002	3000	55,000	1800		1800	3,900			3,900	51,000
2003	3100	58,100	3000	400	3400	7,300			6,900	51,200
2004	3300	61,400	3000	400	3400	10,700			9,900	51,500
2005	3400	64,800	3000	400	3400	14,100			12,900	51,900
2006	3800	68,600	3000	400	3400	17,500	900	900	16,800	51,800
2007	4100	72,700	3000	400	3400	20,900	1800	2,700	21,600	51,100
2008	4700	77,400	3000	400	3400	24,300	1800	4,500	26,400	51,000
2009	4500	81,900	3000	400	3400	27,700	1800	6,300	31,200	50,700
2010	4500	86,400	3000	400	3400	31,100	1800	8,100	36,000	50,400
2011	4000	90,400	3000	400	3400	34,500	2400	10,500	41,400	49,000
2012	4100	94,500	3000	400	3400	37,900	3000	13,500	47,400	47,100
2013	4200	98,700	3000	400	3400	41,300	3000	16,500	53,400	45,300
2014	4200	102,900	3000	400	3400	44,700	3000	19,500	59,400	43,500
2015	4300	107,200	3000	400	3400	48,100	3000	22,500	65,400	41,800
2016	4300	111,500	3000	400	3400	51,500	3000	25,500	71,400	40,100
2017	4500	116,000	3000	400	3400	54,900	3000	28,500	77,400	38,600

Year											
2018	4700	120,700	3000	400		3400	58,300	3000	31,500	83,400	37,300
2019	4700	125,400	3000	400		3400	61,700	3000	34,500	89,400	36,000
2020	4900	130,300	3000	400		3400	65,100	3000	37,500	95,400	34,900
2021[d]			3000	400		3400	68,500	3000	40,500	101,400	28,900
2022			1100	400		1500	70,000	3000	43,500	105,500	24,800
2023								3000	46,500	108,500	21,800
2024								3000	49,500	111,500	18,800
2025								3000	52,500	114,500	15,800
2026					8000[e]			3000	55,500	117,500	12,800
2027								3000	58,500	120,500	9,800
2028								3000	61,500	123,500	6,800
2029								3000	64,500	126,500	3,800
2030								3000	67,500	129,500	800
2031								800	68,300	130,300	

Source: (DOE 1985).

[a] Data from *Commercial Nuclear Power 1984: Prospects for the United States and the World*, DOE/EIA 0438(84), November 1984. Includes discharge from decommissioned reactors.

[b] Approximate waste-acceptance rates for high-level waste from atomic energy defense activities and commercial high-level waste from the West Valley Demonstration Project. Quantities have been "normalized" to metric tons of uranium (MTUs) on a curie-equivalent basis. Direct comparison with spent fuel is not equivalent, because defense high-level waste (DHLW) and commercial high-level waste (CHLW) resulted from the reprocessing of spent fuel. In the example, 400 MTU of defense waste equals 800 canisters. Actual acceptance rates are to be negotiated between Defense Programs and the Office of Civilian Radioactive Waste Management in the DOE.

[c] The first repository currently is designed to begin operation in two phases. This example shows the acceptance of DHLW and CHLW in the first phase when the second phase reaches its maximum receipt rate.

[d] The Energy Information Administration projects spent fuel generation only through 2020. For waste created after 2020, either the capacity of the first two repositories could be increased or additional repositories could be built.

[e] The example shows a total of 8000 MTU of DHLW and CHLW emplaced by 2022. Additional DHLW can be accommodated by extending the operation of the first repository, emplacing DHLW in the second repository, or constructing additional repositories, as indicated in footnote *d*.

TABLE 3-6 Waste-Acceptance Schedule—With MRS [Metric Tons of Uranium (MTU) per Year]

| Year | Spent Fuel Generation[a] | | MRS Acceptance | MRS Inventory[b] | First Repository | | | | Second Repository | | Cumulative Spent Fuel Acceptance | Spent Fuel Backlog |
	Annual	Cumulative			SF from MRS	High-Level Waste[c]	Total	Cumulative Total Waste	Spent Fuel	Cumulative		
Pre-1998		40,100	2200	2,200							2,200	37,900
1998	2900	43,000	3000	4,800	400		400	400			5,200	37,800
1999	3000	46,000	3000	7,400	400		400	800			8,200	37,800
2000	3000	49,000	3000	10,000	400		400	1,200			11,200	37,800
2001	3000	52,000	3000	12,100	900		900	2,100			14,200	37,800
2002	3000	55,000	3000	13,300	1800		1800	3,900			17,200	37,800
2003	3100	58,100	3000	13,300	3000	400	3400	7,300			20,200	37,900
2004	3300	61,400	3000	13,300	3000	400	3400	10,700			23,200	38,200
2005	3400	64,800	3000	13,300	3000	400	3400	14,100			26,200	38,600
2006	3800	68,600	3000	13,300	3000	400	3400	17,500	900	900	30,100	38,500
2007	4100	72,700	3000	13,300	3000	400	3400	20,900	1800	2,700	34,900	37,800
2008	4700	77,400	3000	13,300	3000	400	3400	24,300	1800	4,500	39,700	37,700
2009	4500	81,900	3000	13,300	3000	400	3400	27,700	1800	6,300	44,500	37,400
2010	4500	86,400	3000	13,300	3000	400	3400	31,100	1800	8,100	49,300	37,100
2011	4000	90,400	3000	13,300	3000	400	3400	34,500	2400	10,500	54,700	35,700
2012	4100	94,500	3000	13,300	3000	400	3400	37,900	3000	13,500	60,700	33,800
2013	4200	98,700	3000	13,300	3000	400	3400	41,300	3000	16,500	66,700	32,000
2014	4200	102,900	3000	13,300	3000	400	3400	44,700	3000	19,500	72,700	30,200
2015	4300	107,200	3000	13,300	3000	400	3400	48,100	3000	22,500	78,700	28,500
2016	4300	111,500	3000	13,300	3000	400	3400	51,500	3000	25,500	84,700	26,800

Year												
2017	4500	116,000	2800	13,100	3000	400	3400	54,900	3000	28,500	90,500	25,500
2018	4700	120,700		10,100	3000	400	3400	58,300	3000	31,500	93,500	27,200
2019	4700	125,400		7,100	3000	400	3400	61,700	3000	34,500	96,500	28,900
2020	4900	130,300		4,100	3000	400	3400	65,100	3000	37,500	99,500	30,800
2021				1,100	3000	400	3400	68,500	3000	40,500	102,500	27,800
2022				1,100	1100	400	1500	70,000	3000	43,500	105,500	24,800
2023									3000	46,500	108,500	21,800
2024									3000	49,500	111,500	18,800
2025									3000	52,500	114,500	15,800
2026						8000			3000	55,500	117,500	12,800
2027									3000	58,500	120,500	9,800
2028									3000	61,500	123,500	6,800
2029									3000	64,500	126,500	3,800
2030									3000	67,500	129,500	800
2031									800	68,300	130,300	

Source: (DOE 1985).

[a] Data from *Commercial Nuclear Power 1984: Prospects for the United States and the World*, DOE/EIA 0438(84), November 1984. Includes discharge from decommissioned reactors.

[b] The MRS facility is assumed to reach a constant acceptance rate and discharge to the first repository as fast as the first repository can accept spent fuel. The MRS facility will stop accepting spent fuel when its inventory will fill the first repository.

[c] See footnotes *b* and *c* in Table 3-5.

repository. (As noted above, Congress revised the latter portions of the site selection procedure by essentially providing for sequential rather than concurrent site characterization beginning with the Nevada site.)

DOE is required to submit a license application for construction of the repository within 90 days of site designation. NRC review of the application for a construction authorization under 10 CFR Part 60 must be accomplished within 3 yr. (There is a possible 1 yr extension of the review period, if required.) This licensing review will determine whether the proposed system will provide the level of protection of public health and safety required to conform to EPA's standards in 40 CFR Part 191 as described in Section 3.2. 10 CFR Part 60 further specified criteria for individual components of the system that will enable these standards to be met. The principal criteria are

1. Primary reliance for isolation in the short term is placed on the integrity of the waste package. Substantially complete containment of the waste by the package for 300 to 1000 yr is required.
2. The release rate of each significant radionuclide from the engineered-barrier system must be less than one part in 10,000 per year of the inventory of that radionuclide at 1000 yr after permanent closure. (The engineered-barrier system includes the waste form, packing material, waste container, and repository features such as backfill and seals).
3. The pre-waste-emplacement groundwater travel times from the repository to the accessible environment must be greater than 1000 yr.
4. Waste must be retrievable for a 50-yr period after emplacement. This period of time allows for monitoring and verification of performance in accordance with the requirements of 10 CFR Part 60.111. Once satisfactory performance has been demonstrated, the repository may be sealed.

Regulatory review will occur a number of times during the life cycle of the repository. Interaction between NRC and DOE has already been initiated and will continue. For example, DOE's Siting Guidelines (10 CFR Part 960) were reviewed by NRC and final adoption by DOE followed written concurrence by NRC that took into account changes that had been made in response to its earlier comments on draft guidelines. The requirement for public interaction and publication of documentation such as the environmental impact statement provides additional specific opportunities for interaction between NRC and DOE during the siting process. NRC representatives will be stationed at DOE Operations Offices responsible for the repository program and technical meetings and workshops will also be held by both agencies.

10 CFR Part 960 requires that candidate sites be identified in different geohydrologic settings and different rock types to ensure that a wide range of alternatives is considered. As specified in the NWPA, identification of sites for succeeding repositories must consider the need for, and advantages of, a regional distribution of repositories with respect to such effects as the system transport-

ation requirements. Procedurally, the site screening process was initially designed to identify "potentially acceptable" sites from which several would be nominated as "suitable for characterization." After characterization, a site would be recommended for repository development. The types of information that must be developed at each stage of the siting process are specified in Appendix IV of 10 CFR Part 960 and are listed in Table 3-7 for information.

The guidelines are organized to address the performance of the system as a whole as well as specific technical characteristics. Further detail is provided by identifying the criteria as being of four types:

"Qualifying Conditions"—must be satisfied for a site to be acceptable.

"Favorable Conditions"—are not necessary but enhance confidence that the qualifying condition can be met.

"Potentially Adverse Conditions"—are presumed to detract from expected performance but additional data may indicate that the performance is acceptable.

"Disqualifying Conditions"—eliminate the site from further consideration.

Conditions included in the guidelines are additionally identified with respect to the time frame over which they are particularly important. Preclosure guidelines address characteristics of the site and environs of particular importance during construction and operation of the repository. This includes public and worker exposure during operation, environmental impact, and socioeconomic impacts. Postclosure guidelines address characteristics that determine the long-term behaviour of the site and related effects on public health and safety.

Formal licensing actions will occur with application for construction authorization, application for authorization to operate the repository (two separate authorizations are anticipated reflecting the phased construction and operation of the repository as described below), application to close the repository, and eventual termination of the repository license.

Repository construction is expected to take 6 yr and is planned to occur in two phases:

Phase 1: Construct surface, shaft, and underground facilities to accommodate 400 MTU spent fuel/yr beginning in 1998.

Phase 2: Construct remaining facilities needed for full-scale repository operation. (Full scale operation is 400 MTU/yr in the Phase 1 facilities and 3000 MTU/yr in the Phase 2 facilities). These facilities will include rod consolidation capabilities. As noted in Table 3-4, Phase 2 facilities are expected to be fully operational 5 yr after initial fuel acceptance at the repository.

Development and operation of the first geologic repository represents over a 90-yr commitment to preclosure responsibilities on the part of DOE. (Similar

Table 3-7   Types of Information for the Nomination of Sites as Suitable for Characterization

Geohydrology—To identify current groundwater flow paths, travel times, and local uses.

Geochemistry—To identify potential resources and the interaction of soil with the waste package and effect on radioisotope migration.

Rock characteristics—To identify a baseline for predicting performance under the thermal, mechanical, and radiation effects of the repository.

Climatic changes—To identify projected changes due to climatic conditions that would affect surface groundwater or repository performance.

Erosion—To identify the potential for uncovering the waste in less than one million yr.

Dissolution—To identify the areal extent of any projected rock dissolution.

Tectonics—To identify the probable effects on site integrity and performance from projected seismic activity over the next 10,000 yr.

Human interference—To identify factors that may affect the probability of intrusion affecting waste containment and isolation.

Natural resources—To identify the probability that past or future exploration and recovery of resources—including groundwater—may adversely affect site performance.

Site ownership and control—To determine if DOE can obtain ownership, and control access to, the site.

Population density and distribution—To identify highly populated areas and the nearest 1 mile by 1 mile area having a population greater than 1000 persons.

Meteorology—To identify current weather conditions and predict the occurrence of conditions likely to disperse material into the air either during or after site operation.

Offsite installations and operations—To identify nearby industrial, transportation, and military facilities that might affect construction, operation, or closure of the repository.

Environmental quality—Baseline description of environmental conditions to estimate effects on public health and safety, and the environment.

Socioeconomic impact—Impact of characterization and operation and closure of the repository on local social conditions (such as population, housing), the need for services (such as schools), and revenue sources.

Transportation—To identify costs and risks of shipping waste to the site on existing or new transportation networks.

Surface characteristics—To identify the impact on construction, operation, and closure of site topography, surface water bodies, and unstable soils.

Rock characteristics—To confirm that the rock will be capable of accepting the planned amount of waste and provide the required long-term isolation.

Hydrology—To confirm the compatibility of the hydrologic conditions with repository construction, operation, and closure.

procedures would be followed for the second repository.) The project schedule subsequent to the receipt of a construction authorization is as follows:

- 6 yr construction,
- 28 yr operation,
- 50-yr period of waste retrievability after first waste emplacement for performance confirmation required by 10 CFR Part 60.111(b),
- 34 yr for waste retrieval and relocation, if necessary,
- repository sealing, decommissioning of surface facilities, reclamation of surface area to the extent feasible.
- erection of permanent markers, installation of necessary postclosure monitoring, and surveillance equipment,
- license termination.

Selecting a site for the geologic repository is a multistep process in which the level of detail is inversely proportional to the size of the area being evaluated. Initial screening efforts (at the national, physiographic province and regional survey level) are based on national maps of faults, earthquake epicenters, land use, recent volcanic activity, host rock and mineral resources, geohydrologic conditions, and other information available in the open literature. Surveys at the area and location level will expand on such information by field exploration and testing. This would include drilling boreholes to confirm indicated characteristics such as depth of host rock, geohydrologic conditions at different depths, geochemical characteristics, and the chemistry of the environment that would be in contact with the waste package. Laboratory studies would also be performed to confirm that host rock characteristics such as sorption coefficients, effective porosity, permeability, mineral composition, radionuclide solubility in the groundwater, strength, elastic properties, coefficients of thermal expansion, and thermal conductivity are sufficient to ensure long-term isolation of the radionuclides in the waste.

The site-screening portion of the first repository program was concluded in February 1983 with the identification of nine potentially acceptable sites in six states (see Figure 3-3). The host geologic medium for seven sites is salt: two sites in the bedded salt of the Palo Duro Basin in Deaf Smith and Swisher Counties, Texas, two sites in bedded salt at Davis Canyon and Lavender Canyon in the Paradox Basin, Utah, and three salt domes in the Gulf Interior region of the Gulf Coastal Plain (the Richton and Cypress Creek Domes in Mississippi and the Vacherie Dome in Louisiana). One site is in basalt at the Hanford Reservation in the Pasco Basin, Washington. The final site is in tuff at the Yucca Mountain site in the Southern Great Basin, Nevada (adjacent to the Nevada Test Site). Based on draft environmental assessments prepared for each of the nine potentially acceptable sites and in accordance with the DOE siting Guidelines (10 CFR Part 960) the sites in Deaf Smith County, Texas, Hanford, Washington, Yucca Mountain, Nevada, Davis Canyon, Utah, and Richton, Mississippi were

Figure 3-3   Candidate sites for a geologic repository. Source: (DOE 1985).

proposed for nomination. DOE recommended that the Deaf Smith County, Hanford, and Yucca Mountain sites be fully characterized for consideration as the site of the first repository.

Crystalline rock formations in 17 states are being studied for possible identification of a site for the second repository. Regional characterization reports have been developed compiling information available in the open literature. DOE has also developed a screening methodology document that discusses how the data in the regional characterization reports will be used to identify areas within the regions that will be the subject of further study. As shown in Figure 3-4 three general regions of crystalline rock have been identified. They include the states of Maine, Vermont, New Hampshire, New York, Pennsylvania, Connecticut, Massachusetts, New Jersey, Rhode Island, Michigan, Minnesota, Wisconsin, Maryland, Virginia, North Carolina, South Carolina, and Georgia.

Longer term alternatives being investigated for repository sites are other host rock types and subseabed disposal. DOE is investigating, in cooperation with Belgium, Canada, the Federal Republic of Germany, France, Italy, Japan, The Netherlands, Switzerland, the United Kingdom, and the Commission of the European Communities, the feasibility of isolating radioactive waste within the thick stable beds of sediments under the ocean floor. Three areas in the North Pacific Ocean and two in the North Atlantic Ocean are being studied to determine if they conform to predictive models of chemical, thermal, and mechanical conditions. Studies being performed by the individual members of the Subseabed Working Group described above include site assessments, barrier

Figure 3-4 Crystalline rock deposits potentially useful for a second geologic repository. Source: (DOE 1985).

assessments, and legal and institutional studies. Additional discussion of this methodology is provided in Chapter 6 (Section 6.12).

In addition to the geologic repository, there are two other important components of the HLW management system: storage and transportation.

Storage has been recognized as being primarily the responsibility of the owners and operators of the nuclear power plants at which the spent fuel is generated. They are currently fulfilling this responsibility by storing the fuel onsite in water-filled basins (spent fuel storage pools). Depending on the age of the facility, many reactors were initially designed to store fuel for about 6 months prior to shipment offsite for reprocessing. Utilities have designed and installed spent fuel storage racks that enable more fuel to be stored in less space without the potential for criticality by incorporating neutron poisons (such as boron) into the rack design. Some utilities are also pursuing programs to demonstrate the feasibility and licensability of rod consolidation. Consolidation of spent fuel rods is accomplished by dismantling the fuel assembly and rearranging the rods in a more compact array. It is particularly effective in extending the storage capacity of storage pools limited by space rather than structural strength or seismic loads. Initial demonstration of the technique and equipment was performed in an interagency cooperative demonstration program at the Tennessee Valley Authority's Browns Ferry nuclear power plant. Initial licensing of the rod consolidation option was pursued by Northeast Utilities at its Millstone station.

An additional storage option being developed for use at facilities in which additional reracking or rod consolidation is not feasible (because of weight

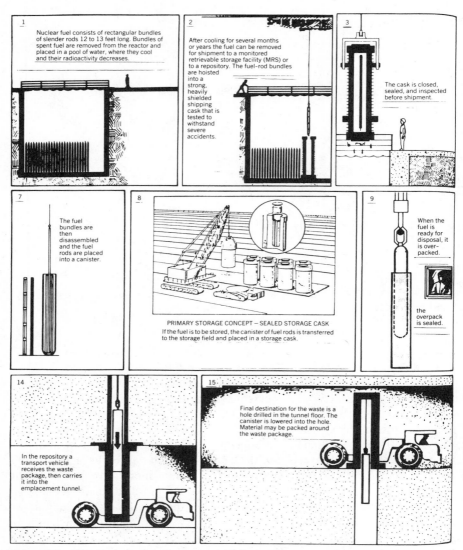

Figure 3-5 Illustration of the integrated high-level waste disposal system (with monitored retrievable storage). Source: (DOE 1985).

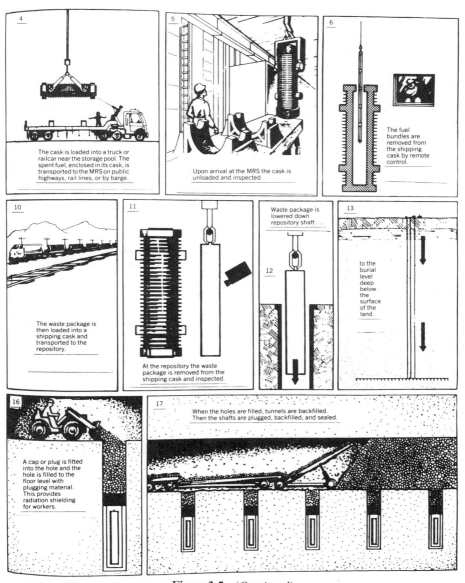

4 The cask is loaded into a truck or railcar near the storage pool. The spent fuel, enclosed in its cask, is transported to the MRS on public highways, rail lines, or by barge.

5 Upon arrival at the MRS the cask is unloaded and inspected.

6 The fuel bundles are removed from the shipping cask by remote control.

10 The waste package is then loaded into a shipping cask and transported to the repository.

11 At the repository the waste package is removed from the shipping cask and inspected.

12 Waste package is lowered down repository shaft . . .

13 to the burial level deep below the surface of the land.

16 A cap or plug is fitted into the hole and the hole is filled to the floor level with plugging material. This provides radiation shielding for workers.

17 When the holes are filled, tunnels are backfilled. Then the shafts are plugged, backfilled, and sealed.

Figure 3-5 (*Continued*)

105

limitations on the spent fuel storage pool, site seismic conditions, or the cost of necessary renovations) is onsite dry storage of spent fuel. Such storage has been performed in casks, vaults, silos, and drywells at DOE facilities. Current programs are building and expanding on the existing data base by verifying the performance of different storage technologies with different commercial fuel at a range of temperatures and using both inert gases and air as cover gases in the dry storage system. Licensing of any onsite dry storage system would be in compliance with NRC's 10 CFR Part 72, "Licensing Requirements for the Storage of Spent Fuel in an Independent Spent Fuel Storage Installation (ISFSI)" and related guidance documents (NRC 1981). It is expected that use of reracking, rod consolidation, and/or onsite dry storage will preclude the need for any federal interim storage prior to the operation of the geologic repository.

The Monitored Retrievable Storage (MRS) facility as currently planned by DOE would provide interim storage ancillary to its primary function of waste preparation for emplacement in the geologic repository. The integrated system including the MRS is illustrated in Figure 3-5. As planned, the MRS would consist primarily of a receiving and handling building in which spent fuel assemblies would be remotely disassembled in concrete "hot cells" that provide shielding to control worker exposure and air filtration systems to minimize release of radioactive materials from the building. The disassembled rods would be consolidated and placed into canisters. The related fuel assembly hardware would be compacted and packaged for eventual disposal. The spent fuel canisters are described in Chapter 2 (Section 2.3). Storage of the sealed canisters pending transfer to the repository would be accomplished using sealed storage casks consisting of steel-lined reinforced concrete. The casks are planned to range from 17 to 22 ft in height and have an inside diameter of 12 ft. The 22-ft high cask will weigh 200 tons empty and hold 20 tons of spent fuel canisters. Casks would be filled in the receiving and handling building and transported outside for storage on a concrete pad. Casks would be periodically tested by collecting gas samples from the storage cavity to determine if there were leakage from any of the waste canisters stored therein. The cask design will provide for sufficient heat transfer from the stored fuel to the surrounding atmosphere to prevent thermal damage to the stored fuel or the cask concrete. An alternative to the concrete storage cask is use of field drywells for storage pending shipment of the containerized fuel to the geologic repository. The drywells are in-ground sealed metal enclosures into which the waste canister would be lowered. A top shield plug would be placed over the waste canister and a closure plate welded to the cavity liner. Drywells, like the storage casks, would be routinely monitored to detect any leakage from the stored containers. Additional monitoring would be performed in the environment of the storage area. Plans are also being made to accept and store transportable metal casks containing fuel previously in dry storage at reactor sites. Current plans provide for approximately 15,000 MTU storage capacity at the MRS. This capacity would allow the MRS to accommodate projected variations in waste generation rates, waste acceptance schedules, and repository emplacement capability. Storage is planned on a modular basis to maximize flexibility and avoid expensive overcapacity.

Under the terms of the spent fuel contracts in effect between DOE and the owners and generators of spent fuel, title to the material transfers to DOE at the reactor site and transportation is a part of the service provided by DOE. Exact plans for shipping spent fuel to the geologic repository (or the MRS) will continue to evolve as the facilities are sited and developed. All transport will be accomplished as a licensed activity in conformance with NRC and DOT regulations as described in Chapter 5. Specific issues still to be resolved include the optimum routes for such shipments, tracking of individual shipments, prenotifications to jurisdictions through which the transport occurs, assistance to local jurisdictions in developing emergency response capabilities, and the relative benefit of fewer, heavier (overweight) shipments rather than more shipments that meet the generally applied weight limits based on wear and tear on roads and bridges.

3.4 TRANSURANIC WASTE MANAGEMENT PROGRAM

The transuranic (TRU) waste management program consists of three principal components: the waste repository, the transportation system, and waste preparation at the generating sites. The program was estimated to cost approximately $5 billion through 2000 with most of that the responsibility of DOE (DOE 1981b).

TRU Waste Repository. Public Law 96–164, the Fiscal Year 1980 DOE Defense Authorization Act, authorized the Waste Isolation Pilot Plant (WIPP) as a defense activity of the DOE. WIPP is a geologic repository in a salt formation in the Permian Basin near Carlsbad, New Mexico. It is intended to dispose of TRU waste produced in defense activities under DOE rules rather than as an NRC licensed facility. It is subject to EPA's standards in 40 CFR Part 191, however.

The WIPP facility is designed to accept almost 6 million ft^3 (168,000 m^3) of contact handled (CH) TRU waste and almost 25,000 ft^3 (700 m^3) of remote handled (RH) TRU waste over a 25-yr period. Provision has also been made to receive a limited amount of defense HLW for experimental purposes. Once the research and development projects related to the HLW are completed, the material will be removed from WIPP. Waste acceptance is scheduled to begin in late 1988. This will be the culmination of a process that begin with initial geologic exploration and facility design in 1975. Environmental reviews and documentation were completed in October 1980. Actual site construction began in April 1981.

As a research and development facility, provision is being made at the WIPP site to conduct experiments designed to test and verify projected performance of the salt and the waste under a variety of conditions. Tests or experiments planned at WIPP include simulations of emplaced defense HLW both with and without external heating, emplacement, and retrieval of a limited amount of defense HLW, rock behavior, package durability, emplacement and retrieval of CH and

RH waste, and plugging and sealing of the excavation (Hunt 1986) Construction completion and operation of the WIPP facility will provide much useful data on waste handling and emplacement procedures applicable to the HLW management program described in Section 3.3. In addition to the technical requirements, another area that has received a great deal of attention at WIPP is the interaction with the neighboring community (Krenz 1986). Whenever possible, program decisions have sought to maximize economic benefit to the local economy. For example, preference was given to local hiring, and support components such as the waste shipment containers (TRUPACT) will be assembled at the site and procurement is likely to be from local businesses on a competitive bid basis. There is a continuing interchange of information between WIPP Project staff and local citizens and legislators. Cooperative agreements have been developed between DOE and the New Mexico State University to establish a health physics technician training program at the University's branch in nearby Carlsbad. Agreements and information exchanges have also taken place with local law enforcement, medical, and governmental agencies regarding the extent of, and access to, emergency response capability that exists at the WIPP site.

One issue that had been addressed in developing the standards for disposal of high-level and transuranic waste was the possibility that water trapped within the salt deposits (when they were formed by the evaporation of an ancient sea) would migrate to the area of waste emplacement due to the construction impact and the decay heat of the emplaced waste. The presence of a significant amount of salty water (brine) around the emplaced waste canisters would present the possibility of corrosion and failure of the canisters at times different from those projected under dry conditions. Such a phenomenon has been observed at the WIPP site and appears to be due to the effects of pressure differences in the salt bed after the emplacement rooms were excavated. DOE has recently responded to recommendations of a committee of the National Academy of Sciences regarding these conditions by reducing the planned rate and amount of material that will be accepted at WIPP pending study of the effects of the brine conditions on the emplaced waste.

Transportation. The process of moving the waste from the generating sites to the WIPP facility will be accomplished by using shielded shipping casks for the relatively small volume of RH waste with high external exposure levels. The larger volume of CH waste does not require shielding but does require ensured confinement because of the α-emitting radionuclides present. Particular care is required to prevent conditions such as a fire in which the material could become dispersed into the atmosphere and inhaled. For those wastes in which ^{239}Pu is a major component, package loading and the configuration of several packages must be performed to prevent any potential for accidental criticality. Protection must also be provided against package failure due to internal gas buildup from the generation of helium from α decay during the extended periods that much of the material has been in storage at the generating sites.

Shipments to date between generating sites have been able to make use of DOE-certified systems for either truck or rail transport. The rail system, the

ATMX-600 series rail cars, were specially designed for this purpose and incorporate such special safety features as roller-bearing wheels, interlocking couplers to prevent uncoupling in a derailment, and locking type center pins to prevent loss of the wheel trucks under accident conditions. The car body is specially designed to withstand accident forces and maintain its integrity. TRU waste within the cars is packaged within multiple confinement barriers. The first packaging is a polyethylene-lined drum or box that is covered with fiberglass-reinforced polyester. These packages are placed within a steel cargo container.

Truck shipments have been made in a container certified for shipping Type B quantities of radioactive materials (see discussion in Section 5.8 for a definition of Type B quantities) either by rail or truck. The container, called the Super Tiger, may be handled, stored, and shipped like any standardized shipping container. It consists of two reinforced rectangular steel shells separated by a layer of fire-retardant polyurethane foam that provides thermal, fire, and shock absorption protection to the contents. Future shipments will be made in a newly designed package, the TRU package transporter (TRUPACT), which will also provide double containment, impact limitation, and fire-retardant capabilities. Current program plans include a fleet of 24 TRUPACTs to transfer waste by truck from the generator site to WIPP. Consideration was given to the receiving rate at WIPP, the generation rate at the sites (including the rate of transfer of material in storage or retrieved), and the time in-transit in developing plans for the size and capacity of the individual TRUPACTs and the total fleet. Studies on optimization of waste-handling efforts have also been performed to determine the benefit of handling several drums as a unit and the optimum packing array. A satellite-based monitoring and communication system will be used to coordinate the movements of the 100 shipments per month of CH waste and 10 shipments per month of RH waste (Weaver 1986). Such a system provides real-time, two-way communication between the transport vehicle and DOE. This improves both the quality and the timeliness of data on the status and location of each shipment. Safety and security of transport benefit from the ability to respond quickly to unexpected mechanical problems, to avoid road conditions that might be dangerous such as extreme rain, snow, or ice, and to reroute around traffic congestion, if necessary. Scheduling and efficiency benefit from being able to anticipate the arrival at the facility and being aware of any conditions that should be corrected upon arrival. Actual movements onsite will be tracked by using bar codes on the equipment and stationing either fixed or portable readers throughout the site.

Waste Processing at Generating Sites.   As described in Section 3.1, TRU waste has been stored retrievably since at least 1970 at several facilities around the country. The major source of TRU waste is the nuclear weapons program and DOE sites at Savannah River, South Carolina, Hanford, Washington, Idaho Falls, Idaho, Los Alamos, New Mexico, and the Nevada Test Site are the primary generators of the waste material. Both newly generated waste and previously stored waste will eventually be shipped from these sites for long-term disposal at WIPP. Management activities involve certifying that waste packages meet the

WIPP acceptance criteria on contents and characteristics such as weight, surface contamination, dispersibility, and liquid content. Volume reduction and improvement of final waste form stability are two driving forces in defining onsite waste treatment systems. Systemwide studies are also providing guidance on whether treatment should be performed at each site or should be centralized at one or more sites with shipment from smaller volume generators or those whose waste may be more difficult to treat or to certify. These programs reflect the use of advanced instrumentation that provides the capability to go beyond the early conservative practice of storing suspect TRU if it could not be definitively shown that the TRU classification was inappropriate. Also, the revision of the definition of TRU from 10 nCi/g to material having no significant economic value which, at the end of institutional control periods, is contaminated with alpha-emitting radionuclides of atomic number greater than 92 and half-lives greater than 20 years in concentrations greater than 100 nCi/g, or has smearable alpha contamination greater than 4000 dpm/cm^2 averaged over the accessible surface (as specified in DOE Order 5820.1) will involve the reevaluation of all the material currently in storage to determine what meets the new criterion, how that can be certified, what further processing is necessary and appropriate, and what the costs and schedules will be for achieving this processing for both newly generated and stored waste. Beyond the material in storage, there are areas at these same sites in which TRU wastes were previously buried. At some of these areas it has been determined that the original containers have failed and that the surrounding soil is contaminated with TRU. Additional studies are being performed to identify the extent of the contamination, the feasibility of exhumation of the waste, and processes for accomplishing its stabilization, where necessary. The types of equipment being evaluated include shredders to achieve volume reduction, incinerators to achieve further volume reduction of combustibles and provide a less reactive waste form because the ash would be immobilized in a matrix such as borosilicate glass, and surface decontamination facilities that will result in substantial volumes of noncombustible equipment no longer being classified as TRU.

3.5   LOW-LEVEL WASTE MANAGEMENT PROGRAM

The LLW program is an amalgam of responsibilities at all governmental levels. As previously described, early experience with LLW disposal was at government facilities. Commercial facilities were first licensed to receive LLW in 1963. The learning curve from operation of these facilities led to the standardization of regulatory practices in 10 CFR Part 61 that became effective in January 1983. The provisions of the regulation were intended to preclude use of sites and waste forms that had demonstrated problems in earlier facilities.

10 CFR Part 61 combines requirements on site characteristics, waste form, site design, operating procedures, and closure and postclosure care to increase the assurance that the following performance objectives will be met:

- Exposure to any member of the public from concentrations of radionuclides released to the general environment will be limited to 25 mrem whole body;

75 mrem thyroid; and 25 mrem to any other organ. Further, such releases should be as low as reasonably achievable (ALARA) (Section 61.41).

- Protection must be provided for someone who inadvertently comes in contact with the waste at some time after institutional controls are removed (generally evaluated at 100 yr postclosure) (Section 61.42).
- All operations must be performed in compliance with the requirements of 10 CFR Part 20. Radiation exposures due to operation must be maintained ALARA (Section 61.43).
- The site and technology characteristics must minimize the need for ongoing active maintenance after the site is closed (Section 61.44).

One of the common problems encountered with the closed commercial LLW disposal sites was inadequate funding for monitoring and maintenance. Subpart E of the regulation defines necessary financial assurances that must be provided before a license may be issued. Section 61.63 requires that funds be collected to support monitoring and maintenance during the required 100-yr postclosure institutional period. The adequacy of these funds must be reviewed periodically to ensure that cost increases, inflation effects, or changes in site technology or operating conditions have not adversely affected the ability to perform the necessary postclosure care.

Specific failure mechanisms and contributors are also addressed in the regulation. One major change from earlier practice was to define three classes of waste based on differences in concentration of both short- and long-lived radionuclides. Incremental protection is then required for successively higher concentrations to provide additional protection.

Class A   waste represents the lowest potential hazard level because of a combination of limited concentration and short half-life characteristics. External radiation levels for packages containing Class A waste are generally low enough to allow direct handling. Protection of operating personnel and the public during operation is the main objective with Class A waste.

Class B   waste contains higher concentrations of radioactivity. Class B waste is generally handled using shielded casks or remote handling. The waste form requirements are more stringent for Class B waste because the higher concentration and longer hazardous life mean that protection must be continued beyond the operating lifetime of the facility.

Class C   waste is the limit for near-surface disposal because of the concentrations involved and/or the length of time the material will persist. Class C waste is usually heavily shielded during handling and barriers are required to reduce the probability and consequences of inadvertent intrusion into the waste through excavation, agriculture, or plant or animal uptake. The "inadvertent intruder" risk is evaluated at 500 yr postclosure.

Section 61.56 imposes minimum waste form characteristics that must be met for any LLW disposed in near surface facilities. These include using packaging that is resistant to biodegradation (that is, no cardboard boxes) and limitations on liquid content (less than 1% by volume). Liquids must have absorbent material capable of absorbing twice the liquid volume. The waste may contain no explosive material or material that may react explosively with water. Waste must not produce toxic gases, vapors, or fumes. Waste must not be pyrophoric. Gaseous waste must contain less than 100 Ci per package and have a pressure less than 1.5 atmospheres at 20°C. Any hazardous, biological, pathogenic, or infectious material must be treated prior to disposal.

Class B and C waste is required, in addition, to be structurally stable to reduce the likelihood of slumping and trench cover failure leading to contact of infiltrating water with the waste. It is intended that structurally stable waste remain recognizable for 300 yr. Structural stability can be provided through waste solidification, through use of a waste container or through the disposal structure into which the waste is placed. Class C waste must be further isolated by emplacement within an "intruder barrier" that may be an engineered structure (such as a concrete slab) or the Class A and B waste that is preferentially disposed above the Class C waste.

The distinction between the waste classes is made on the basis of both long- and short-lived isotopes contained in the waste as shown in Tables 3-8 and 3-9. If Class A waste is segregated from other waste there are no supplemental stability requirements.

Section 61.50 specifically addresses site suitability requirements that will minimize problems related to incomplete understanding of the long-term interaction of the site, the contained radionuclides, and the local environment. A fundamental premise is that long-term integrity is a paramount goal. To achieve that, the first requirement is that the site can be characterized and modeled before operation and monitored during and after operation. Population characteristics must be considered to minimize the potential public exposure and the probability of inadvertent intrusion. Natural resources exploitation must be considered with respect to the possibility for compromising the integrity of the facility. Drainage is important since groundwater transport is considered to be the prime mechanism by which contained radionuclides may come in contact with the biosphere. Therefore, 100 yr flood plains must be avoided and the site must be free of areas of flooding and frequent ponding. To minimize contact between the waste and groundwater, there must be a sufficient depth to the water table that normal fluctuations will not intrude into the waste disposal unit. Long-term processes that might uncover the waste include erosion, landsliding, and weathering. Each of these factors must be considered in selecting a site as well as local characteristics such as how site soils behave and whether landsliding or slumping would be anticipated. The operation of the proposed LLW disposal facility must also be considered in context with other facilities nearby. For example, the operation of another facility must not mask the ability to monitor the performance of the LLW disposal facility.

Individual compacts or states can and are adopting supplemental siting criteria that reflect concerns or sensitive areas that are unique to that locality. For example, state lands may be removed from consideration as potential sites,

TABLE 3-8   Classification of LLW by Concentrations of Long-lived Nuclides[a]

Radionuclide	Concentration (Ci/m^3)
^{14}C	8
^{14}C in activated metal	80
^{59}Ni an activated metal	220
^{94}Nb in activated metal	0.2
^{99}Tc	3
^{29}I	0.08
α-emitting transuranic nuclides with half-life greater than 5 yr	100[b]
^{241}Pu	3,500[b]
^{242}Cm	20,000[b]

Source: (NRC 1987).

[a] 10 CFR Part 61.55(a):

(3) Classification determined by long-lived radionuclides. If radioactive waste contains only radionuclides listed in Table 3-8, classification shall be determined as follows:

(i) If the concentration does not exceed 0.1 times the value in Table 3-8 the waste is Class A.

(ii) If the concentration exceeds 0.1 times the value in Table 3-8 but does not exceed the value in Table 3-8, the waste is Class C.

(iii) If the concentration exceeds the value in Table 3-8, the waste is not generally acceptable for near-surface disposal.

(iv) For wastes containing mixtures of radionuclides listed in Table 3-8, the total concentration shall be determined by the sum of fractions rule described in paragraph (a) (7) of this section.

[b] Units are nanocuries per gram.

environmental effects on endangered species may rule out other areas, and the specific formula for considering population effects varies among the several jurisdictions. Some use a direct population density criterion as a firm cutoff to exclude areas from further consideration. Others use population centers as a secondary preference criterion for choosing among areas delineated by application of other criteria.

The process of selecting sites for a new LLW disposal facility also varies among the different jurisdictions engaged in it. The multistep approach described for the HLW program is the basic method being used, but the criteria and restrictions are different because of the near surface location of the facilities and the difference in the degree and length of isolation required. A variation on the basic site selection methodology that is being considered is to solicit volunteer proposals from candidate areas since this may reduce the public resistance to the facility's being located in their community. The selection agent may be the state itself or a contractor retained to perform site characterization and/or to develop

TABLE 3-9   Classification of LLW by Concentrations of Short-Lived Nuclides[a]

Radionuclide	Concentration (Ci/m³)		
	Col. 1	Col. 2	Col. 3
Total of all nuclides with less than 5 yr half life	700	[b]	[b]
^{3}H	40	[b]	[b]
^{60}Co	700	[b]	[b]
^{63}Ni	.5	70	700
^{63}Ni in activated metal	35	700	7000
^{90}Sr	0.04	150	7000
^{137}Cs	1	44	4600

Source: (NRC 1982).

[a]10 CFR Part 61.55(a):

(4) Classification determined by shortlived radionuclides. If radioactive waste does not contain any of the radionuclides listed in Table 3-8 classification shall be determined based on the concentrations shown in Table 3-9. However, as specified in paragraph (a) (6) of this section, if radioactive waste does not contain any nuclides listed in either Table 3-8 or 3-9, it is Class A.

(i) If the concentration does not exceed the value in Column 1, the waste is Class A.

(ii) If the concentration exceeds the value in Column 1, but does not exceed the value in Column 2, the waste is Class B.

(iii) If the concentration exceeds the value in Column 2, but does not exceed the value in Column 3, the waste is Class C.

(iv) If the concentration exceeds the value in Column 3, the waste is not generally acceptable for near-surface disposal.

(v) For wastes containing mixtures of the nuclides listed in Table 3-9, the total concentration shall be determined by the sum of fractions rule described in paragraph (a) (7) of this section.

(5) Classification determined by both long- and short-lived radionuclides. If radioactive waste contains a mixture of radionuclides, some of which are listed in Table 3-8, and some of which are listed in Table 3-9, classification shall be determined as follows:

(i) If the concentration of a nuclide listed in Table 3-8 does not exceed 0.1 times the value listed in Table 3-8, the class shall be that determined by the concentration of nuclides listed in Table 3-9.

(ii) If the concentration of a nuclide listed in Table 3-8 exceeds 0.1 times the value listed in Table 3-8 but does not exceed the value in Table 3-8, the waste shall be Class C, provided the concentration of nuclides listed in Table 3-9 does not exceed the value shown in Column 3 of Table 3-9.

(6) Classification of wastes with radionuclides other than those listed in Tables 3-8 and 3-9. If radioactive waste does not contain any nuclides listed in either Table 3-8 or 3-9, it is Class A.

[b]There are no limits established for these radionuclides in Class B or C wastes. Practical considerations such as the effects of external radiation and internal heat generation on transportation, handling, and disposal will limit the concentrations for these wastes. These wastes shall be Class B unless the concentrations of other nuclides in Table 3-9 determine the waste to be Class C independent of these nuclides.

the facility. The current status of the several compacts and state programs is summarized in Appendix D.

In addition to selecting the site for new disposal facilities, evaluations are underway to determine the technology that will be used for disposal. Several states' legislation includes a prohibition on use of "shallow land burial" for disposal of LLW. The legislation may or may not provide any further description of exactly what is meant by that term but generally it has been interpreted to mean disposal without engineered enhancements to improve the site's natural capability to meet the performance objectives of 10 CFR Part 61. No facility has yet been sited and developed in accordance with all the requirements of 10 CFR Part 61. (That is, all the operating facilities were in existence before the regulation became effective.) Operating facilities have, however, made changes to their procedures that result in their meeting all of the performance objectives and most of the other requirements discussed in Section 3.1.

There are a variety of alternative technologies that may be used to dispose of LLW in compliance with 10 CFR Part 61. The choice among these technologies, like the choice among sites, is unique to the state performing the selection. It entails a balancing of factors and reflects the relative importance of such considerations as relative exposure of members of the public, relative occupational exposure, relative ease of postclosure care and/or monitoring, compatibility with site-specific environmental parameters, public perception, and cost. Developing a methodology for choosing among the alternatives is a fundamental requirement for the Compact Commissions, states, and/or operators of the new facilities. Additional information on several alternative technologies is provided in Chapter 6 (Section 6.8).

3.6 URANIUM MILL TAILINGS MANAGEMENT PROGRAM

Management of tailings at currently licensed uranium mills is performed in accordance with the facility's NRC (or Agreement State) license under the regulations in 10 CFR Part 40 relating to source material. The provisions of EPA's 40 CFR Part 190 standards on uranium fuel cycle facilities described in Section 3.1 apply to these facilities and are implemented through NRC regulations. In addition, EPA final rules on radon emissions from tailings piles, issued as one section of the regulations known as National Environmental Standards for Hazardous Air Pollutants (NESHAPS-40 CFR Part 61, Section W) (EPA 1986) require that mill owners discontinue use of large tailing piles within 6 yr. Small impoundments ($<$ than 40 acres) would be allowed for collection and retention of tailings during facility operation. Alternative practice includes continuous dewatering and burial of tailings essentially as produced ($<$ 10 acres could be uncovered at any one time).

The basic standard that such facilities must meet is that dose commitment to a member of the general public in the nearest residence will not exceed

25 mrem/yr whole body or 75 mrem/yr to the critical organ. Compliance is determined using the AIRDOS-EPA code to model and evaluate the facility.

Achievement of this change in tailings management practice (from a major task at the time of facility shutdown to essentially a "clean as you go" basis) is likely to be affected by the economic condition of the uranium production industry. As noted in Chapter 2, at the end of 1986 only six mills, representing about 9% of domestic capacity, were in active operation. Slower than anticipated nuclear power growth and competition from overseas sources of production combined to severely limit the amount of domestic uranium purchased in the last several years. A determination by DOE in 1987 that the industry is "nonviable" may require more domestic uranium to be purchased and alter these figures somewhat in future years. Some efforts have been made to stabilize existing tailings storage areas, but the vast majority of these materials will require additional stabilization to comply with the EPA standards.

Responsibility for management of uranium mill tailings has been delegated by Congress to the Department of Energy for those inactive sites at which production was performed to supply the national defense program. As described in Chapter 2, there are 25 sites that been designated as part of this program. Funding for the necessary remedial actions is split 90–10% between the U.S. DOE and the state in which the site is located in accordance with the terms of the Uranium Mill Tailings Radiation Control Act of 1978 (UMTRCA-P.L.95–604). The need for and extent of remedial action are defined by the ability of the site, and, in many cases, surrounding areas (known as "vicinity properties") to comply with standards developed by the U.S. EPA. The trigger values requiring remediation under the uranium mill tailings program are an external γ exposure rate > 0.02 mR/hr, a radon daughter concentration of 0.015 WL (including background), and a concentration of ^{226}Ra on open land of > 5 pCi/g.

Eligible sites were classified on a priority basis reflecting the amount and type of material present, spread of material beyond the original site and potential for further movement, and potentially affected population. Ten sites were identified as high priority, seven as medium priority, and eight as low priority. Actions taken—or to be taken—are determined on a case-by-case basis and reflect engineering, economic, and environmental impact evaluations. They may include stabilization on site, removal of waste for stabilization elsewhere, and extraction of residual uranium content where practical. Three sites have been remediated. Work was originally scheduled to be completed in 1990 under the terms of the UMTRCA. DOE is hoping to extend the project schedule until 1993.

Remedial actions in the Grand Junction, Colorado area have been essentially completed and the current schedule calls for the program to be completed in the Fall of 1988. The Grand Junction Remedial Action Program began in 1974. Over the course of the project almost 600 structures and properties in Grand Junction, Colorado were subject to removal of uranium mill tailings and contaminated material. About 140,000 tons of such material was removed for disposal elsewhere.

3.7 PHOSPHOGYPSUM WASTE MANAGEMENT PROGRAM

Control of the release of radioactive materials from phosphogypsum deposits must be accomplished in accordance with EPA's regulations in Subpart K of the NESHAPS regulations (EPA 1985b). The fundamental standards for radiation protection are the same as those for uranium mill tailings as described in the previous section. In particular, evolution rates for ^{210}Po must be limited for such facilities. Promulgation of these EPA standards eliminated one uncertainty concerning jurisdiction over this material since the NRC was not given responsibility for naturally occurring radioactive material until after its removal from its occurrence in nature. Earlier efforts to regulate phosphogypsum wastes were undertaken by individual states such as Florida and Montana in which large-scale processing had occurred and in which structures had been built on reclaimed land that had not been adequately stabilized to minimize release of radioactive materials.

3.8 DECONTAMINATION AND DECOMMISSIONING WASTES

3.8.1 Planned Decontamination and Decommissioning

As described in Chapter 2, wastes arising from the decontamination and decommissioning (D&D) of nuclear facilities will be varying combinations of high-level waste, transuranic waste, low-level waste, and tailings, depending on the operations conducted at the specific facility. As such, they will generally be managed under the regulations of the agency that licensed the facility or in accordance with DOE rules established for facilities operated by its contractors. The major difference from operating conditions is expected to be the larger volumes involved and the rate of production. Actual responsibility for accomplishing the D&D will rest with the licensee or the DOE contractor.

Several relatively small reactor facilities have been shut down or dismantled after having reached the end of their design life. This includes both research and test facilities as well as several early commercial power reactors. Industrial, educational, and medical facilities at which radioactive material was used have also been decommissioned. Experience with D&D at these facilities has provided information needed to estimate the amounts and types of material expected from D&D of currently operating facilities. The actual volumes and rate of production will vary with the time at which decommissioning is initiated and the procedures used (see discussion Section 2.4.2). Another factor that may alter the volume of material requiring disposal is the radioactivity level at which no further cleanup is required. The below regulatory concern (BRC) rule-making actions at EPA and NRC discussed in Section 3.1 can also provide systematic rules for decisions that have been made on a case-by-case basis to date.

There have been several cases identified in which facilities decontaminated to standards existing at shutdown have been determined not to meet current requirements. Remedial actions have been mandated to permit unrestricted access to and use of the facilities. The two primary programs established to accomplish such remediation are the Formerly Utilized Sites Remedial Action Program (FUSRAP) and the Surplus Facilities Management Program (SFMP). Both programs are the responsibility of the Department of Energy.

FUSRAP, begun in 1974, is addressing conditions at 29 sites in 12 states that were used in conjunction with Manhattan Engineering District and early Atomic Energy Commission efforts or designated by Congress for other reasons. The sites processed uranium and thorium ores, concentrates, and residues. Actions taken at the individual sites were determined in accordance with guidelines for radiological surveys performed (ORO-831) and pathways analyses (ORO-832) published by the Department of Energy in 1983. The radiological surveys performed to identify material to be removed or otherwise treated included measurements of

- fixed and removable radiation on surfaces;
- $\alpha, \beta/\gamma$ and γ exposure rates;
- radionuclide contamination in surface water, groundwater, and drains;
- surface and subsurface deposits of radioactive materials; and
- radionuclide concentrations in air, water, and vegetation.

Most of the waste produced by FUSRAP operations consists of soil, building materials (flooring and piping), and rubble. Remedial action has included removal for disposal as LLW at commercial and government disposal facilities, *in situ* stabilization, and, in several cases, interim storage pending final disposal.

SFMP was established in 1978 and is currently scheduled to continue through 2005. Over 300 radioactively contaminated facilities are subject to decontamination to levels that will enable them to be released for unrestricted use. The cleanup will be sponsored either by the Defense Programs or Nuclear Energy Office of the Department of Energy, depending on whether the site was originally used for defense or civilian programs. Waste involved may be LLW, TRU, or mill tailings. Plans call for most of the site wastes to be disposed at existing DOE disposal facilities including the Hanford Reservation, Idaho National Engineering Laboratory, Los Alamos National Laboratory, Nevada Test Site, Oak Ridge National Laboratory, and the Grand Junction Project Office Site. In most cases this corresponds to disposing of the material at the site at which it was produced. Final disposition has yet to be determined for waste produced in the cleanup of the Weldon Spring site in Missouri, the Niagara Falls Storage Site and Knolls Atomic Power Laboratory Separations Process Research Unit in New York, and the Monticello Mill Tailings and vicinity properties in Utah.

3.8.2   Decontamination and Decommissioning Wastes from Accident Recovery

Unplanned events producing radioactive waste at nuclear facilities may range from occurrences with limited consequences that produce amounts and types of waste that are manageable within the capabilities of the existing waste treatment systems, to infrequent accident conditions that require development of new procedures and/or equipment.

An example of the former would be the additional liquid and trash produced incident to cleanup of a spill in a nuclear medicine laboratory. The classic examples of the latter are the accidents at the nuclear power reactors at Three Mile Island and Chernobyl. The limited events are essentially indistinguishable from normal conditions. This section will discuss the licensing issues related to the Three Mile Island recovery and contrast them to recovery operations at Chernobyl.

The events involved in the Three Mile Island accident are briefly described in Section 1.6 of this text. A detailed chronology of the accident and subsequent events can be found in the report of the investigative commission appointed by the President after the accident (TMI 1979). Many actions taken to achieve recovery from the accident were actually waste management decisions, although they were not necessarily addressed as such at the time they occurred. The major portions of the program include

- treatment of the water in the containment and auxiliary buildings;
- removal of surface contamination from walls floor, and other structures;
- reduction of airborne radionuclide concentrations;
- removal of the fuel from the reactor vessel; and
- shipment of the waste material produced for offsite disposal.

Planning and performance of the recovery operations involved cooperative actions among the NRC, DOE, the reactor operator (General Public Utilities Nuclear), and the system manufacturer (Babcock and Wilcox). Liquid waste cleanup employed a submerged demineralizer system (SDS) and an organic ion-exchange resin system known as EPICOR II. Remote-controlled robots were employed to apply high-pressure water sprays to walls, ceilings, and other surfaces to remove transferable contamination. Vacuum attachments on the robots facilitated sludge removal from floors and transfer to treatment facilities. Dewatered sludge was solidified in cement for disposal as LLW at commercial facilities.

The most fundamental difference between the recovery action at Three Mile Island and planned decontamination and decommissioning at other power reactors was the need to remove spent fuel and core structural material from the reactor vessel when the fuel was no longer intact. A substantial amount of research and development effort was required before this could begin. There was a need to determine what the in-core conditions were, what the fuel material inventory was in various parts of the primary system, and feasible methods for removing the material from the vessel. Removal began in January 1986 and was scheduled to take about 2 yr to complete. This material, placed in shielded shipping casks after removal from the vessel, was shipped to the DOE's Idaho National Engineering Laboratory (INEL) for research and development studies and for storage thereafter. By the end of 1986 about 466 ft^3 of spent fuel and core debris containing 706,000 Ci of radioactive material had been shipped to INEL (DOE 1987). Permanent disposal is expected to occur in the geologic repository being developed to accept spent fuel and HLW as described in Sections 3.3 and 6.10. The extensive planning, procedure, and equipment development performed for this recovery operation have yielded substantial benefits in terms of occupational radiation protection. Worker exposure in the first year of the core recovery operation averaged 700 mrems. This is a small fraction of the regulatory limit of 5000 mrem/yr. The collective occupational dose (the product of the individual exposures and the number of people exposed) through mid-1988 was 5,100 person-rem. The operator currently estimates that about 6,000 person-rem will be incurred by the time the plant is placed in the Post De fueling, Monitoring, and Storage (PDMS) status (GPUN 1988). This compares very favorably with the NRC's 1983 projection of 13,000 to 46,000 person-rem (EPRI 1987).

The accident that occurred at Unit 4 of the Chernobyl power station in the U.S.S.R. on April 26, 1986 differed fundamentally from the Three Mile Island accident in almost all respects. At Chernobyl the accident was initiated during an experimental procedure for which protective systems had been deliberately disconnected by the plant operators. The reactor design itself differs from those used in the United States, which are essentially self-limiting in cases where power levels increase rapidly. Further, the second United States line of defense against release of radioactive materials, the pressure containment structure, is not used on these Soviet reactors. Details of the reactor accident and the resulting environmental transport and deposition and projected dose commitments have been reported by the Soviets and by study teams from agencies around the world (DOE/NE-0076; USSR 1986; IAEA 1986; BNL-38550). The comparison to Three Mile Island can probably be most effectively appreciated by considering that the spent fuel and core debris that was removed from the Three Mile Island pressure vessel under controlled conditions following almost 7 yr of planning and preparatory decontamination had been explosively released and widely dispersed by the accident at Chernobyl.

Waste management needs following the accident included a variety of actions related to decontamination of structures at the other three reactors onsite (two of which resumed operation in October 1986), at residences within about a 20-mile radius of the site [about 60,000 buildings had been decontaminated by spring 1987 (Wilson 1987)], and disposal of foodstuffs confiscated because radioactivity concentrations exceeded levels designated by individual countries based on potential exposure of consumers. With respect to the confiscation of foodstuffs, the World Health Organization concluded that most of the restrictions were unwarranted in terms of protecting public health but that lack of measurements on levels of contained radioactivity and uncertainties in the exact area of origin of the foodstuffs resulted in administrative decisions to err on the side of overprotection (WHO 1986). Initial concentrations were identified as due to ^{131}I whereas subsequent peaks have been identified from ^{137}Cs. Because of the cesium's 28.8 yr half-life, areas subject to elevated deposition may have to be removed from agricultural production for extended time periods, or surface soil in which the material is most highly concentrated may be removed to minimize crop uptake and resuspension and further dispersion of radioactivity.

It is expected that a great deal of further information will become available on the effectiveness of methods used for identifying areas (localized and dispersed) requiring decontamination, methods for performing the decontamination and long-term isolation of the radioactivity, and the transport and dispersion of radionuclides in the environment.

REFERENCES

(APS 1978)   "Report of the Study Group on Nuclear Fuel Cycles and Waste Management." *Rev. Mod. Phys.* **50**: 1, (1978).

(Blanchard 1978)   "Supplementary Radiological Measurements at the Maxey Flats Radioactive Waste Burial Site—1976 to 1977." U.S. EPA, EPA-520/5-78/001.

(BNL-38550)   "Preliminary Dose Assessment of the Chernobyl Accident." A. P. Hull, Safety and Environmental Protection Division, Brookhaven National Laboratory, March 1987.

(DOE 1979)   *Progress Report on the Grand Junction Uranium Mill Tailings Remedial Action Program.* U.S. DOE, Washington, D.C., 1979.

(DOE 1980a)   *Project Review: Uranium Mill Tailings Remedial Action Program.* U.S. DOE, Washington, D.C., April 25, 1980.

(DOE 1980b)   Final Environmental Impact Statement on *Management of Commercially Generated Radioactive Waste.* U.S. DOE, Washington, D.C., DOE/EIS-0046F, October 1980.

(DOE 1981a)   "Decision Regarding Use of Geologic Repositories for Disposal of Radioactive Waste." 46 FR 26677, May 14, 1981.

(DOE 1981b)   "The National Plan for Radioactive Waste Management." Working Draft No. 4, U.S. DOE Office of Nuclear Waste Programs, Washington, D.C., January 1981.

(DOE 1983)   *Standard Contract for Disposal of Spent Nuclear Fuel and/or High Level Radioactive Waste.* 10 CFR Part 961, April 1983.

(DOE 1984)   10 CFR Part 960 "Nuclear Waste Policy Act of 1982; General Guidelines for the Recommendation of Sites for the Nuclear Waste Repositories." 49 FR 47714, December 6, 1984.

(DOE 1985)   DOE/RW-0005, "Mission Plan for the Civilian Radioactive Waste Management Program." U.S. DOE Office of Civilian Radioactive Waste Management, Washington, D.C., June 1985.

(DOE 1987)   "Integrated Data Base for 1987: Spent Fuel and Radioactive Waste Inventories, Projections and Characteristics." DOE/RW-0006, Rev. 3, U.S. Department of Energy, Washington, D.C., September 1987.

(DOE/NE-0076)   "Report of the U.S. Department of Energy's Team Analyses of the Chernobyl-4 Atomic Energy Station Accident Sequence." U.S. DOE, Washington, D.C., November 1986.

(EPA 1982)   Proposed 40 CFR Part 191 "Environmental Standards for the Management and Disposal of Spent Nuclear Fuel, High-Level and Transuranic Radioactive Wastes." U.S. EPA 47 FR 58196, December 29, 1982.

(EPA 1985a)   40 CFR Part 191 "Environmental Standards for the Management and Disposal of Spent Nuclear Fuel, High-Level and Transuranic Radioactive Wastes." Final Rule 50 FR 38066, September 19, 1985.

(EPA 1985b)   National Environmental Standards for Hazardous Air Pollutants, Subpart K, Elemental Phosphorous Production Facilities. U.S. EPA, *Federal Register*, 50, pp. 1586 ff, February 6, 1985.

(EPA 1986)   National Environmental Standards for Hazardous Air Pollutants, Subpart W, Uranium Mills. U.S. EPA, *Federal Register*, 51, pp. 34056 ff, September 24, 1986.

(EPRI 1987)   "Radiation Protection Management Program at TMI-2: Noteworthy Practices and Accomplishments." Electric Power Research Institute, Palo Alto, California, 1987.

(GPUN 1988) Kintner, E.E. "Three Mile Island: A Nine Year Perspective." Presented at ANS Topical Conference on Radiological Effects on the Environment Due to Electricity Generation, July 1988.

(Hunt 1986)   "The Operational Status of WIPP." A. E. Hunt, et al. Presented at Waste Management '86, Tucson, Arizona, March 1986.

(IAEA 1986)   "Summary Report on the Post-Accident Review Meeting on the Chernobyl Accident." IAEA Safety Series No. 75-INSAG-1, International Nuclear Safety Advisory Group, Vienna, Austria, 1986.

(IRG 1979)   *Report to the President by the Interagency Review Group on Nuclear Waste Management.* U.S. DOE Washington, D.C., TID-29442, March 1979.

(ITF 1979)   "Interagency Task Force Report on the Proposed Decommissioning of the Sheffield Nuclear Waste Disposal Site." K. Dragonette, J. Blackburn, and K. Cartwright, September 1979.

(KDHR 1976)   "Radiation Concentrations at the Maxey Flats Area of Fleming County, Kentucky, January 1, 1975 to December 31, 1975." Kentucky Department for Human Resources, June 1976.

(Krenz 1986)   "WIPP and the Local Communities." D. L. Krenz and C. A. Sankey, U.S. DOE, Albuquerque, New Mexico, Presented at Waste Management '86, Tucson, Arizona, March 1986.

(Matuszek 1982)   Matuszek, J. M. "Radiochemical Measurements for Evaluating Air Quality in the Vicinity of Low-Level Waste Burial Sites—The West Valley Experience." Presented at the NRC Symposium on Low-Level Waste Disposal, June 1982, NUREG/CP-0028 CONF-820676, Volume 2.

(NAS 1979)   "Solidification of High-Level Radioactive Wastes," Report prepared by the National Academy of Engineering and National Academy of Sciences for the U.S. NRC (NUREG/CR-0895), June 1979.

(NRC 1975)   "Report of the Nuclear Regulatory Commission Review Group Regarding Maxey Flats, Kentucky Commercial Waste Burial Ground, July 7, 1975." Transmitted by J. R. Chapman, Director, Office of Nuclear Materials Safety and Safeguards to Hon. J. M. Carroll, Governor of Kentucky, July 14, 1975.

(NRC 1979)   10 CFR Part 60 "Proposed Procedures for Siting and Licensing a High Level Waste Geologic Repository." U.S. NRC, 44 FR 70408, December 1979.

(NRC 1980)   10 CFR Part 60 "Draft Technical Criteria for Licensing a High Level Waste Geologic Repository." U.S. NRC, 45 FR 31393, May 13, 1980.

(NRC 1981)   Regulatory Guide 3.48, "Standard Format and Content for the Safety Analysis Report for an Independent Spent Fuel Storage Installation (Dry Storage)." U.S. NRC Office of Standards Development, Washington, D.C., November 1981.

(NRC 1982)   10 CFR Part 61, "Licensing Requirements for Land Disposal of Radioactive Waste." 47 FR 57446, December 30, 1982.

(NUREG-0217) "NRC Task Force Report on Review of the Federal/State Program for Regulation of Commercial Low-Level Radioactive Waste Burial Grounds." March 1977.

(NUREG/CR-1759)   "Data Base for Radioactive Waste Management." Prepared by J. J. Clancy, D. F. Gray, and O. I. Oztunali, of Dames & Moore for the U.S. NRC, November 1981. Volume 1 Review of Low-Level Radioactive Waste Disposal History.

(ORO-831)   "Radiological Guidelines for Application to DOE's Formerly Utilized Sites Remedial Action Program." U.S. Department of Energy, Oak Ridge, Tennessee, March 1983.

(ORO-832)   "Pathways Analysis and Radiation Dose Estimates for Radioactive Residues at Formerly Utilized MED/AEC Sites." U.S. Department of Energy, Oak Ridge, Tennessee, March 1983.

(TMI 1979)   "Report of the President's Commission on the Accident at Three Mile Island." J. G. Kemeny, Chairman. Washington, D.C., October 1979.

(USSR 1986)   "The Accident at the Chernobyl Atomic Energy Station and Its Consequences." State Committee for Using Atomic Energy in the U.S.S.R., Data prepared for the IAEA Expert Conference, Vienna, Austria, August 25–29, 1986.

(Weaver 1986)   "Real-Time Tracking and Scheduling for WIPP Waste Transportation." J. M. Weaver, et al. Presented at Waste Management '86, Tucson, Arizona, March 1986.

(WHO 1986)   Initial assessment of the Chernobyl reactor accident by World Health Organization Scientists as reported in "Response to Chernobyl." IAEA Bulletin Vol. 28, No. 2, Summer 1986, IAEA, Vienna, Austria.

(Wilson 1987)   "A Visit to Chernobyl," R. Wilson, *Science*, **236**, June 26 (1987).

CHAPTER 4

PUBLIC HEALTH, ENVIRONMENTAL, AND SAFETY ASPECTS OF RADIOACTIVE WASTE MANAGEMENT PRACTICES

The siting and operation of new radioactive waste treatment, storage, and disposal facilities and the stabilization and closure of active and inactive facilities are governed by the need to ensure short- and long-term protection of the public health and the environment. This protection is required by the provisions incorporated in the federal laws and regulations governing the licensing of radioactive waste management operations, and is obtained through use of design concepts and operating practices that minimize the potential for dispersion of the radioactive contaminants in the environment. This chapter will first describe the mechanisms that result in dispersion of the radioactive waste into the environment and potentially cause impacts on biota and humans, and then enumerate the regulatory controls and operational programs at radioactive waste management facilities that are designed to minimize the potential for this environmental dispersion to occur. Potential accident conditions and the measures taken to control and minimize the effects of accidents are described.

The sequential steps in the process by which a radioactive waste source ultimately can cause health impacts on humans or environmental degradation are illustrated in the following schematic and discussed in Sections 4.1–4.4.

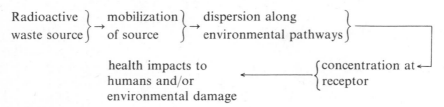

4.1 CHARACTERIZATION OF THE RADIOACTIVE WASTE SOURCE

The hazardous nature of a radioactive waste stream, as measured by both activity level and potential mobility, is a function of a number of factors. The fundamental radiological nature of the waste stream is established by the individual radionuclides present in the stream, generally in terms of the type and intensity of emissions during decay (α, β, or γ) and the persistence of the radionuclides as measured by the half-life. The activity level of these constituents will also be a function of their concentration in the waste stream. In general, the total hazardous nature of a waste stream is established by the mixture of materials present in this stream rather than the individual constituents. Thus, it is also necessary to consider the presence of other than radiological materials such as metals, process chemicals, and organics that, either acting individually or through interactions such as chemical reactions, complex formations, or synergisms, create a source that is not only radioactive, but may be hazardous in other ways. In addition, these interactions may change the physical or chemical form of the radioactive material or its matrix, thus altering its potential for mobilization.

Radioactive waste streams include source, special nuclear, or by-product material. The NRC classification system for low-level waste (LLW) developed in conjunction with the Low Level Waste Regulation (10 CFR 61) divides the waste streams into A, B, and C wastes as a function of type and concentrations of radionuclide constituents. Although these designations and classifications assist in establishing uniform requirements for handling, packaging, transportation, and disposal of the wastes, they do not characterize the source from the standpoint of determining its strength for potential environmental effects; the above-described characteristics do.

4.2 MOBILIZATION OF THE SOURCE

The dispersion of the radioactive source cannot occur unless a mechanism exists to mobilize the source and separate it from the waste stream in the form of an emission. The mobilized emission is measured in terms of a concentration and/or release rate to a dispersion pathway. The emission is a function of a combination of factors including

1. The initial concentration of radionuclides in the source. The greater the concentration, the more material is available for mobilization.

2. The physical form of the matrix in the waste stream the radionuclides are bound in. This factor is important in establishing the emission rate. Considerable variation in emission rate occurs as a function of parameters such as moisture content, density, permeability, diffusion coefficient, surface particle size, and surface area to volume ratio. Calculations of release rate must take these factors into consideration.

3. The nature and intensity of mobilization mechanisms. The mobilization mechanisms, which are the natural or man-made actions that release the material into the environment, can be categorized as air and liquid release mechanisms.

Air mobilization mechanisms include

- Fugitive particulate emissions from soil disturbance due to erosion or excavation activities, movement and placement of waste, and improperly packaged wastes. The particulates are dispersed directly into the air, or cling to vehicle or package surfaces and are subsequently released during handling or movement.
- Fugitive gaseous emissions from diffusion through the waste and soil, decomposition of organic materials, volatilization, and chemical reaction. The decay of radium to form radon gas and its subsequent diffusion through the waste or soil is an example of a fugitive gaseous emission.
- Emissions resulting from accidents, spills, fires, or explosions. Improper packaging and handling of wastes have resulted in past releases from this mechanism, as has the less common occurrence of fires or explosions in waste management operations.
- Residuals emitted from waste incineration or other treatment activities. Although the emission control systems are capable of removing the large majority of radioactive particulates and gases (in excess of 99%) from the process exhaust stream, there is always some small amount released to the air.
- Emissions from the generation process that forms the waste. In each process, whether it be the generation of tailings in the production of uranium or of LLW in fuel fabrication, a certain small fraction of the airborne emissions passes through the emission control system into the environment.

Liquid mobilization mechanisms can be further divided into those that result in releases to the surface waters and those that cause contamination of the groundwater. Surface water mobilization mechanisms include

- Spills during handling and movement of waste such as has occurred from improperly packaged wastes containing liquids, or from rupture of pipes moving radioactive waste slurries (e.g., tailings).
- Surface runoff from waste storage, treatment, or disposal areas in which precipitation or drainage has eroded contaminated material from the surface and carried it to nearby surface water, or where the precipitation has percolated through surface waste and formed leachate that has run over the surface into nearby water bodies. A less common but more damaging occurrence of contaminated surface runoff results from flooding or rupture of impoundments (e.g., Churchrock, New Mexico tailings dam) with the subsequent dispersion of radioactive wastes into local surface water.

- Direct discharge of liquid or semisolid wastes into surface water, a practice that is not permitted at nuclear facilities in this country, but that has occurred in the disposal of material such as tailings in other countries.
- Cross-connection with a contaminated aquifer resulting in inflow of contaminated groundwater.

Groundwater mobilization mechanisms include

- Infiltration of leachate from percolation of water through waste in near surface burial units and into an underlying aquifer. This release mechanism has been of primary concern in the operation of LLW disposal facilities and is a major factor in the shutdown of the West Valley and Maxey Flats commercial burial sites.
- Infiltration of spilled contaminated liquid from the surface or from water passing through wastes stored on permeable surfaces.
- Leakage of liquid from the bottom and sides of poorly sealed surface impoundments.

4. Effect of depletion of the source. As the radioactive constituents in the source are removed, particularly at or close to the surface of the matrix, the release rate of these constituents is reduced unless the source continues to be replenished. Thus, in the case of a nonreplenished in-place waste source the release rate can be expected to decrease with time. In addition, in certain cases in which radionuclides with relatively short half-lives are the isotopes of concern, natural decay will reduce the concentration of this material with time.

5. Extent of in-place control mechanisms. Where either emission control equipment is used to entrap particulates and/or gases from a mobilized waste stream or operational controls are employed to prevent releases at the source, the quantity of mobilized material that is dispersed can be substantially reduced. Examples of effective uses of controls to minimize releases of the waste source include filter systems, precipitators, and scrubbers incorporated into airborne emission streams from incinerators, grinders, and crushers in mills, and ore and fuel process areas, and water sprays and crusting agents used to reduce fugitive emissions from roadways, tailings, and waste facility surfaces. Control technology is discussed in Chapter 5.

4.3 DISPERSION ALONG ENVIRONMENTAL PATHWAYS

The radioactive material in the waste source that is mobilized and eludes the control mechanisms can then be dispersed along various pathways into the environment. Figure 4-1 illustrates the potential environmental pathways to biota and ultimately humans from a mobilized radioactive waste source. The primary pathways to humans include

- Direct γ radiation of individuals in general proximity to the source or disposed material.
- Inhalation of emissions dispersed directly into the air.
- Direct ingestion of ground and/or surface water.
- Ingestion of contaminated vegetation on which particulates have settled, or where gaseous exchange has occurred, or which have concentrated radionuclides absorbed through the root systems from contaminated soils.
- Ingestion of fauna (e.g., livestock, fish) in the food chain that have ingested and concentrated radionuclides from a lower species in the chain.

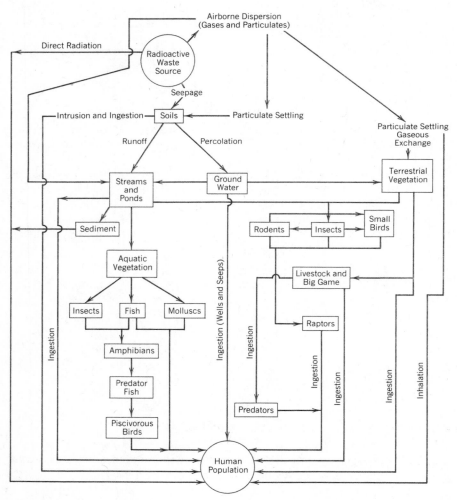

Figure 4-1 Pathway analysis to biota and man: generation and disposal locations on common site.

To evaluate the concentration at various locations along these environmental pathways, the mobilized source concentration and release rate at the point of origin are determined, and the subsequent movement through air, biota, soil, and ground and surface water is mapped until the dispersed concentrations at the receptor locations are evaluated. Because the resultant dispersion is a function of many interacting variables, computer models have been developed to assist in evaluating the pathway dispersion and to convert the concentration at receptors into doses.

The significant parameters effecting airborne dispersion of a mobilized waste in addition to the concentration and release rate of the material are the local climatological conditions including wind direction, speed, and stability class, soil moisture conditions, and the local topographic profile. A number of computer models have been developed, of successively greater sophistication, to evaluate the dispersed airborne and ground concentrations of the radionuclides, and the resultant doses they produce. These are described in Appendix B. Of particular utility is the MILDOS model that was developed for use in evaluating airborne releases from uranium processing and disposal facilities and is relevant to any releases containing constituents of the uranium decay chain. The MILDOS model simulates emissions of radioactive materials from fixed and areal sources using a sector-averaged Gaussian plume dispersion model, which utilizes user-provided wind frequency data. Mechanisms such as deposition of particulates, resuspension, radioactive decay, and ingrowth of daughter radionuclides are included in the transport model. Annual average air concentrations are compiled, from which subsequent doses to humans through various inhalation and ingestion pathways are determined. Ground surface concentrations are estimated from deposition buildup and ingrowth of radioactive daughters. The surface concentrations are modified by radioactive decay, weathering, and other environmental processes. The MILDOS code allows the user to vary the emission sources as a step function of time by adjusting the emission rates, which includes shutting them off completely. Thus, the results of a computer run can be made to reflect changing processes throughout the facility's operational lifetime. The pathways considered to obtain individual dose commitments and population doses are inhalation, external exposure from ground concentrations, external exposure from cloud immersion, and ingestion of vegetables, meat, and milk. Dose commitments are calculated using dose conversion factors that are based on recommendations of the International Commission on Radiological Protection (ICRP). These factors are fixed internally in the code, and are not part of the input option.

Groundwater dispersion of the mobilized source is dependent on (1) its concentration and release rate in the leachate, (2) the ease of access of the contaminant to the underlying aquifer as influenced by the ion-exchange capacity of the intervening soil layers, the depth to the groundwater from point of mobilization (travel distance), and the extent fracturing of those layers opens up direct conduits for the leachate, and (3) the driving force (head) that establishes the extent of water infiltration into the site as measured by the site's recharge capability. The recharge capability is determined by the balance created by the extent of precipitation on the surface, the evapotranspiration rate from the surface, the permeability of the underlying soil layers, and the surface slope and roughness, which determines the division between precipitation runoff and infiltration. In addition to these factors, the variation in characteristics between saturated and unsaturated zones through which the radiologically contaminated material passes effects leachate dispersion.

Much work has been done to develop groundwater dispersion models that will consider the interacting effects of these variables and predict, with an acceptable level of accuracy, the two- and three-dimensional movement of the contaminants in groundwater. The NRC considers the ability to model groundwater dispersion accurately to be so critical to the satisfactory characterization of a new LLW site that they have made this a basic criterion in 10 CFR 61. Although effort continues to improve the predictive nature of the models, particularly when treating a nonhomogeneous site containing fractured layers, a number of models are available to provide satisfactory accuracy in predicting concentrations as a function of distance from the source for most relatively homogeneous sites (see Appendix B). Among the most commonly used models are

- FEMWATER—A finite element model defining water flow through saturated – unsaturated porous media. The model evaluates moisture content within the bounded region.
- FEMWASTE—A two-dimensional transient model for estimating the transport of dissolved radioactive constituents through porus media. The transport mechanisms considered include conversion, hydrodynamic dispersion, chemical sorption, and radioactive decay.
- CRWATRR—A model that assesses impacts from groundwater migration of radionuclides with emphasis on waste form and packaging parameters and site selection and design parameters. The model permits the evaluation of the effect of incorporating engineered barriers into the disposal unit onsite.

Dispersion of a waste source into surface water is influenced by a number of site-related characteristics between the point of mobilization of the source and the point of entry into the surface water body. As with air and groundwater pathway dispersion, the starting point is the concentration and release rate of the mobilized source. Then the movement of the contaminant is controlled by the local topography in terms of the surface slope and roughness, and the degree of underlying soil saturation. These conditions will control the extent of groundwater infiltration and thus establish the amount of runoff into adjoining surface water. Once the radiological contaminant enters a flowing surface water body as a point source, the dispersion of the contaminant can be assessed through established predictive models, including

- LADTAP II—Determines the radiation dose to humans from the pathways in the aquatic environment such as ingestion of potable water and aquatic foods, direct radiation from shoreline deposits, swimming and boating, and ingestion of terrestrial foods irrigated with contaminated surface water.
- LPGS—Calculates radiological impact resulting from the release to the hydrosphere (i.e., estuary, small river, well, lake). Determines radiation dose to humans as a function of time for various environmental pathways.

4.4 IMPACT ON BIOTA AND HUMANS

The fundamental concern about the radioactive contaminants that are dispersed into the environment is that these materials will reach the regional biota and the human population at concentrations that result in an elevated dose to the receptor.

The regional biota are impacted when uptake of radionuclides by plants occurs through the root system that has penetrated contaminated soil or by grazing animals consuming foliage on which contaminants have been deposited. Consumption by animals of surface water that has been contaminated by runoff or deposition is another means by which elevated radionuclide concentrations may be established in the local biota. These concentrations in the biota have potential impact in two ways: through entry into the food chain to higher life forms and through elevated doses to the receptor, which may result in genetic or somatic changes.

The concentrations of radionuclides that are developed in plants and animals that are exposed to these environmental pathways are a function of the concentration of the individual radionuclide, and the balance of acceptance (bioaccumulation) and elimination (biorejection) that are inherent in the species. Each species will establish different internal concentrations of a radionuclide for the same pathway exposure because the bioaccumulation and biorejection factors will vary from species to species. Furthermore, individual parts of plants (i.e., roots, leaves, stems) and the organs in animals concentrate radionuclides at different rates. This variation will continue up the food chain, as animals consuming the plants tend to establish different concentrations in various organs in the body.

The variation in ability of individual plant species to concentrate radionu-clides is of particular concern in establishing a vegetative cover over an LLW disposal facility or a stabilized site containing radioactive constituents. Indigen-ous plant species are usually selected that do not have root systems that penetrate into the waste zone, and that do not tend to excessively concentrate the radionuclides that are present in the underlying soil.

A comparable interaction of bioaccumulation and biorejection factors exists in human receptors, producing a concentration within the whole body and individual organs (skin, kidney, bone, lung) that converts to a dose. The ratio of radiation dose produced to concentration is the dose conversion factor, which is experimentally determined for each radionuclide, pathway to the receptor (i.e., inhalation, ingestion), and organ. The unit of measurement for the radiation dose is the rem, and the low doses generally produced in the management of radioactive wastes are measured in millirem (mrem).

The effect of exposure to radiation has generally been deduced from epidemiological studies correlating radiation dose and cancer incidence. These correlations have been difficult to obtain for the doses typically produced by exposure to releases from radioactive wastes because of the existence of other impacting variables, the long period required for the onset of the induced cancer, and the inability to establish a safe or "threshold" level of exposure below which no elevated cancer risk exists. The consensus of current thinking in the field is that there is no clearly definable threshold level of exposure. As scientific opinion has moved to accept this concept, regulatory agencies and advisory bodies have successively lowered allowable exposure levels of workers and individuals in the exposed population. The ALARA (as low as reasonably achievable) concept represents the imposition of a further set of standards that requires that exposure levels be reduced below regulatory limits if the incremental reduction can be justified by the increased incremental costs associated with additional engineering barriers, emission controls, operating constraints, or other dose-reducing mechanisms.

4.5   PROGRAM TO MITIGATE IMPACT OF WASTE DISPERSION

The approach to ensuring that a facility is designed and constructed in a manner that minimizes potential contaminant releases, and is then operated and closed in a manner that minimizes short- and long-term releases, is multifaceted. It depends on the interaction of the following programs:

- Regulatory controls imposed during the period of site selection, facility design, and construction incorporated into licensing procedures and regulatory requirements.
- Use of engineered controls to minimize effluent releases.
- Monitoring programs at the facility designed to provide an early warning of unplanned releases to the environment.
- Radiation protection programs, consisting of administrative and operating controls designed to minimize worker and public exposure to contaminant sources.
- Procedures to mitigate the effects of accidents and natural catastrophes.

The success of these programs in ensuring minimal releases and exposures requires that the facility workers and management are sensitive to meeting their requirements and are educated as to the effects of lax adherence, and that the regulations impose consistent and demanding compliance activities to validate adherence to license requirements.

These programs are described in Sections 4.6–4.10.

4.6 REGULATORY PROGRAM TO MITIGATE IMPACT AT RADIOACTIVE WASTE MANAGEMENT FACILITIES

The initial and most effective method by which the government ensures that a facility managing and/or disposing of radioactive wastes successfully contains the waste and does not pose an environmental or public health risk is a comprehensive regulatory program and strict enforcement of compliance. The regulatory program that has evolved in the United States, as described in Chapter 3, has the protection of the environment and public health as a major focus. This program, which is now being applied to the handling and subsequent land burial and storage of LLW and mill tailings, will also be applied to the construction and operation of geologic repositories for high-level waste (HLW), transuranics, and spent fuel. This section will focus on the current regulatory program for packaging, handling, and land burial of radioactive wastes.

4.6.1 Licensing and Regulating Waste Management Facilities

To obtain a license to construct and operate a facility to receive, possess, and dispose of radioactive waste containing source, special nuclear, or by-product material, the potential licensee must submit an application to the NRC. This application consists of general information, specific technical data, and institutional and financial information and is accompanied by an environmental report (Section 4.6.2) prepared by the applicant or their agent. Among the information included in the application are data and evaluations to demonstrate that stipulated performance objectives, in terms of permissible exposure levels (Section 4.6.3), are achieved. The information required includes a description of the site, the design of the facility, the construction and operation of the facility, the site closure plan, the radioactive materials to be received, possessed, and disposed of at the facility, the quality control program, the radiation safety program, the environmental monitoring program, and the administrative procedures that the applicant will apply to control activities at the facility. The emphasis for each item will be the parameters and/or characteristics that relate to containment capabilities and controls on potential releases.

Several technical analyses must also be submitted to demonstrate that the performance objectives will be met. These include pathway analyses to validate protection of the general population from releases of radioactivity, protection of individuals from inadvertent intrusion, protection of individuals during operation of the facility, and analyses of the long-term stability of the disposal site and the need for ongoing active maintenance after closure.

Institutional information is also required that, in the case of an LLW disposal site or tailings impoundment, includes a certification by the federal or state government that owns the site that it will assume responsibility for custodial care after site closure and postclosure observation and maintenance. Where the proposed disposal site is on land not owned by the federal or state government, the applicant must submit evidence that arrangements have been made for assumption of ownership in fee by the federal or state government. Financial information that is submitted must, among other things, be sufficient to demonstrate that the financial qualifications of the applicant are adequate to carry out the activities for which the license is sought, and to meet financial assurance requirements relative to site closure and remediation.

The provisions of 10 CFR Part 51 "Licensing and Regulatory Policies and Procedures for Environmental Protection" and the relevant governing regulations require that applicants for NRC permits and licenses for radioactive waste management facilities submit an Environmental Report with the application. The report is discussed in Section 4.6.2.

After receipt of the application, the NRC has 60 days to determine the acceptability of the application from the standpoint of completeness and adherence to regulatory requirements. If the application is accepted for action, the NRC conducts an in-house review, culminating in the preparation of an Environmental Impact Statement (EIS) by the NRC staff or a contractor (Section 4.7.2). The review process will involve

- Distribution of the Environmental Report and supporting material for review to other involved government agencies (e.g., EPA), and private parties (e.g., public interest groups) with an interest in the proceedings. Comments are obtained and weighed.
- Conducting public hearings to obtain input from interested parties on the proposed project.
- Independent analysis of the viability of the proposed site and alternatives, and in verification of the applicant's studies, conducted by NRC staff or consultants.
- Preparation of a Draft Environmental Impact Statement (DEIS) summarizing the parameters of the proposed action and the NRC's position on the application. This may include stipulation as to additional actions or constraints.
- Circulation of the DEIS to the involved parties for comment, resolution of the comments, and revision of the DEIS in accordance with the resolution.
- Publication of a Final Environmental Impact Statement (FEIS).

The review process will be accompanied by the preparation of documentation reflecting the analyses and decisions made, and culminating in the development of a "Record of Decision."

Upon completion of the above-defined process, authorization would be issued by the NRC to develop the site in accordance with the parameters specified in the application, as modified by the conditions established (if any) in the FEIS. The operating license would be issued upon verification that, in fact, the facility had been developed in accordance with the approved conditions.

During the operational phase, the licensee carries out the disposal activities in accordance with the requirements of the governing regulation (i.e., 10 CFR 61 for LLW) and any conditions of the license. Periodically, during the operational phase, at least 30 days prior to the license expiration date, the licensee must file an application to renew their license to conduct the above ground operations and dispose of waste. This application is accompanied by a report, generally following the format of the original Environmental Report, that summarizes the operational history, and describes changes in facility characteristics and operating conditions. At this time, the Commission will review the operating history and make a decision to permit or deny continued operation. When disposal operations are to cease, the licensee applies for an amendment to their license to permit site closure. After final review of the licensee's final site closure and stabilization plan, the Commission may approve the final activities necessary to prepare the disposal site so that ongoing active maintenance of the site is not required during the period of institutional control.

The disposal site closure phase is the period when the final site closure and stabilization activities are being carried out. The licensee must then remain at the disposal site for a period for postclosure observation and maintenance (5 yr for an LLW site) to ensure that the disposal site is stable and ready for institutional control. The Commission may alter the duration of the period if conditions warrant. At the end of this period, the licensee applies for a license transfer to the disposal site owner. After a finding of satisfactory disposal site closure, the Commission will transfer the license to the state or federal government agency that owns the disposal site.

During the institutional control phase and under the conditions of the transferred license, the owner will carry out a program of monitoring to ensure continued satisfactory disposal site performance and physical surveillance to restrict access to the site and carry out minor custodial activities. At the end of the prescribed period of institutional control, the license will be terminated by the Commission.

While the above-described process will vary in details for the licensing and regulation of HLW geologic repositories, the approach will be similar.

4.6.2   Application of NEPA Process to Radioactive Waste Management

Inherent in the licensing process for radioactive waste management facilities described in Section 4.6.1 is adherence to the requirements of the National Environmental Policy Act (NEPA) of 1969 (Public Law 91-190 83 Stat. 852). To ensure that issuance of a facility operating license meets the national environmental goals expressed in NEPA, the NRC is required to assess the potential environmental effects of the proposed activities prior to issuance of the license. In order to obtain the information essential for performance of this assessment, the Commission requires that each applicant for a license submit a report on the potential environmental impact of the proposed facility and related activities.

The Environmental Report must emphasize the environmental impact of the proposed action, any adverse effects that cannot be avoided if the proposed action were implemented, alternatives to the proposed action, the relationship between local short-term uses of man's environment and the maintenance and enhancement of long-term productivity, and any irreversible and irretrievable commitments of resources that would be involved in the proposed action if it were implemented. The Environmental Report should include a benefit–cost analysis that considers and balances the environmental effects of the facility and the alternatives available for reducing or avoiding adverse environmental effects, as well as the environmental, economic, technical, and other benefits of the facility. It should also include a discussion of the status of compliance of the facility with applicable environmental quality standards and requirements that have been imposed by federal, state, and regional agencies having responsibility for environmental protection.

The NRC's regulations in 10 CFR Part 51 provide only general information concerning the contents of the applicant's environmental report. Specific and detailed guidance for the preparation of Environmental Reports is provided in Regulatory Guides (NRC 1978; NRC 1983), which define information requirements and the format for its presentation. The content of the Environmental Report for a radioactive waste disposal facility, presented in accordance with a representative format, is provided in Table 4-1.

The components of the environmental report that are of particular import to the evaluation of the license application based on environmental considerations are

- Environmental impact of the proposed action. The impact evaluation provides a means of (1) analyzing and assessing the effects of the proposed disposal facility and its operation on the environment and the public, and (2) determining the proper technological or administrative controls required to ensure the disposal of the waste in accordance with the facility performance objectives (Section 4.6.3). The methodology for assessing environmental impact involves the use of predictive analyses of doses from exposure of receptors to dispersed releases under natural and accident conditions, and comparison of these doses with established exposure limits.
- Any adverse environmental effects that cannot be avoided. A determination must be made of those environmental effects that are unavoidable and subject to later amelioration and those that are unavoidable and irreversible. Where amelioration is to be accomplished, the reduction in impact is projected and the cost of achieving the reduction becomes a factor in weighing costs and benefits. Where the effect cannot be ameliorated, the

TABLE 4-1   Contents of Environmental Reports for Land-Based Radioactive Waste Disposal Facilities

1.0   Purpose and need for proposed project—Necessity for, and basis of, new disposal site. Summary of wastes to be disposed of and their current status.

2.0   Alternatives to proposed action—Description of procedures used to evaluate alternatives, in sufficient detail to permit independent analyses of final selection. Alternatives evaluated include

- Siting Alternatives—Objectives and limitations of the site selection process, and documentation of the criteria and data used to comparatively weigh siting alternatives.

- Alternative Facility Design—Comparative assessment of proposed facility design and alternatives considering variables associated with (1) receiving, classifying, and processing waste, (2) planned location and configuration of waste disposal units, (3) construction of disposal units, (4) onsite transport of waste and placement in disposal units, and (5) construction of disposal unit cover. Environmental impact is to be documented and quantified to the extent practicable.

3.0   Characteristics of the proposed site—Description of site geography, demography, ecology, meteorology and air quality, hydrology, geology and seismology, radiological characteristics, socioeconomics, and regional historic, archaeologic, architectural, scenic, cultural, and natural landmarks.

4.0   Design of the proposed facility—Description of the wastes to be accepted in terms of types and average annual volumes, and of the principal features of the facility and how they are to be utilized. Construction plans are also described. The types and design of support facilities are provided.

5.0   Environmental effects of proposed action—Both short- and long-term effects are described. Short-term effects include those deriving from site preparation and construction, facility operations (i.e., radiological impacts on biota and man, and effects of chemical and sanitary waste discharges), and facility closure activities. Long-term effects include impact of long-term containment and potential radionuclide releases.

6.0   Environmental effects of accidents—Environmental and health effects of accidents that may occur during the operating life of the facility. Severity and probability of such accidents, as well as precautionary measures to lower the probability of their occurrence, are addressed.

7.0   Summary evaluation of proposed project—Summary of the important adverse environmental impacts and overall cost—benefit analysis for the proposed project. Description distinguishes between the unavoidable adverse impacts and irreversible and irretrievable commitment of resources. The benefit—cost balance summarizes and categorizes project benefits and costs.

8.0   Environmental measurements and monitoring programs—Description of the means by which the baseline data at the site were collected, as well as the plan and programs for monitoring the impact of the proposed activities on the environment. Preoperational baseline monitoring programs cover meteorology, hydrology and water quality, terrestrial environment, and radiological parameters. Operational monitoring programs cover the meteorological, hydrological, ecological, and radiological parameters. Conceptual designs of postoperational monitoring programs are developed.

9.0   Status of compliance—Listing of all permits, licenses, approvals, and other entitlements required by federal, state, local, and regional authorities.

impact should be measured, preferably in quantifiable terms, and weighed in the cost–benefit analysis.

- Alternatives to the proposed action. To the extent practicable, an applicant is expected to discuss all alternative parameters. The NRC wants the option of considering all available alternatives that "may reduce or avoid adverse environmental, social and economic effects expected to result from continuation and operations" (NRC 1978) of the proposed facility. The selection of alternatives is left to the applicant, but the basis and rationale for the selection in regard to number, availability, and suitability of the alternative must be specified.
- The relationship between local short-term uses of man's environment and the maintenance and enhancement of long-term productivity. This relationship is reflected in the benefit–cost balance.
- Any irreversible and irretrievable commitment of resources that would be involved in the proposed action should it be implemented. There are two categories of resources to be considered: environmental and material resources involved in facility construction, operation, and closure.
- The benefit–cost analysis that considers and balances the environmental effects of the facility and the alternatives available for reducing or avoiding adverse environmental effects, as well as the environmental, economic, technical, and other benefits of the facility. Every attempt should be made to quantify the various factors involved. Where these factors cannot be quantified, discussion should be used to present the "argument" in qualitative terms.

The summary of benefits should include services provided by the waste management facility in terms of factors such as removal of the waste from societal control and long-term isolation capabilities. Other benefits are increases in temporary and permanent employment, tax revenues, recreational, environmental, or aesthetic enhancements, and creation or improvement of roads or other facilities.

The summary of costs should include the capital costs of land acquisition and facility construction, transportation, operating and maintenance costs, facility decommissioning, site stabilization, and site reclamation costs. External social, ecological, and economic costs need also be considered. These may include short-term, external costs such as shortage of housing, inflationary rentals or prices, congestion of local streets and highways, overloading of water supply and sewage treatment facilities, crowding of local schools, hospitals, or other public facilities, overtaxing of community services, and disruption of people's lives or the local community caused by acquisition of land for the proposed site (NRC 1978).

Long-term external costs may include decreased real estate values in areas adjacent to the proposed facility, increased costs to local governments for the additional services required by the permanently employed workers and their families, lost income from environmental degradation and/or from recreation that may be impaired by environmental disturbances, reduced availability of desired (for recreation) species of wildlife and sport animals, restrictions on access to land or water access preferred for recreational use, deterioration of esthetic and scenic values, removal of land from alternative uses, and restriction on access to areas of scenic, historic, or cultural interest or degradation of these areas.

- The status of compliance of the facility with "applicable environmental quality standards and requirements that have been imposed by Federal, State and regional agencies having responsibility for environmental protection." In Section 5.0 of the Environmental Report compliance with NRC radiological concentration and exposure criteria is to be demonstrated as a necessary requirement to meet the licensing standards of the Atomic Energy Act. In the benefit–cost analysis of Section 7.0, both the radiological and other environmental impact of the facility construction, operation, and closure must be considered. Section 9.0 provides a summary of the status of compliance with all the regulatory requirements imposed by governmental authorities. These requirements include numerous other pieces of legislation and regulations imposed by federal and state bodies in addition to NEPA, the Atomic Energy Act of 1954, and the Energy Reorganization Act of 1974.

4.6.3 Performance Objective/Prescriptive Requirements

The approach of combining minimal prescriptive technical requirements with performance objectives has become an integral component of the waste management regulations, as exemplified by the provisions of 10 CFR Part 61, the LLW Regulation. Prescriptive requirements generally take the form of prohibitions against siting in certain areas (e.g., floodplain, regions of high seismic activity), definition of limiting values for certain design or operational parameters, and/or the requirement that certain health and safety-related practices be performed. Performance objectives take the form of limiting dose constraints to the public and onsite workers.

Adherence to the prescribed (specified) practices is intended to provide the supporting framework to ensure that the performance objectives are met, but, for the most part, the specific facility design characteristics, and the parameters of the operational programs within the framework of these practices are left to each facility operator. Thus, the operator retains the flexibility of determining how the performance objectives would best be achieved. For example, the design of engineered barriers such as a trench liner or trench cap is determined by the applicant based on their analysis of the requirements to achieve the necessary isolation to keep doses below the limiting performance value. Table 4-2 provides the performance objectives for commercial LLW disposal facilities as defined in 10 CFR Part 61, Supart C, and also adopted by the DOE for Federal sites.

The approach to validating whether a waste management facility meets the

TABLE 4-2 Performance Objectives

The following performance objectives, which are paraphrased from 10 CFR 61, Subpart C, represent the limiting dose constraints applicable to LLW disposal facilities:

General Requirement

Land disposal facilities must be sited, designed, operated, closed, and controlled after closure so that reasonable assurance exists that exposures to humans are within the limits established in the following performance objectives.

Protection of the General Population from Releases of Radioactivity

Concentrations of radioactive material that may be released to the general environment in groundwater, surface water, air, soil, plants, and animals must not result in an annual dose exceeding an equivalent of 25 mrem to the whole body, 75 mrem to the thyroid, and 25 mrem to any other organ of any member of the public. Reasonable effort should be made to maintain releases of radioactivity in effluents to the general environment as low as is reasonably achievable.

Protection of Individuals From Inadvertent Intrusion

Design, operation, and closure of the land disposal facility must ensure protection of any individual inadvertently intruding into the disposal site and occupying the site or contacting the waste at any time after active institutional controls over the disposal site are removed.

Protection of Individuals During Operation

Operations at the land disposal facility must be conducted in compliance with the standards for radiation protection set out in 10 CFR Part 20.101, except for releases of radioactivity in effluents from the land disposal facility, which shall be governed by the above-described objective for Protection of the General Population. Every reasonable effort shall be made to maintain radiation exposures as low as reasonably achievable.

Stability of the Disposal Site After Closure

The disposal facility must be sited, designed, operated, and closed to achieve long-term stability of the disposal site and to eliminate to the extent practicable the need for ongoing active maintenance of the disposal site following closure so that only surveillance, monitoring, or minor custodial care is required.

performance objectives varies with the phase in its life cycle that the facility is in. The following approach applies to each phase:

- *Preoperational*—The activities during this phase are associated with the selection and qualification of a new site. Compliance with site selection and qualification practices and requirements, in conjunction with segregation of the waste based on a classification system, should ensure that the performance objectives are met. Predictive analysis of potential long-term dispersion along viable pathways and assessment of its probable impact are required as part of the basis for site qualification.

- *Operational*—The operational phase for the land disposal facility is relatively short in the context of the entire span of concern about potential impact. For those facilities that are operational, adherence to the performance objectives is validated primarily through conducting environmental and occupational monitoring programs. Examination of the trends in the monitoring data provides a basis for determining whether performance objectives are being adhered to.

- *Closure/Postclosure Observation/Institutional Control*—These three phases are considered together because the actions taken in the closure phase to achieve the performance objectives are related to the long-term isolation capabilities of the facility during the subsequent postclosure observation and institutional control phases. The approach to accomplishing this for operating sites would be as follows:

  During the operational phase, as a component in the planning of the closure actions, an analysis should be made of the capability of the facility of achieving the performance objectives in the long term. This is done by evaluating the source terms, dispersion pathways and mechanisms with time, and doses at the potential receptor locations using the site and locality parameters.

  If the analysis shows that the site and locality objectives will not be achieved over the long term (institutional control phase) with the site as it stands, additional engineered barriers will have to be incorporated into the closure until the performance objectives are achieved. This may take the form of actions such as the reworking of the trench cap to decrease permeability and water infiltration or, at the extreme, the removal of buried waste for segregation and repacking.

  If the analysis indicates that the performance objectives can be achieved over the long term with sufficient margin of safety, the closure plan can proceed as originally defined.

If the facility is in the postclosure observation or institutional control phase, a similar analysis of the ability to achieve performance objectives over the long term should be performed with the objective of introducing new engineered barriers or enhanced monitoring and maintenance if required to achieve the objectives. In some instances it would be anticipated that the need for additional site remedial work would be required, even after the site is in a closed status, to achieve the performance objectives.

4.6.4 Application of the ALARA Concept

The concept of maintaining occupational radiation exposures ALARA ("as low as reasonably achievable") grew out of a 1954 recommendation of the National

Council on Radiation Protection (NCRP) (NCRP 1954). This recommendation, which was to keep radiation levels "as low as practicable" below the recommended maximum permissible dose equivalent, presumed that any radiation exposure might carry some risk. The NCRP continued to press this recommendation over the years (NCRP 1975), and was joined by other prestigious governmental, scientific, and professional organizations in espousing this approach (NAS 1972; FRC 1960). As a result, this basic radiation philosophy has been incorporated in regulations and guides of the NRC, including those related to radioactive waste management, and in recommendations to member countries by the Nuclear Energy Agency of the Organization for Economic Cooperation and Development. The ALARA concept has also been broadened to apply to environmental exposures in addition to occupational exposures.

In application, the ALARA concept requires that the cost of achieving an incremental reduction below regulatory limits be weighed against the benefit received in terms of reduced occupation or population exposures. Since regulatory limits are being achieved, the cost of the additional administrative or system controls or incorporation of an engineered barrier must be justifiable by the sustained reduction in dose achieved.

The application of the ALARA concept has had a profound impact on the philosophy behind the design and operation of waste management facilities. It has played a role in causing changes in three broad areas: (1) imposition of engineered controls and barriers to limit effluent releases, (2) improved measuring instrumentation to validate lower objectives for allowable concentrations in conjunction with enhanced monitoring of the workplace and the surrounding regions, and (3) evolution of radiation protection programs in the facilities designed specifically to achieve ALARA conditions.

The ALARA concept has been incorporated by reference into the regulations and regulatory guides issued by the NRC (e.g., the LLW regulation 10 CFR Part 61) and has been specifically defined and required by regulatory guides directed at facility management's program for ensuring that employee exposures and effluent releases are ALARA (NRC 1980).

4.7 USE OF ENGINEERED CONTROLS TO MINIMIZE EFFLUENT RELEASE AND/OR PERSONNEL EXPOSURE

Significant capability to control exposures from a nuclear material processing facility or a radioactive waste management facility, and thus achieve ALARA operation, exists during the design phase. Processing equipment and ventilation and dust control system design is based not only on maximizing process efficiency, but also on consideration of the potential for exposure to radiological and chemical contaminants. Thus, in planning the design of the facility the following major control aspects are considered:

1. Design the nuclear material processing operations to minimize potential

radiological sources. Reduction of the radiological source concentration is most effective in controlling emissions since it eliminates or reduces the need for imposition of subsequent effluent controls on the waste stream. An example of this approach is the use of wet (semiautogenous) grinding instead of dry crushing and grinding in the milling of uranium ores to suppress dust and radon gas release.

2. Control airborne releases of radioactive particulates through the use of properly designed ventilation and dust control systems. The techniques to accomplish this include the use of systems to increase the moisture content of ore and process material, dust suppressant systems for road beds and tailings areas (sprinklers or foam ejectors), exhaust filtering systems or scrubbers designed to minimize particulate emissions, with efficiencies in excess of 99%, ventilation systems with a ventilation rate adequate to maintain the concentration of radionuclides to less than 10% of the concentration limits in Appendix B to 10 CFR Part 20, maintenance of enclosed work areas in which potentially high airborne concentrations may exist under a negative pressure to prevent dispersion, use of individual air suction devices at work stations involving handling of radioactive materials, design of facility exhaust stacks to prevent exhausted air from entering intakes that service other areas of the facility, and location of equipment exhaust vents to minimize effluent releases to unrestricted areas.

3. Site burial units and waste impoundment based on a combination of geohydrologic and topographic features that will minimize the potential for radionuclide dispersion. Further, make use of engineered barriers to achieve ALARA releases over the lifetime of the facility. (See Chapter 3 for further discussion of siting radioactive waste disposal facilities.)

4. Layout the facility and equipment locations in a manner that maintains employee exposure ALARA while at the same time ensures that exposure to others is not thereby increased. To accomplish this, the facility layout must provide ready access to process equipment for operation and routine maintenance, isolate areas of high radiation and high potential levels of airborne concentrations from other processing areas, control access to the facility and further restrict entry to any airborne radioactivity area, locate emergency personnel decontamination facilities (e.g., showers) adjacent to equipment that could cause major contamination of a worker if an accident should occur, and control access to airborne radioactivity areas by use of signs, security locks, or administrative procedures.

5. Incorporate fire control and chemical hazard protection systems into the facility design that will minimize the potential for a release. The relevant design features include provision for isolation of facility areas in which there is a high potential for fire and resultant dispersion of radioactivity, use of automatic fire suppression and detection equipment in high fire-potential areas, provision for drainage of flammable materials (e.g., solvents) to sumps or to outside lined ponds or impoundments, use of dikes and/or cribs around process and storage tanks to

confine hazardous chemicals in the event of a spill or a leak, and use of high-level alarms or tanks containing hazardous chemicals to minimize possibility of spillage from a process circuit malfunction.

6. Design radioactive waste-handling equipment to optimize the ease of carrying out procedures, especially routine maintenance, in a manner that minimizes working time and exposure consistent with the requirements of the procedure.

7. Separate and isolate laboratory facilities, such as metallurgical or health physics laboratories, to avoid contamination from processing or waste-handling operations. Laboratory equipment and surfaces should be readily decontaminated, and the laboratory should be equipped with exhaust fume hoods. Radioactive wastes generated from sample analysis must be handled, stored, and disposed in a safe manner.

8. Conduct closure and postclosure operations in a manner that ensures that releases from the facility and site are maintained ALARA for an extended period. To accomplish this, measures are employed such as the incorporation of a stabilization cover of sufficient thickness to reduce and hold radionuclide releases below regulatory standards (e.g., clay and topsoil cover over uranium mill tailings areas designed to keep radon emanation below 20 pCi/m^2 for a period in excess of 1000 yr) and decontamination and demolition of facility buildings after operations cease, with the rubble disposed of in a controlled on– or offsite disposal facility.

4.8   RADIOLOGICAL MONITORING PROGRAMS

The primary objective of a multimedia operational monitoring program at a waste management facility is to verify that there is no unacceptable migration of pollutants through pathways that could lead to man, and thus to achieve and maintain safe working conditions for onsite employees and safe environmental conditions for members of the general public. In addition, monitoring programs provide data needed to assess the level of impact of site operations on the environment and the public and to determine if the site is in compliance with applicable regulations, standards, and performance objectives, to detect any unacceptable migration early enough to permit corrective actions to be taken to prevent or minimize any significant adverse impacts, for long-term predictions of waste isolation capabilities of the site that will be important in the postclosure period, to establish a data base for use in the design of future waste disposal sites and monitoring programs, and to evaluate the effectiveness of effluent control measures and equipment. Thus, a well-designed and conducted monitoring program is a key component in ensuring that the operation of the facility will not pose an environmental or public health threat.

A facility radiological monitoring program has three components:

1. Baseline monitoring prior to facility construction.

2. Environmental monitoring at the site boundary and in the region around the site.

3. Effluent monitoring at plant release points.

4.8.1 Baseline Radiological Monitoring Program

Baseline radiological monitoring programs are conducted to establish the concentrations of radionuclides in the various transfer media at the site (i.e., soil, water, air, biota) prior to disturbance of the land. The radionuclides to be analyzed are those whose presence in the waste is anticipated. If the site has not been previously disturbed, the baseline measurements are background levels of the site. The baseline monitoring program is generally conducted for a 1-yr period, with measurements taken, at a minimum, during all four seasons. Prior analysis of site characteristics (local climatology, topography, planned facility layout, and site boundaries) must be done to enable sampling and measurement locations to be selected where elevated levels of constituents might be expected during operation. A preliminary modeling of potential effluent dispersion is frequently performed to better predict these locations. Accurate establishment of baseline levels permits a direct comparison of the changes (if any) occurring at the same locations during operation.

4.8.2 Operational Environmental Monitoring Program

The operational environmental monitoring program will measure the concentrations in the environment of those radioactive materials, chemically toxic substances, and leachate indicators that are found in the buried (or stored) wastes. Direct penetrating radiation from the waste source will also be measured. The environmental surveillance program will be routinely performed at the same locations and use the same instrumentation and collection procedures as the preoperational program. This program, which is generically described in Table 4-3 for an LLW site, is focused on the site boundary locations and in the unrestricted areas around the proposed facility. Table 4-4 describes the typical radionuclide analyses for this site.

The results of the radiological monitoring program are routinely interpreted to assess existing and potential health and environmental impacts and are compared with regulatory exposure limits and performance objectives to validate that the facility performance is as predicted and within limits.

The performance of the radiological monitoring program is in accordance with an approved monitoring plan specific to this site. The monitoring plan will include an overall description of the proposed monitoring program, and specifications of sample collection and analysis frequency, method of collection, laboratory analysis procedures, and instrumentation requirements including minimum sensitivities. In addition to the radiological monitoring program, supplemental meteorologic, hydrologic, and ecologic data are collected to permit a

TABLE 4-3 Operational Environmental Sampling Program for An LLW Site

Medium	Type	Frequency	Sample size	Locations
Air	Particulate	Continuous—changed weekly	500 m³/sample 6	6 perimeter—4 offsite
	Tritiated water vapor	Continuous—changed weekly	—	4 perimeter—2 offsite
	Gases and iodine	Continuous—changed weekly	—	2 perimeter—2 offsite
	Precipitation	Monthly	Total	1 perimeter—4 offsite
Direct radiation	TLD (or other)	Bimonthly	—	6 perimeter—4 offsite
Water	Surface	Continuous collection or weekly grab	4 L	Water that drains the site and at that location that is downstream of the site and drains the area
	Subsurface—offsite	Quarterly	4 L	Up to 10 locations within 10km
	Subsurface—onsite	Monthly	4 L	12 perimeter monitoring wells and any wells into aquifers
	Subsurface—onsite	Monthly	1 L	Trench sumps and trench monitoring line wells from modeling results
Soil	Subsurface—onsite		1 kg	Collect a representative number of cores from each borehole as dug
Soil	Surface	Annually	4-5 kg	10 onsite—10 offsite
Bottom sediment	Offsite	Annually	Several kg	Above and below the site
Vegetation and farm crops	Offsite	Annually	1 kg each	Representative samples of the dominant species of the area of the site
Small mammals	Onsite	Annually	1 kg each	Representative samples of the common species that inhabit the site
Game birds	Offsite	Annually	1 kg each	In-season species at convenient locations within 10 km of the site
Fish	Offsite	Annually	1 kg each	Upstream and downstream of the site
Milk	Offsite	Quarterly	4 L	Upwind and downwind of the site

TABLE 4-4 Typical Radionuclide Analysis Schedule: Environmental Sampling Program for an LLW Site

Sample Type	Analysis	Conditions[a]
Air		All (combine samples by
Particulate	Total α, β γ-ray spec.	location monthly)
	TRU[b]	5%
Water vapor	^{3}H	All
Gases	^{3}H, ^{14}C	All
	^{129}I, ^{85}Kr, ^{222}Rn	10%
Precipitation	γ-Ray spec.	All
Water	γ-Ray spec.	25%
	^{3}H	All
	^{14}C, ^{63}Ni, ^{90}Sr, ^{99}Tc, ^{129}I, U, ^{226}Ra, TRU[b]	5–25%
Soil	γ-Ray spec.	All
	^{3}H	25%
	^{90}Sr, U	5%
Plants, crops	γ-Ray spec.	25%
	^{3}H	All
Animals	γ-Ray spec.	All
Milk	γ-Ray spec.	50%
	^{3}H	All
	^{90}Sr	50%

[a]Fraction of collected samples selected for each analysis indicated.
[b]Transuranic elements, especially α-emitting isotopes of Np, Pu, and Am.

comparison with preoperational data in these media and provide parameters for relating the measured concentrations of contaminants to health impacts.

4.8.3 Effluent Monitoring Program

The effluent monitoring program is designed to determine the releases, in terms of concentrations and release rates, of the airborne and liquid effluents that are emitted from the facility. Airborne effluents are released from stacks and vents in the facility; liquid effluents may be released directly into surface or groundwater or occur as a result of seepage from surface impoundments, or burial units. Effluent monitoring programs provide an indication of potential environmental concentrations of the released contaminants, and data to assess the effectiveness of control systems and engineered barriers in preventing releases.

Stack sampling of concentrations consists of taking an isokinetic sample of each stack discharge by drawing sample air through filter paper. Stack flow rates are measured to determine volume discharge rates. Planned releases of process water must be in compliance with the monitoring and quality requirements of the facility's National Pollutant Discharge Elimination System (NPDES) discharge permit. Any unusual releases of liquid wastes are sampled. The release rates and concentrations of the contaminants in the liquid seepage from impoundments or burial units are monitored by onsite wells located close to and around the unit, with the majority of the wells placed hydrologically downgradient of the unit.

4.9 RADIATION PROTECTION PROGRAMS

The radiation protection programs at waste management facilities consist of a number of administrative and operating components and controls that are designed to create a safe operating environment and thus minimize worker and public exposure to contaminant sources. The radiation protection program is a cornerstone of the ALARA program at the facility. The implementation of a successful radiation protection program becomes the responsibility of everyone involved in the operation of the facility with emphasis on the role of facility management and the onsite personnel responsible for radiation protection.

A radiation protection program at a waste management facility is composed of the following components, which were enunciated in a series of Regulatory Guides issued in the late 1970–early 1980 time period (NRC 1979; NRC 1980; NRC 1981).

- Appointment of an onsite Radiation Safety Officer (RSO)
- Management commitment to achieving ALARA.
- Management audit and inspection program.
- Radiation safety monitoring covering external radiation surveys, airborne radiation surveys, bioassay, and contamination surveys.
- Training Program.
- Emergency procedures.

4.9.1 Radiation Safety Officer

The Radiation Safety Officer (RSO) is responsible for developing and carrying out the radiation safety program. The RSO serves as an advisor to key project personnel with respect to all phases of the program and the implementation and adherence to the facility ALARA policy. The RSO has the authority to halt unsafe operations or to remove personnel from an unsafe work environment. Principal routine responsibilities of the RSO include the performance of the facility radiological health surveillance program, improving procedures to perform tasks with less exposure, reviewing and maintaining personnel exposure records, and processing, calibrating, and maintaining radiation monitoring instrumentation.

4.9.2 Management Commitment to Achieving ALARA

Management's commitment is expressed in the form of policy statements, information documents, and instructions to personnel that convey facility-specific mechanisms for implementing ALARA on the job. Management should require that all principal work assignments are conducted in accordance with standard written operating procedures that include consideration of relevant safety practices. Nonroutine work or maintenance activity that may result in exposure to radioactive materials is to be carried out in accordance with "Special Work Permit Procedures" that require review and approval by the RSO prior to initiation of work. The work is then performed under the supervision of a member of the radiation safety staff.

4.9.3 Management Audit and Inspection Program

Periodic management audits of procedural and operational efforts to maintain exposures ALARA encompass reviews of exposure records, plant inspections, and consultation with the radiation protection staff to ensure that all steps have been taken to reduce exposures. Table 4-5 lists the items that should be reviewed in an audit.

In addition to the periodic audits conducted by corporate or plant management, the RSO or designee would be expected to conduct audits and inspection of facility operations in accordance with the following typical schedule.

- Daily walkthrough (visual) inspection of all areas of the facility to ensure proper implementation of good safety practices. Daily work order and shift logs are also reviewed to ensure that jobs having radiation exposure were approved in writing prior to initiation of work.
- Weekly inspection of all work and storage areas to validate compliance with operating procedures, license requirements, and safety practices.
- Monthly inspection of all work and storage areas, and a review of all monitoring and exposure data for the month. The monthly inspection summary addresses any deviations from the ALARA program and any unsolved problems and proposed corrective measures.

TABLE 4-5 Material to Be Reviewed in an Audit of a Radioactive Waste Management Facility

1.	Employee exposure records (external and time-weighted calculations) showing trends for categories of workers
2.	Bioassay results
3.	Inspection log entries
4.	Documented training program activities
5.	Safety meeting reports
6.	Radiological survey and sampling data
7.	Radioactive effluent and environmental monitoring data showing trends
8.	Reports on overexposure of workers that were submitted to regulators
9.	Operating procedures that were reviewed during this time period
10.	Records on control equipment use, maintenance, and inspection

- A formal semiannual audit of the ALARA program to evaluate the overall effectiveness of the program. A written report is provided to the plant manager summarizing the relevant plant and personnel radiological safety records and discussing the trends in personnel exposures. Recommendations to further reduce personnel exposures and effluent releases are made consistent with ALARA standards.

Violations noted in audit reports are corrected and the response is documented. Recommendations are evaluated and their disposition documented.

4.9.4   Security Program

Access control at a radioactive waste management facility is achieved through posting, fencing, and use of security personnel. Radioactive material caution signs are posted at all public entrances to the property as required by law. The facility is fenced and posted as a controlled area. Caution signs, labels, and signals are posted in the area in compliance with the requirements of 10 CFR Part 20.203, "Rules and Regulations Pertaining to Radiation Control." Security personnel are on the property at all times. A gate is established, typically adjacent to the administrative office at the facility to provide access for personnel and authorized visitors. Visitors are required to register at the office and are escorted while within the plant area. Contractors having work assignments, such as equipment repair, are given security, safety, and radiation protection orientation prior to performing their duties.

4.9.5   Radiation Safety Monitoring Programs

Radiation safety monitoring programs are conducted to validate that occupational radiation exposures and effluent releases within the facility meet regulatory criteria and are further consistent with the ALARA philosophy.

External Radiation Surveys.   External radiation (β/γ) measurements are initially made at a number of selected sites both within and external to the buildings. These measurements are generally made with hand-held portable survey instruments. Sampling is performed frequently (e.g., monthly) during the initial period after operation commences and may be reduced in frequency subsequently in "low-radiation areas" if levels remain consistent and well below action levels. If action levels are exceeded, the initial step is to conduct a survey to determine the probable source of the increased levels. Follow-up action is then taken to reduce employee exposure to acceptable levels.

Film badges are used to determine individual radiation exposures to external radiation. Employees working in the project source, processing, or handling areas wear film badges during all working hours. Personnel working in "high-radiation areas," as defined in 10 CFR 20.202(b), are issued pocket dosimeters. If total calculated annual exposure exceeds 80% of the 10 CFR 20 limits, personnel involved are assigned to duties in areas of known lower radioactivity. Outside personnel (contractors, drivers, visitors) are also monitored.

Airborne Radiation Surveys. Airborne radiation surveys include particulate sampling performed at specified locations within the facility, breathing zone sampling, and surveys performed during nonscheduled maintenance. Where the facility is handling material containing uranium or radium, radon concentrations are measured at a number of locations on the site.

Airborne particulate sampling is performed to determine the radionuclide concentration on airborne dust. These data can be used to estimate personnel exposure to airborne contaminants and also serve as a check on proper operation of pollution control equipment, plant processes, ventilation systems, and housekeeping procedures.

Breathing zone samples are taken at employee work stations, generally in proximity to the airborne particulate sampling stations. The information provided by this sampling is used to determine an employee's direct exposure through the inhalation pathway. Airborne dust is sampled in the breathing zone whenever nonscheduled work or maintenance is performed in the areas of the facility where personnel may be exposed to sufficient radioactive material to cause a significant increase in their exposure. Respirator use is typically required by the RSO under these circumstances.

When radon measurements are required, they are made at sites within the buildings where the uranium or radium-bearing material is handled (e.g., waste-packaging stations), around and over external areas containing the material (e.g., tailings impoundments). Measurements are made using radon gas monitors, passive radon monitors, or equivalent instruments.

The success of the external exposure monitoring airborne radiation surveys, and other external measurements, as tools to assess internal radiation is highly dependent on the accuracy of the exposure records kept for each individual and of calculations made to convert these measurements to equivalent internal doses. Time spent by each individual at a specific area, the average concentration measured in that area, and the whole body exposure levels are recorded. Time-weighted average exposure levels and concentrations are then determined on a weekly or monthly basis and evaluated to detect any trends that may indicate a potential problem.

Internal Radiation Monitoring Program (Bioassay). The bioassay program includes analysis and *in vivo* measurements. Urinalysis is used as a check on an individual's inhalation and exposure to airborne dust concentrations, and also to assess the employee's general exposure and uptake of radionuclides. Generally individuals routinely working in a high radiation area will participate in regularly scheduled urinalysis testing. Urinalysis is also performed on mainten-ance personnel required to wear respirators during nonscheduled repair activities. The program is performed by a qualified laboratory using approved quality assurance procedures. *In vivo* measurements are generally performed on a periodic (i.e., annual) basis for all facility employees to confirm the analysis of body and organ exposure from radionuclide uptake in the workplace.

Contamination Surveys. Radioactive waste materials being handled in the facility may cause contamination of personnel, enclosed handling areas, exterior of packages, and equipment. The monitoring and control of this contamination is an important component of a Radiation Safety Monitoring Program.

Personnel protection against contamination is achieved through the imposition and enforcement of radiation safety procedures by the RSO. Table 4-6 describes representative safety instructions furnished employees in a facility processing, or handling radioactive materials and wastes. Dust generated in waste or product handling and packaging areas can be controlled by use of dust pickup hoods at designated points of release. The particulate laden air is then exhausted into a dry-type dust collector. Contaminated equipment is decontaminated onsite using water, steam, decontamination solutions, and specialized equipment and tools. The waste generated by the decontamination operation is immobilized and packaged for disposal as LLW.

Surface contamination levels are evaluated through periodic smear tests that involve taking wipes of the area and assessing the amount of radioactive material on the wipe. Decontamination procedures are used, if necessary, to achieve

TABLE 4-6   Representative Radiation Safety Instructions

1.	Do not store or eat food in any area in which the food may be contaminated by radioactive materials.
2.	Wash your face and hands thoroughly before eating.
3.	Do not enter areas of "high" dust generation unless wearing a properly fitted respirator. Clean respirators are available.
4.	Cover skin eruptions, cuts, abrasions, or other open wounds with a sterile bandage.
5.	Change clothes before leaving the facility. Work clothing that is contaminated must not be worn outside the work area or in designated lunchroom areas.
6.	Shower or have a personal radiation survey conducted prior to leaving the plant.
7.	Keep protective clothing, such as coveralls, as clean as possible. In the event protective clothing becomes contaminated, it must be changed.
8.	Do not allow dust to collect on equipment, pipes, ledges, and floors.
9.	Remove dust with a vacuum cleaner or with a hose and water. Do not use a broom or compressed air.
10.	Clean up spills as soon as possible after they occur.
11.	Wear personnel monitoring devices as directed by the Radiation Safety Officer.
12.	Conduct contamination surveys as directed by the Radiation Safety Officer.
13.	Provide urine samples for bioassay purposes as directed by the Radiation Safety Officer. Be careful not to contaminate the sample with dirt from clothing.
14.	Obey all warning signs. Do not remove or deface these signs.

TABLE 4-7 Acceptable Surface Contamination Levels

Nuclide[a]	Average[b,c]	Maximum[b,d]	Removable[b,e]
U-Nat, ^{235}U, ^{238}U, and associated decay products	5,000 dpm/100 cm²	15,000 dpm/100 cm²	1,000 dpm/100 cm²
Transuranics, ^{226}Ra, ^{228}Ra, ^{230}Th, ^{228}Th, ^{231}Pa, ^{227}Ac, ^{125}I, ^{129}I	100 dpm/100 cm²	300 dpm/100 cm²	20 dpm/100 cm²
Th-nat, ^{232}Th, ^{90}Sr, ^{223}Ra, ^{224}Ra, ^{232}U, ^{126}I, ^{131}I, ^{133}I	1,000 dpm/100 cm²	3,000 dpm/100 cm²	200 dpm/100 cm²
β/γ emitters (nuclides with decay modes other than α emission or spontaneous fission) except ^{90}Sr and others noted above	5,000 dpm/100 cm²	15,000 dpm/100 cm²	1,000 dpm/100 cm²

[a] Where surface contamination by both α and β/γ-emitting nuclides exists, the limits established for α- and β/γ-emitting nuclides should apply independently.

[b] As used in this table, dpm (disintegration per minute) means the rate of emission by radioactive material as determined by correcting the counts per minute observed by an appropriate detector for background, efficiency, and geometric factors associated with the instrumentation.

[c] Measurements of average contaminant should not be averaged over more than 1 m². For objects of less surface area, the average should be derived for each such object.

[d] The maximum contamination level applies to an area of not more than 100 cm².

[e] The amount of removable radioactive material per 100 cm² surface area should be determined by wiping that area with dry filter or soft absorbent paper, applying moderate pressure, and assessing the amount of radioactive material on the wipe with an appropriate instrument of known efficiency. When removable contamination on objects of less surface area is determined, the pertinent levels should be reduced proportionally and the entire surface should be wiped.

153

acceptable surface contamination levels. Table 4-7 lists the acceptable levels specified by the NRC (NRC 1982). Radioactive waste packages prepared for shipment or received at a storage or disposal facility are subjected to smear tests prior to handling.

4.9.6 Training Program

The purpose of an in-house radiation safety training program is to accurately inform employees of potential short- and long-term radiation hazards associated with their jobs, acquaint them with practices instituted by management to limit occupational exposures, and ensure that each has an understanding of radiation safety procedures that must be followed. New employees are required to complete the training program prior to starting work and receive refresher training annually, and are tested to validate their knowledge of safe practices as it relates to their work assignment.

The radiation protection training program sessions cover the areas of basic radiation safety procedures, health effects, and proper decontamination proce-

---

TABLE 4-8   Topics to Be Covered in a Radiation Protection Training Program

1. Fundamentals of health protection
   a. Radiologic and toxic hazards of exposure to radionuclides handled in the facility
   b. Inhalation and ingestion pathways for radionuclides of concern to enter the body
   c. ALARA concept
2. Personal hygiene at waste management facilities
   a. Protective clothing
   b. Respirators
   c. Designated areas for eating, drinking, and smoking
   d. Personal decontamination (i.e., showers)
3. Facility-provided protection
   a. Cleanliness of the work place
   b. Safety-designed features for process equipment
   c. Ventilation systems and effluent controls
   d. Standard operating procedures
   e. Security and access control to designated areas
4. Health protection measurements
   a. Measurements of airborne radioactive materials
   b. Bioassays to detect radionuclides
   c. Surveys to detect contamination of personnel and equipment
   d. Personnel dosimetry
5. Radiation protection regulations
   a. Regulatory authority of Nuclear Regulatory Commission (NRC), and Agreement States
   b. Emergency procedures

---

dures and include presentations based on situations that could arise within the facility. The course material to be covered in the training program is outlined in Table 4-8.

4.10 PROGRAMS TO MITIGATE EFFECTS OF ACCIDENTS AND NATURAL CATASTROPHES

4.10.1 Potential Accidents at Waste Management Facilities

Facilities that generate, handle, store, or dispose of radioactive waste are potentially subject to a range of accident situations that must be assessed during the planning and licensing phase to permit the incorporation of design features, operating practices, and emergency plans that minimize the environmental and public health impact of releases resulting from the accidents. These accidents would occur as a result of natural phenomena or actions of humans. Natural phenomena that could cause impact-resulting events to occur including earthquakes, tornadoes, high winds, heavy precipitation, or floods. Events resulting from human error or equipment failure include system failures such as tank or pipe leaks, electrical outages, liquid waste retention system rupture, procedural failures or operator errors, aircraft crashes into the facility, and train derailment or truck accidents.

The consequences of these events could range from no impact to a range of impacts of varying degree inclusive of

- Spillage or release of waste material.
- Fire or chemical reaction.
- Potential criticality event if special nuclear material is being handled.
- Facility, or system and equipment within the facility, loses its function and becomes a potential source of danger.

The impact from these consequences results from (1) dispersion of the waste material with the potential for movement along various pathways to humans and/or (2) immediate acute physical danger from bodily injury.

4.10.2 Accident Analysis

In order to assess the extent of impact that could arise from an accident, the potential spectrum of accidents is first defined from available information about the site and the facility design, and from the potential natural phenomena that could occur in the region. The accident spectrum will range from trivial, where no release or impact to the environment occurs, to very serious, where large releases occur to the environment and major local and/or regional impact would be anticipated. The next step is to assign accident probabilities based on a detailed examination of the site and the facility design and relevant probability data

TABLE 4-9 Representative Probability Classification System

Class	Description	Probability of Occurrence
High	Event is likely to occur several times during the life of the facility	$10^{-1} < p < 1.0$
Moderate	Event is likely to occur some time during the life of the facility	$10^{-2} < p < 10^{-1}$
Low	Event is unlikely to occur during the life of the facility	$10^{-4} < p < 10^{-2}$
Extremely low	Event is extremely unlikely to occur during the life of the facility	$p < 10^{-4}$

derived from comparable situations. While there are a number of ways to classify the probability of accident occurrence, the system in Table 4-9 has proven to be both accurate, and widely accepted.

Potential onsite and offsite impact to humans resulting from injury and from release of radioactive and other hazardous materials is estimated using analytic techniques such as source release calculations for gradual or rapid release, followed by dispersion and dose modeling for the affected pathways. Design and operational features of the facility that reduce accident effects are considered in estimating the impact. Once the consequences of the accidents are assessed, they

TABLE 4-10 Representative Hazards Classification System

Class	Criteria
Low (trivial)	Meets all of the following criteria: 1. Potential for personnel radiation exposure or effluent concentrations less than 10 CFR Part 20 limits 2. No potential for nuclear criticality 3. No potential for exposure to toxic material in excess of one threshold limit value (TLV) 4. No property loss over $250,000
Moderate	Meets *any* of the following criteria: 1. Could cause personnel injury on- or offsite 2. Could result in releases and/or exposures exceeding applicable limits in 10 CFR Part 20 3. Could cause offsite exposures to toxic materials exceeding TLVs 4. Could cause more than $250,000 property damage
High (very serious)	Meets *any* of the following criteria: 1. Could cause injuries to more than five persons on- or offsite 2. Could cause one or more fatalities on- or offsite 3. Could cause unplanned nuclear criticality

are classified according to the hazards they represent. Table 4-10 provides a representative system for classifying accident hazards.

Frequently, after the potential facility accidents are classified by probability and consequence, numerical values are assigned to these classes, and the spectrum of accidents is tabulated to provide a readily compared summary of the results. Whether or not this tool is used, the results of the risk assessment provide a basis for (1) modifying site or design characteristics to reduce the risk, if required, (2) defining precautionary measures, in terms of operational constraints, to lower the probability of accident occurrence, and/or (3) developing emergency procedures to be implemented in case the accident should occur. Thus, the accident analysis leads to a safety analysis.

4.10.3 Transportation Accidents

When radioactive waste or other nuclear material is transported on or off a site, the probability and potential consequences of accidents along the transport route must be evaluated to assess the risk of these shipments. The results of these analyses are used to (1) modify waste packages or transport vehicles and/or (2) impose procedural constraints and/or (3) develop emergency procedures to clean up the results of vehicle accidents. A comparable analytic procedure is followed as was described for facility accidents. Worst-case accident scenarios are defined for the waste transport, generally involving the breaching of the waste containers, and release of the material. The accident probability for the transport mode (i.e., truck, train, plane) and route is determined from accident statistics, and converted into a total accident probability for all shipments. The presumption is made that a significant fraction of the waste becomes airborne and is dispersed, and airborne dispersion and dose models are then used to calculate impact on the transport workers and population along the transport route. Additionally, the accident is conservatively assumed to occur adjacent to a surface water body that is a drinking water source. A significant fraction of the waste enters the water body, is dispersed downstream from the source, and is drawn into the drinking water inlet. Generally, even with cumulative conservative assumptions along both the airborne and surface water pathways, the calculated doses are quite low.

4.10.4 Safety Analysis Report

The Safety Analysis Report (SAR) is the vehicle for documenting the results of the accident and safety analyses. The SAR identifies and evaluates the onsite and offsite effects of accidents on occupational and public health and facility integrity and enumerates the safety measures to mitigate against the impact of the accident. The SAR is prepared by the license applicant for review by the NRC, other federal and state agencies, and the public, and focuses on the following items.

- A description of the site operations and the measures incorporated into the

TABLE 4-11   Representative Format of Safety Analysis Report (SAR)

1.0   Introduction and general description of plant
2.0   Site characteristics (focus is on characteristics of significance in an accident situation such as surface faulting, stability of slopes, flood potential, ice effect)
3.0   Design of structures, components, equipment, and systems (conformance with regulatory design criteria, classification of structures, components, and systems, seismic design, flood design, missile protection, protection against dynamic effects, environmental design of equipment)
4.0   Engineered safety features (e.g., engineered barriers in waste disposal sites, double walled waste storage tanks)
5.0   Instrumentation and controls
6.0   Radioactive waste management (source terms, liquid, gaseous, and solid waste management systems design bases and system description, monitoring and sampling systems)
7.0   Radiation protection (ALARA program, radiation sources, dose assessments, health physics program, radiation protection design feature)
8.0   Conduct of operations (organization, training, emergency planning, review and audit, security, plant procedures)
9.0   Startup test program (scope, schedule, startup procedures, records)
10.0   Accident analyses (risk assessment of probabilities and hazards of each accident scenario)
11.0   Quality assurance plan

facility design and operation to ensure the protection of public health and safety.

- A probability analysis of potential accidents based on the site and facility characteristics, and data from other comparable sites, to assess the likelihood of such accidents occurring at this site.
- Precautionary measures to be followed to lower the probability of occurrence of these accidents based on the results of the risk assessment.
- Discussion of emergency procedures to be implemented if an accident occurs.

Table 4-11 provides a representative format of an SAR for a radioactive waste management facility. The environmental report submitted with the license application will also discuss the environmental effect of possible accidents at the facility and during transportation of waste material to and from the facility.

REFERENCES

(FRC 1960)   Federal Radiation Council, "Background Material for the Development of Radiation Protection Standards." Report No. 1, Washington, D.C., 1960.
(ICRP 1974)   International Commission on Radiological Protection (ICRP), "Imple-

mentation of Commission Recommendations That Doses Be Kept As Low As Readily Achievable." Report No. 22, Pergamon Press, Elmsford, New York, 1974.

(NAS 1972)   National Academy of Sciences—National Research Council, "The Effects on Population of Exposures to Low Levels of Ionizing Radiation." Washington, D.C., 1972.

(NCRP 1954)   National Council on Radiation Protection (NCRP), "Recommendations of the NCRP—Permissible Dose From External Sources of Ionizing Radiation." NCRP Report No. 17, Washington, D.C., September 1954.

(NCRP 1975)   National Council on Radiation Protection and Measurements, "Review of the Current State of Radiation Protection Philosophy." Report No. 43, Washington, D.C., January 1975.

(NRC 1978)   U.S. Nuclear Regulatory Commission, Regulatory Guide 3.8, "Preparation of Environmental Reports for Uranium Mills." Revision 1, Washington D.C., September 1978.

(NRC 1979)   U.S. Nuclear Regulatory Commission, "Acceptable Programs for Respiratory Protection." Regulatory Guide 8.15, Washington, D.C., 1979.

(NRC 1980)   U.S. Nuclear Regulatory Commission, "Health Physics Surveys In Uranium Mills." Regulatory Guide and Value/Impact Statement, Washington, D.C., August 1980.

(NRC 1981)   U.S. Nuclear Regulatory Commission, "Information Relevant to Ensuring That Occupational Radiation Exposures at Uranium Mills Are As Low As Reasonably Achievable." Draft Regulatory Guide OH941–4, Washington, D.C., 1981.

(NRC 1982)   U.S. Nuclear Regulatory Commission, "Guidelines for Decontamination of Facilities and Equipment Prior to Release for Unrestricted Use or Termination of License for By-product, Source or Special Nuclear Material," Washington, D.C., 1982.

(NRC 1983)   U.S. Nuclear Regulatory Commission, "Standard Format and Content of Environmental Reports for Near-Surface Disposal of Radioactive Waste." Regulatory Guide Guide 4.18, Washington, D.C., June, 1983.

CHAPTER 5

LOW-LEVEL RADIOACTIVE WASTE MANAGEMENT TECHNOLOGY

The management of low-level radioactive waste encompasses each step in the system from the generation of the waste source to final disposal (Figure 5-1). Because disposal is in itself a broad and predominant aspect of the waste management process, it is treated separately in Chapter 6. However, the action taken at each step in the waste management process influences the type and volume of waste material shipped to the disposal site; in turn the design and operating constraints at the disposal site influence the other steps in the waste management process. Thus, it is necessary to consider the entire waste management system and the interactions between the various steps.

With the increasing pressures to reduce the volume of low-level waste (LLW) shipped to disposal sites, the ability to minimize waste generation and to optimize subsequent processing to achieve a high degree of cost-effective volume reduction becomes a more significant aspect of LLW management. Volume reduction pressures derive primarily from the increasing costs of transporting and disposing LLW, which in turn are driven by evolving federal and state regulations for parameters such as waste form and classification. These pressures will continue to intensify as the individual LLW compacts impose further constraints on LLW received at their disposal facilities.

The determination as to what path these generated and collected wastes take enroute to the disposal facility is a function of the following factors, which are considered sequentially:

- Whether the as-generated LLW form can be packaged and shipped to a disposal facility.
- The acceptability of the packaged waste for disposal.

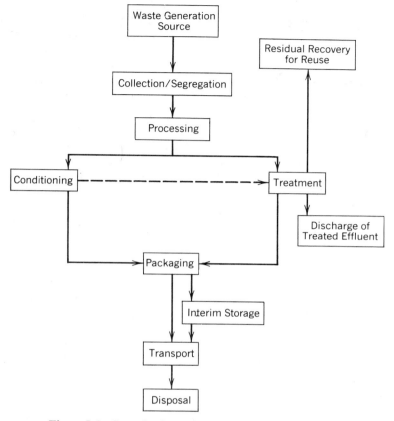

Figure 5-1   Steps in the management of low-level waste.

• The cost effectiveness of disposing of the waste in the regenerated form.

The assessment of these factors leads either to the selection of treatment and/or conditioning steps prior to handling and ultimate disposal, or to the direct shipment of the as-generated and packaged waste to the disposal facility.

5.1   LOW-LEVEL WASTE GENERATION SOURCES AND FORM

5.1.1   Major Low-Level Waste Generation Sources

The LLW generation sources, which may exist wherever radioactive materials are handled and/or processed, can be categorized into three broad classes:

1. Industrial and institutional sources—manufacturers, test and research laboratories, hospitals, and academic institutions.

2. Commercial nuclear fuel cycle sources—production facilities for uranium hexaflouride, isotope enrichment, and fuel fabrication, nuclear power plants, spent fuel storage facilities, and fuel cycle facilities undergoing decontamination and decommissioning.

3. Government sources—Department of Energy (DOE) laboratories, Department of Defense (DOD) processing and production facilities, and DOE and DOD facilities undergoing decontamination and decommissioning. The radionuclides present in the LLW from these sources cover a wide spectrum. Typically waste from conversion, enrichment, and fuel-fabrication facilities will contain the radionuclides in the uranium decay chain as the predominant constituents. Industrial and institutional waste will contain a broader spectrum of radionuclides in small quantities, with ^{3}H, ^{14}C, ^{60}Co, ^{99m}Tc, and ^{137}Cs representative of constituents from medical, manufacturing, or test operations. Nuclear power plants and spent fuel storage facilities will produce LLW containing primarily activation products, fission products, and transuranic elements.

Because of the growing trend toward reducing volumes of LLW shipped for disposal, the generation source has come under increasing attention as a place to prevent or minimize the production of the waste. Among the steps being taken to achieve this end are processing of liquid radwaste more effectively, including plants operating their liquid radwaste systems as designed to ensure they do not produce excess waste. In addition, facilities are actively working to reduce the liquid input to radwaste by taking actions such as minimizing valve leaks through early detection and accelerated repair. LLW volumes can also be reduced by preventing the materials being handled from becoming radioactive by maintaining good housekeeping practices, and reducing the number of items brought into contaminated areas (e.g., packing materials, disposables such as plastic bags).

5.1.2 Low-Level Waste Forms

The physical form of the LLW is the primary consideration in the selection of the treatment alternative for the material, and in the choice of packaging approach. The physical forms of the waste are liquids, wet solids (semisolids), dry solids, and gases. Radioactive gases are treated to remove contaminants and the resultant waste generated by the treatment process (e.g., filtration, scrubbing, absorption) falls into one of the other three categories. The decontaminated gas is then usually discharged directly to the environment.

Liquid LLW streams are fluids with relatively low concentration of solids, generally less than 1% suspended solids. The "liquid" designation, however, may be applied to fluids containing up to 10% suspended solids, which would then be categorized as slurries. The predominant liquid LLW form are

Decontamination solutions generated from decontamination of building and equipment surfaces. These solutions are either water based, containing suspended solids and small quantities of chemicals and detergents, or chemical based, containing strong detergent agents that are usually also chelating agents.

Chemical regenerative solutions produced from the regeneration of organic ion-exchange media such as demineralizer beds and resin beds.

Contaminated oils arise from the oil/water separators used to treat floor and equipment drains, from hydraulic scrubbers used in system piping, and from used oil from reactor coolant pumps.

Liquid scintillation fluids generated as organic solvents (e.g., toluene, xylene) in liquid scintillation counting operations.

Miscellaneous liquid wastes produced from varying sources such as laundry waste, discharges from equipment and floor drains, process and steam condensate, steam-condensate cooling water discharge, and rinse water from cleaning and purging operations. Semisolids often are produced from the treatment of liquid waste, or as thick (high solids content) slurries (e.g., tailings) from a materials processing operating. This waste form covers evaporator bottoms, expended filter cartridges, biological waste (i.e., animal carcasses, animal bedding, excreta, vegetation, culture media), and miscellaneous slurries.

Dry solid LLW streams can be categorized as

Contaminated equipment and materials, inclusive of major plant equipment (e.g., piping sections, pumps, valves, and motors) that has been taken out of service, building material from decommissioning and remediation operations, and radiologically contaminated soil from remediation operations.

Bulk trash, categorized as compactible or noncompactible, and combustible or noncombustible. For example, clothing, rags, plastic, and paper products are both compactible and combustible, whereas construction material, tools, and piping are generally noncompactible and noncombustible.

Irradiated hardware consists of plant components activated by exposure to a neutron flux and then removed from service.

5.2 COLLECTION AND SEGREGATION OF LOW-LEVEL WASTE/WASTE CLASSIFICATION

The collection and segregation of LLW prior to either processing or packaging of the waste are driven by the acceptance criteria established by the NRC and DOE at commercial or government LLW disposal facilities, respectively. These criteria, which are essentially comparable, are in turn based on a waste classification system and a set of corresponding waste form acceptance criteria for land burial of LLW initially defined in 10 CFR Part 61 and described in Chapter 3. This classification system defines three categories of waste: Class A,

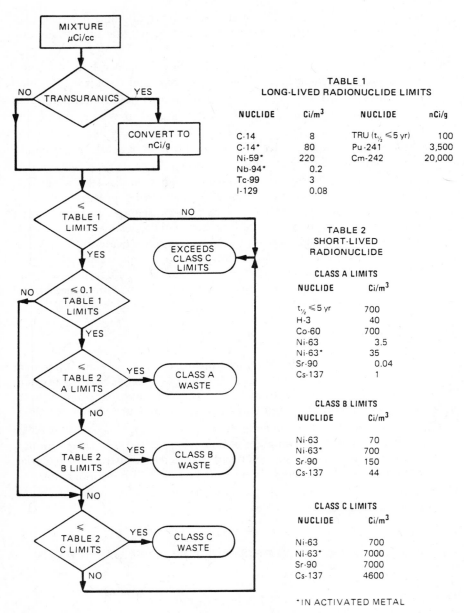

Figure 5-2 Classification chart for radioactive waste based on 10 CFR Part 61. Source: (DOE 1984).

The figure contains the following text and tables:

Flowchart:
- MIXTURE μCi/cc
- TRANSURANICS — NO / YES
- CONVERT TO nCi/g
- ≤ TABLE 1 LIMITS — NO / YES
- EXCEEDS CLASS C LIMITS
- ≤ 0.1 TABLE 1 LIMITS — NO / YES
- ≤ TABLE 2 A LIMITS — YES → CLASS A WASTE / NO
- ≤ TABLE 2 B LIMITS — YES → CLASS B WASTE / NO
- ≤ TABLE 2 C LIMITS — YES → CLASS C WASTE / NO

TABLE 1
LONG-LIVED RADIONUCLIDE LIMITS

NUCLIDE	Ci/m³	NUCLIDE	nCi/g
C-14	8	TRU (t½ ≤ 5 yr)	100
C-14*	80	Pu-241	3,500
Ni-59*	220	Cm-242	20,000
Nb-94*	0.2		
Tc-99	3		
I-129	0.08		

TABLE 2
SHORT-LIVED RADIONUCLIDE

CLASS A LIMITS

NUCLIDE	Ci/m³
t½ ≤ 5 yr	700
H-3	40
Co-60	700
Ni-63	3.5
Ni-63*	35
Sr-90	0.04
Cs-137	1

CLASS B LIMITS

NUCLIDE	Ci/m³
Ni-63	70
Ni-63*	700
Sr-90	150
Cs-137	44

CLASS C LIMITS

NUCLIDE	Ci/m³
Ni-63	700
Ni-63*	7000
Sr-90	7000
Cs-137	4600

*IN ACTIVATED METAL

Class B, and Class C. In each class, maximum concentration limits are set for individual radionuclides, with the limits increasing from Class A to C. Class A places only minimum requirements on waste form and characteristics; Classes B and C set more stringent requirements on waste form and characteristics to ensure physical stability after disposal, mainly requiring that the waste must remain stable and recognizable for 300 yr. This is accomplished by measures such as stipulating that less than 1% of the volume as free water is permitted for waste packaged in a disposal container designed to ensure stability. Where radionuclide concentrations exceed the Class C values, the waste would not be accepted for land disposal unless specific approval is granted by the NRC for an exception. Figure 5-2 is a schematic representation of the classification decision-making process that leads to segregation of the waste on the basis of class.

In addition to the classification system, 10 CFR Part 61 also imposes other minimal prescriptive acceptance criteria on LLW disposers for all waste classes, which in turn result in actions on the part of generators and processers to segregate the waste prior to packaging. These include (1) waste must contain less than 1% of the volume as free liquid, (2) wastes must not be capable of deterioration, explosive decomposition or reaction, or explosive reaction with water, (3) wastes must not contain or generate toxic gases, vapors, or fumes that might be harmful, (4) wastes must not be pyrophoric, and (5) hazardous, biological, pathological or infectious waste material must be extracted to minimize the nonradiological hazard.

5.3 WASTE-PROCESSING TECHNOLOGIES

LLW is generally subjected to processing to reduce the volume of waste requiring disposal, to change the waste form, and to separate the radioactive components from the nonradioactive component in the waste stream. LLW processing generally takes place when it is necessary to satisfy regulatory requirements or if processing improves the overall cost effectiveness of disposal. Waste processing takes the form of treatment and/or conditioning.

Treatment involves processing of the wastes to produce a waste stream that has a smaller volume and higher concentration of radioactivity than the original waste (DOE 1984), and a second stream of sufficiently lower radionuclide concentrations than the original waste to permit release or reuse. Conditioning includes those operations that transform the LLW, with or without prior treatment, into forms acceptable for transportation and disposal.

Treatment can be further subdivided into transfer, concentration, and transformation technologies. Transfer technologies are employed to remove the radioactivity from the waste stream and transfer it to another medium, permitting reuse or discharge of the original waste stream or use of other treatment technologies that are more cost effective (DOE 1984). These techniques are usually applied to liquid wastes, but may also be applied to regeneration of ion-exchange resins. Concentration technologies achieve a smaller waste volume by concentrating the radioactivity in the original waste form matrix. Transformation technologies also result in concentration of the radioactivity in the waste, but require changing the physical form of the waste. They are used for processing liquids, and both wet and dry solids. The effectiveness of each of these technologies is generally enhanced by a pretreatment step involving the removal of excess water in liquids and semisolids or such activities as sorting and/or

TABLE 5-1   Matrix of Treatment/Conditioning Technologies and Waste Forms

	Liquids	Wet Solids	Dry Solids
Transfer Technologies			
Decontamination			×
Filtration	×	×	
Ion exchange	×		
Chemical regeneration		×	
Ultrafiltration	×		
Reverse osmosis	×		
Concentration Technologies			
Evaporation	×		
Distillation	×		
Crystallization	×		
Flocculation	×		
Precipitation	×		
Sedimentation	×	×	
Centrifugation	×	×	
Drying		×	
Dewatering		×	
Dehydration		×	
Compaction			×
Baling			×
Shredding			×
Integrated systems	×	×	
Transformation Technologies			
Incineration	×	×	×
Calcination	×	×	
Conditioning Technologies			
High-integrity containers		×	×
Solidification	×	×	×
Absorption	×	×	

Source: (DOE 1984).

shredding for dry solids. Table 5-1 provides a listing of predominant treatment and conditioning technologies applicable to each type of waste form. These are discussed further in Sections 5.4 and 5.5.

The U.S. nuclear industry, particularly the utilities that generate large volumes of LLW, are increasingly using volume reduction treatment techniques to reduce processing and disposal costs. However, there are significant uncertainties for industry as to what techniques to develop because of the inability to predict evolving federal regulations and state LLW compact disposal requirements for waste parameters. At the same time current legislative and regulatory requirements are imposing increased LLW disposal costs requiring that the industry now employ volume reduction techniques.

Under the Low-Level Radioactive Waste Policy Amendments Act of 1985, in 1993 the compacts containing the three commercial disposal sites will be able to exclude waste from generators in other compact regions and independent states. Until then, the amount and cost of LLW sent to these sites from outside their compact region are regulated by a volume limit and imposed surcharge. In addition to a cumulative limit for each of the existing sites, there is an allocation system limiting the volumes that may be shipped by individual power reactors. The current allocation limits will be further decreased in 1990, and the surcharge progressively scaled up from $10 in 1986–1987 to $40 in 1990–92 for waste shipped from nonregional generators.

There is, however, a development that could act to reduce to some extent the amount of volume reduction performed by nuclear facilities. If the NRC enacts a generic rule for some LLW to be considered as "below regulatory concern (BRC)" and disposed as routine industrial waste, less waste would have to be processed.

5.4   LIQUID AND SEMISOLID TREATMENT TECHNOLOGIES

5.4.1   Liquid Treatment Technologies

Liquid LLW treatment is designed to concentrate the radioactivity into a much smaller volume, usually as a semisolid. The semisolid may be further treated or conditioned to remove or stabilize any remaining free liquid. Table 5-2 is a matrix relating the technologies frequently used to process liquid LLW with the predominant LLW streams. The major treatment technologies are discussed below:

Filtration. A transfer process in which the undissolved particulate solids in a liquid waste stream are separated from the liquid by forcing the stream through a porous body. There are a number of commercially available filters used at nuclear facilities with cartridge filters, edge filters, and precoat filters being the most commonly used. In cartridge filters the flow is directed through the filter element. Multiple disposable elements are used. In edge filters, the construction is similar to the cartridge filter, but the filters are reusable and can often be cleaned in

TABLE 5-2 Technologies Often Used for Processing the Various Types of Liquid Low-Level Waste

	Regenerative Solutions	Decontamination Solutions	Oils	Other Organics	Scintillation Vials	Other Inorganics	Biologic Wastes
Transfer Technologies							
Filtration	X					X	
Ion exchange	X	X	X			X	
Ultrafiltration	X					X	
Reverse osmosis	X					X	
Concentration Technologies							
Evaporation	X	X				X	
Distillation	X	X				X	
Crystallization	X	X				X	
Flocculation	X	X				X	
Precipitation	X	X				X	
Sedimentation	X	X				X	
Centrifugation	X	X				X	
Integrated systems	X	X				X	
Transformation Technologies							
Incineration			X	X	X		X
Calcination			X	X	X		X
Conditioning Technologies							
Solidification			X	X	X	X	
Absorption			X	X	X	X	X

Source: (DOE 1984).

place. In precoat filters, porous vertical tubes coated with a filter medium are used, then removed and cleaned, and precoated for reinstallation in the filter. Figure 5-3 shows a common disposable cartridge filter. Table 5-3 provides a comparison of the advantages and disadvantages of filters used in nuclear facilities.

In the generic filtration process, the liquid feed passes through the porous media where the solids are either trapped in the pores of the filter media and/or build up on the surface of the media. This buildup is called filter cake. The filter (see Figure 5-3) typically is composed of the housing or vessel containing inlet and outlet ports, a volume of space to permit cake buildup, holdup space for the feed and filtrate, filtrate drainage channels, and instrumentation and controls.

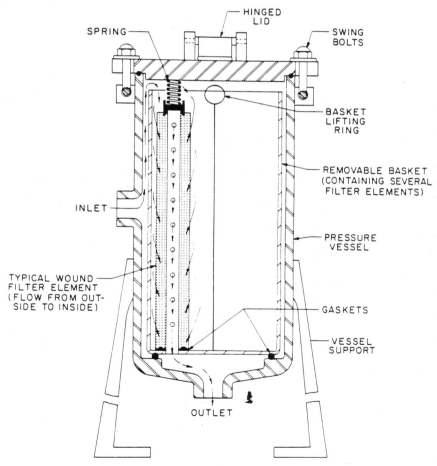

Figure 5-3 Typical disposable cartridge filter illustrating liquid flow from outside to inside of element. Source: (DOE 1984).

TABLE 5-3 Potential Advantages and Disadvantages of Filters for Liquids in Light-Water Reactor Nuclear Power Plants

Type of Filter	Advantages	Disadvantages
Disposable		
Wound cartridge	Compact, low solid waste volume, no backflush gas or liquid to treat, and good solids removal	Remote or automatic changeout difficult because of nonuniformity and poor arrangement, changeout frequently done on radiation level rather than pressure drop, and media migration may occur
Pleated paper cartridge	Compact, low solid-waste volume, no backflush gas or liquid to treat, and good solids removal	Same as above
Pleated wire screen	Can operate at elevated temperatures, good solids removal, and little or no media migration	Fair mechanical strength when adequately supported and plugging may cause uneven flow and nonuniform cake buildup
Woven mesh bag	Rapid changeout and good roughing filter	Manual changeout has limited use to treating low activity streams at low temperature and pressure
Reusable without precoat		
Partially cleanable metallic	Fair mechanical strength and may be used at temperatures up to 540°C	Eventual permanent plugging may force replacement, filter life unknown or limited, and filter medium subject to chemical attack by cleaning solutions
Porous ceramic	Fair mechanical strength and resistant to attack by cleaning solutions	Eventual permanent plugging may force replacement, and filter life unknown
Stacked etched-disc	Short backflush time with thorough cleaning, expected to last for plant life, amenable to automatic or remote operation, low solid waste	Low crud-holding capability, corrosion characteristics unknown, backwash waste to treat, and low oil-holding capacity
Reusable with precoat		
Backflushable tubular bundle	Amenable to automatic or remote operation, powdered resin or diatomaceous earth precoat can be used, and relatively compact	Precoat loss upon loss of flow or fluctuation in pressure, excess or uneven cake can cause strain and possible collapse of supporting screen, and incomplete backflushing causes uneven precoat

Dry cake discharge		
Centrifugal discharge	High crud-holding capacity, can handle automatically and remotely all plant wastes with same filter, low maintenance requirements, and no precoat loss caused by loss of flow, pressure, or power	Relatively high headroom, cake overloading can cause distortion, generates large sludge volume, and some cake difficulty with Solka Floc or resins alone
Flat bed	High crud-holding capacity, can handle automatically and remotely all plant wastes with same filter, and no precoat loss caused by loss of flow, pressure, or power	Relatively high headroom, cake overloading can cause belt wear, generates large sludge volume, some cake difficulty with resin alone, and may require fairly high belt maintenance
Clamshell	High crud-holding capacity, can handle automatically and remotely all types of plant waste, no precoat loss caused by loss of flow, pressure, or power, and relatively small floor space	Relatively high headroom generates large sludge volume, maintenance requirements unknown, and no nuclear operation data to date
Reusable deep bed		
Ground walnut shells	Low sludge discharge and amenable to automatic or remote operation	Requires addition of flocculating agent, which may interfere with water recycle, and unsatisfactory performance with detergent
Reusable magnetic		
Magnetite bed	Can operate effectively at temperatures up to 260°C, effective in removing heavy crud load of corrosion products, and some soluble activity removal	Deep bed plugs rapidly and poor performance at low-crud loads, therefore not applicable to high pH chemical shim at PWRs
Electromagnetic	Operates at elevated temperature, should remove corrosion products effectively, and clean by flushing with magnet shut off	Ineffective for nonferromagnetic cruds and no operating experience with BWRs and PWRs

Source: (Godbee (1978)).

The selection of a filter type is dependent on (1) the degree of filtration required, (2) chemical compatibility of the filter media with the liquid being processed, (3) the concentration and particle size distribution of the solids to be removed, and (4) the volume, flow rates, temperature, and pressure of the liquid waste stream. Significant information is available in the open literature on in-use and developmental filter technology (Kibbey 1978; Kibbey 1980; McCabe, 1976).

Ion Exchange. Ion exchange is a transfer process involving the selective removal of contaminants from liquids through the reversible interchange of similarly charged ions between an electrolyte solution and a solid phase. The contaminants are accumulated on to the exchange medium, which is typically a resin. Ion exchangers are either of the cation or anion type. A further categorization is between separate-bed systems (demineralizers) consisting of either cation or anion resins and the mixed-bed system consisting of a stationary bed containing fixed anion and cation resins. Figure 5-4 is a schematic diagram showing both mixed- and separate-bed systems. The mixed-bed systems have been the predominant ion-exchange choice for treatment of liquid wastes.

When the ion-exchange resins become fully loaded they are either (1) regenerated by use of acidic or basic solutions that permit recycling of the residue and produce a high salt content concentrated liquid waste, or (2) removed from service and disposed of as radioactive waste.

Ion-exchange techniques are used to treat primary coolant, steam condensate,

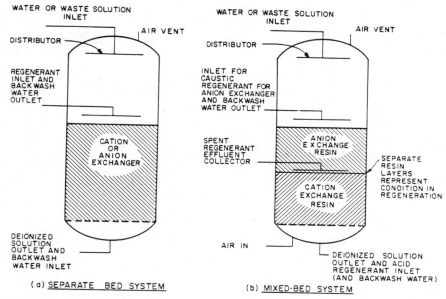

Figure 5-4   Schematic diagram of separate-bed (a) and mixed-bed (b) bed ion-exchange resins. Source: (DOE 1984).

and liquid radwaste streams at BWR power plants. In addition, ion-exchange technology is used to process large volumes of low-concentration liquids to reduce the activity level to within allowable limits for direct release to the environment and for purification of low-to-intermediate level liquid waste solutions generated in processing operations. Among the comparative advantages of ion-exchange systems are

- Portable and disposable units are available permitting the use of this technique at facilities with a low generation volume of liquid waste.
- The technique can be used as an alternative to evaporation to process high-volume, low-solids content liquids.
- Ion exchange is a well-developed, reliable technique usable for a range of soluble and insoluble radionuclides.
- Significant information is available in the open literature on in-use and developmental ion-exchange technology (Liu 1973; Clark 1978).

Evaporation. Evaporation is a widely applied concentration method used in nuclear facilities to reduce volume, remove water from a solution, produce relatively pure water from impure water, and remove soluble and insoluble materials from liquid radwaste streams.

Evaporation can be used on a wide variety of liquid wastes and slurries, particularly those with high dissolved solids content and relatively high conductivity, and requiring a high degree of decontamination (DOE 1984). Since energy requirements for evaporation are high, other liquid waste treatment techniques are typically used for processing large volumes of liquids.

In an evaporator, heat is transferred to the liquid waste that in turn boils. The vapor produced is then separated from the liquid and a concentrated solution (bottoms) or semisolid containing the radioactivity is produced. The cleansed vapor is condensed and released or reused. The major evaporator types used at nuclear facilities can be categorized as natural circulation, forced circulation, and submerged U-tube.

Table 5-4 summarizes the advantages and disadvantages of these evaporator types and describes the best application for each type. Natural-circulation evaporators are of the horizontal tube type, rising-film, long-tube vertical type, and falling-film long-tube vertical type. Figure 5-5 depicts the schematic of a natural-circulation, rising-film, long-tube vertical evaporator with an internal heater. Because of the tendency for excessive deposition to occur in natural-circulation evaporators leading to reduced heat transfer across the tube walls and resultant loss of capacity, their use has been somewhat curtailed at nuclear power plants. In the forced-circulation evaporator (Figure 5-6), the liquid waste feed is pumped through the circuit at relatively high velocities. The heat exchanger is normally a separate unit from the flash chamber to facilitate cleaning, maintenance, and replacement. While the forced-circulation evaporator does have some disadvantages (see Table 5-4), its inherent advantages make it suitable for the widest range of evaporator applications.

TABLE 5-4 Advantages and Disadvantages of the Evaporator Types Used in Light-Water Reactor Power Plants

	Evaporator Type		
	Natural Circulation	Forced Circulation	Submerged U-Tube
Advantages	Low-cost, large heating surface, in one body, low hold-up, small floor space, good heat-transfer coefficients at reasonable temperature differences (rising film), and good heat-transfer coefficients at all temperature differences (falling film)	High heat-transfer coefficients, positive circulation, and relative freedom from salting, scaling, and fouling	Very low headroom, large vapor–liquid disengaging area, good heat-transfer coefficients, and easy semiautomatic descaling
Disadvantages	High headroom, generally unsuitable for salting and severely scaling liquids, poor heat transfer coefficients of rising-film version at low-temperature differences, and recirculation usually required for falling-film version	High cost, power required for circulating pump, and relatively high hold-up or residence time	Unsuitable for salting liquids, high cost, and relatively high hold-up or residence time
Best applications	Clear liquids; foaming liquids; corrosive solutions; large evaporation loads; high-temperature differences—rising film, low-temperature differences—falling film; and low-temperature operation—falling film	Crystalline product, corrosive solutions, and viscous solutions	Limited headroom, small capacity, and severely scaling liquids
Frequent difficulties	Sensitivity of rising-film units to changes in operating conditions and poor feed distribution to falling-film units	Plugging of the tube inlets by salt deposits detached from walls of equipment, poor circulation because of higher than expected head losses, salting because of boiling in tubes, and corrosion–erosion	Slow response to changes in control settings and poor level control in vacuum units

Source: (Godbee 1978).

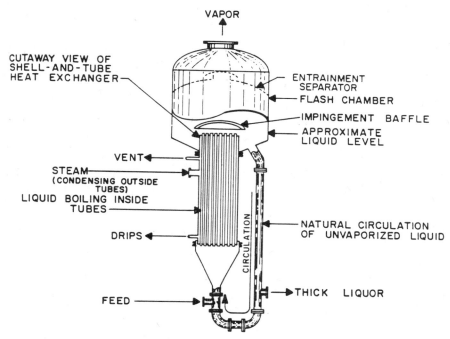

Figure 5-5 Natural-circulation, rising-film, long-tube vertical evaporator with an internal heater. Source: (DOE 1984).

Two specialized types of forced-circulation evaporators are the wiped-film and the spray-film evaporators. In the wiped-film design, the heating surface consists of a single large-diameter cylindrical or tapered tube. The feed liquid is spread into a thin, turbulent film by the blades of a rotor that accelerates the evaporation process. In the spray-film evaporator the liquid feed is atomized into small droplets in spray nozzles and the droplets impinge on the evaporation surface. The submerged U-tube evaporator is similar in design to the spray-film evaporator and is being extensively used in the nuclear industry for treatment of liquid radioactive waste streams.

Crystallization and distillation are concentration technologies that are related to evaporation. Distillation is the process of heating a liquid solution in one vessel to generate a vapor, and collecting and condensing the vapor into liquids in another vessel, leaving the solid materials in the first vessel. Evaporative crystallization involves the generation of a supersaturated solution by evaporation of the water from the solution and subsequent production of crystals and concentrated liquor. Both distillation and crystallization are proven techniques, but do not have wide application in treatment of radioactive waste. Additional details on application of evaporation techniques to liquid radwaste can be found in the current literature (Kibbey 1978; Kibbey 1980; Perry 1973; Trigilio 1981; Yamamoto 1968).

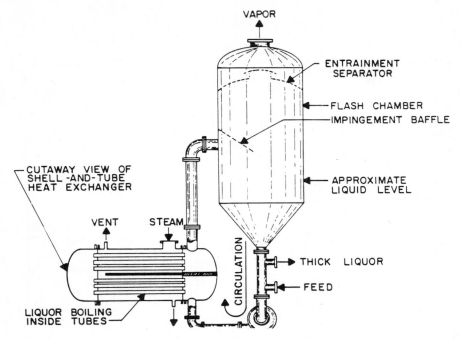

Figure 5-6   Forced-circulation evaporator with an external, horizontal, submerged-tube, and two-pass heater. Source: (DOE 1984).

Centrifugation. Centrifugation is a treatment process based on transfer technology that is used to remove suspended solids from LLW streams. In a centrifuge, the rapid rotation of a perforated basket or bowl containing the waste stream causes the solids to separate from the liquids by centrifugal action. The liquids are forced out through the basket wall and the solids collect and are removed by mechanical action or sprays. Figure 5-7 shows two typical centrifuge types, the perforated-basket and solid-bowl centrifuges. Centrifuges are used to dewater resins and filter sludges and concentrate dilute sludges of 2–5% solids to a sludge of 15–50% solids.

Flocculation/Sedimentation/Precipitation. Grouped together because they are interrelated and often combined into a single treatment process, these processes are used to concentrate the radioactivity in a liquid LLW stream into a small volume of wet solids that is easily separated from the bulk liquid component. Flocculation is a transfer process in which small, suspended particles are agglomerated into large particles that settle out of the liquid with the radioactivity adhering to the particles. Precipitation involves the removal of a substance in solution containing radioactivity from the solution and transformance of it into a solid. Both precipitation and flocculation are enhanced by

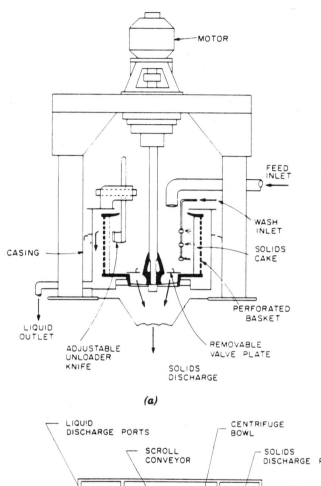

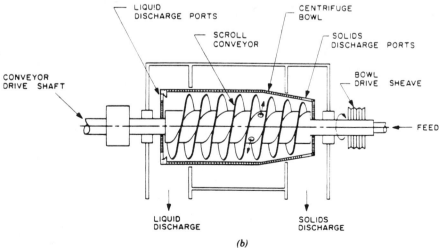

Figure 5-7 Typical centrifuge types (a) top suspended perforated-basket centrifuge and (b) helical conveyor solid-bowl centrifuge. Source: (DOE 1984).

addition of chemicals to the liquid waste stream. In sedimentation, suspended particles settle out of the liquid through the action of gravity.

Solidification. Solidification is a conditioning technique in which small volumes of liquid wastes are immobilized within an inert matrix or binder. Solidification of large volumes of liquids is generally not cost effective compared to treatment methods to remove the radionuclides from the waste stream and produce a smaller volume of waste. As a result of current LLW regulations that restrict free water content in a waste package and then require that the LLW be mixed with a material that can form a monolithic solid, the use of solidification has become widespread at nuclear facilities. Solidification is also used to immobilize

TABLE 5-5   Advantages and Disadvantages of Cement Solidification

Advantages	Disadvantages
Material and technology well known and available	Some wastes affect settling or otherwise produce poor waste forms
Compatible with many wastes	
Most aqueous wastes chemically bound to matrix	pH sensitive—pretreatment of waste may be necessary
Low cost of cement	Swelling and cracking occurs with some products under exposure to water
Good self-shielding	
No vapor problems	Volume increase and high density in shipping and disposal
Long shelf life of cement powder	Excessive setting exotherms may develop with certain cement and waste combinations
Good impact and compressive strengths	
Low leachability for some radionuclides	Dust problems with some systems
No free water if properly formulated	Equipment for powder feeding difficult to maintain
Rapid controllable setting— no differential setting	Potential maintenance problems resulting from premature cement setting, especially in-line mixers
In-container and in-line mixing processes available	May require heating or pressing equipment for some processes

Source: (Furhmann 1981).

semisolids and dry solids at these facilities. The optimal solid incorporating radioactive waste should

1. not fracture under load and have high mechanical strength,
2. be resistant to radiation damage for a long time period,
3. have good thermal resistance,
4. resist leaching from water infiltrating the package, and
5. demonstrate chemical compatibility between the binder and the incorporated waste. Since this combination of characteristics is difficult to achieve in a binder, cement, asphalt (bitumen), and plastic-based materials have evolved as the most suitable materials.

Hydraulic cement is commonly used to immobilize liquids and sludges in the United States. These cement varieties, when mixed with water, form a rigid interlocking matrix of hydration products that binds the cement mass together. Depending on the specific application, portland cement, high alumina cement, or masonry cement may be employed. Additives may be used to adjust the properties of the cement to accommodate individual waste constituents and/or aid in the solidification process. Table 5-5 provides a summary of the major advantages and disadvantages of using cement as a binder in waste immobilization.

Asphalt (or bitumen) is principally used in Europe as an immobilization agent, where it has been popular for several decades. The material is a mixture of asphaltene and malthene hydrocarbons obtained generally from the "bottoms" from petroleum refineries and/or coking plants. Although there are a number of forms of asphalt used as an immobilizing agent, the straight run distillation and oxidized asphalts are most commonly employed. Additives may also be employed to aid in fixing specific constituents in the matrix. The use of an asphalt solidification system requires that the asphalt be heated to over 150°C to soften the material, and vaporize and drive off the water in wet wastes. This heating process may result in a fire hazard during processing, requiring use of special equipment and protection of workers, and has been the primary reason that the use of asphalt solidification has not grown rapidly in the United States. Table 5-6 provides a summary of the major advantages and disadvantages of using asphalt as an immobilization agent.

Even though urea-formaldehyde was one of the earliest materials used to solidify liquid-bearing LLW, and has been generally displaced because of its inability to immobilize water, other plastic-based organic polymers are growing in popularity for use as solidification agents. Among this class of polymers, polyester-type resins, which form a hard permanent solid that can immobilize free water in the cell structure of the matrix, are particularly attractive. Table 5-7 describes the advantages and disadvantages of the "Dow Process," which is an advanced process using polyesters. The polymer-concrete process, used to solidify dry waste, also offers promise for future growth as a solidification agent.

TABLE 5-6 Advantages and Disadvantages of Bitumen Solidification

Advantages	Disadvantages
Technology and material are well known and available	Bitumen is flammable and burns spontaneously at temperatures as low as 390°C
Compatible with a wide range of wastes	Limited loading of salts because of hardening effects
Concurrent volume reduction of aqueous wastes	May swell in water leading to increased leachability and product degradation
No free-standing water	Potential for radiolytic gas generation
Individual waste particles are coated	Exposure to heat may cause melting or phase separation of waste form
Low cost of bitumen	Process requires elevated temperatures
No difficulty with improper setting because it is not a chemical process	Heating must be well controlled and spread evenly
Typically low leachability	Storage of asphalt before use requires elevated temperatures to maintain fluidity of material
	Capital equipment costs are relatively high
	Generation of off-gas during processing oil evaporate may clog filters

Source: (Furhmann 1981).

The advantages and disadvantages of polymer-concrete solidification are described in Table 5-8.

Addition details on solidification can be found in the current literature (Furhmann 1981; Kibbey 1980).

Absorption. Absorption is a conditioning technique, used primarily for institutional liquid LLW, in which the liquid is stored within a porous material such as vermiculite and diatomaceous earth. The liquid, which substantially increases the total volume when absorbed, can be subsequently extracted. Generally, these sorbants can be used by themselves because they do not immobilize the free liquid and thus prevent the waste package from meeting NRC and DOT regulations. However, they may also be used as additions to the solidification agents previously discussed to produce the necessary monoliths. Vermiculite, silica, and selected

TABLE 5-7 Advantages and Disadvantages of Vinyl Ester Styrene—DOW Process Solidification

Advantages	Disadvantages
Adaptable to many wastes—both liquid and solid	Limited binder shelf-life
No free-standing water	Hazards associated with the monomer, catalyst, and promoter handling
Relatively low leachability	
High compressive and impact strength	Some wastes may interact chemically and prevent or affect polymerization
Good radiation stability	
Ease of working with liquid components	Relatively expensive materials
In-container mixing available	
Available in mobile systems	Mixing method important

Source: (Furhmann 1981).

TABLE 5-8 Advantages and Disadvantages of Polymer-Concrete Solidification

Advantages	Disadvantages
High waste loading and negligible volume increase	Applicable only to dry wastes
Low leachability	Limited binder shelf-life
High compressive and impact strength	Hazards associated with monomer handling
No free-standing water	Relatively expensive materials
Relatively insensitive to chemical nature of waste	Some wastes may interact chemically and prevent or affect polymerization
Simple process for combining waste and binder	Not commercialized to full-scale use

Source: (Furhmann 1981).

clays are inorganic materials that have low leachability and long-term stability, but have a number of disadvantages including being usually limited to alkaline wastes and not providing volume-reducing capability. Organic sorbents, which generally have larger surface areas and significant absorptive and ion-exchanging properties, are also being developed. These materials can be used to absorb residual free water, oils, and solvents. One class of organic polymers, referred to as "Superslurpers," has been demonstrated to be useful in cleaning up radioactive spills because of its high absorption capacity with cement to solidify tritium waste. Additional details on absorption can be found in the current literature (Kibbey 1980; Furhmann 1981).

5.4.2 Semisolid Treatment Technologies

Semisolids (wet solids) require treatment and/or conditioning to reduce the free water content to less than 1% and thus make them acceptable for disposal. The selection of the appropriate technology for processing semisolids is dependent upon a number of characteristics of the waste, the most important of which are waste generation rate, free-water content, chemical characteristics, and concentration and specific activities of the radionuclides in the waste. Most of the liquid HLW treatment techniques discussed in Section 5.4.1 are also applicable to wet solids. Conditioning technologies (solidification and/or sorption) may also be more cost effective than treatment technologies in achieving the stipulated low free-water content in the as-shipped waste packages. Table 5-9 provides a summary of the treatment and conditioning technologies for various semisolids, and Table 5-10 summarizes the characteristics of the types of semisolids generated at LWRs that are also representative of the range of semisolids generated at other nuclear facilities. The following discussion references previ-

TABLE 5-9 Technologies Often Used for Processing Various Wet Solids

	Evaporator Bottoms	Spent Ion-Exchange Resins	Filter Sludges	Filter Cartridges	Miscellaneous Sludges
Transfer Technologies					
Filtration	×				×
Chemical regeneration		×			
Concentration Technologies					
Sedimentation	×				×
Centrifugation	×				×
Drying	×		×		×
Dewatering	×	×	×		×
Dehydration	×	×	×		×
Integrated systems	×				×
Transformation Technologies					
Incineration		×	×		×
Calcination	×	×	×	×	×
Conditioning Technologies					
Solidification	×	×	×	×	×
High-integrity containers		×	×	×	
Absorption	×				×

Source: (DOE 1984).

TABLE 5-10   Typical Light-Water Reactor Wet Solid Wastes and Characteristics

Reactor Type and Location	Waste Description	Annual Volume (m^3)	Gross Specific Activity (Ci/m^3)	Major Nuclide
BWR A[a]	Dewatered powdered resin and ion-exchange resin	255	7.1	^{51}Cr, ^{60}Co, and ^{65}Zn
BWR B[b]	Spent resins, concentrates, and filter sludges	269	7.1	^{60}Co and ^{137}Cs
BWR C[c]	Spent resins and concentrates	368	2.5	^{54}Mn and ^{60}Co
PWR D[d]	Resins, sludges, and evaporator bottoms	65	2.1	^{57}Co and ^{60}Co
PWR E[e]	Solidified concentrates	283	3.5	^{58}Co, ^{60}Co, and ^{134}Cs
	Dewatered resins, solidified resins, and filter cartridges	142 340 17		^{137}Cs
PWR F[f]	Spent resins	9	0.10	^{54}Mn, ^{58}Co, ^{60}Co, and ^{137}Cs
	Solidified concentrates and filter cartridges	283 71	0.18 1.06	^{137}Cs, ^{134}Cs, ^{58}Co, and ^{60}Co

Source: (DOE 1984).
[a]Typical 1000-MWe unit (6 yr old).
[b]Typical 800-MWe unit (10 yr old).
[c]Typical 800-MWe unit (5 yr old).
[d]Westinghouse 800-MWe unit (2 yr old).
[e]Babcock & Wilcox 800-MWe unit (5 yr old).
[f]Babcock & Wilcox 900–MWe unit (6 yr old).

ously described technologies applicable to this waste form, and elaborates on technologies specifically relevant to semisolids.

Filtration, as described in Section 5.4.1 for liquid LLW, could be used to treat semisolids if the liquid content of the waste is greater than 90%. Chemical regeneration is another transfer technology that can be used for the removal of radionuclides from loaded ion-exchange resins. The resultant liquid waste stream will require further treatment. Among the available concentration technologies, centrifugation is a technique that can be applied to semisolids as well as liquids, with the primary application being dewatering of resins and filter sludges. Sedimentation is used to treat high water-content wet solids (e.g., slurries, floor drains) to remove solids as a prestep before use of ion-exchange techniques.

Liquid removal techniques using concentration technologies that are particularly applicable to treatment of semisolids are dewatering and drying. In dewatering, either pumping and/or gravitational drainage is used to remove the water from a semisolid. A commonly employed approach for treating ion-exchange resins, called "in-container dewatering," involves the use of multiple filter elements placed in a disposable container and connected to a pump. Drying uses heat to drive off the water to produce a dry solid. Drying has been used at DOE facilities and will undoubtedly be more frequently applied at commercial facilities in the future to meet the 1% free water requirement at competitive costs.

The conditioning technologies of solidification and absorption (see Section 5.4.1) are applied to the treatment of semisolids. In applying solidification technology, the choice of the binder material based on relative economics to yield a monolith that meets regulatory criteria is generally the governing factor. This choice is dependent on factors such as the extent the semisolid must be dewatered prior to solidification, the chemical compatibility of the waste and binder, the degree of mixing needed to yield an acceptable product, the packaging efficiency of the solidified material (i.e., the bulk volume of the dewatered wet solids relative to the volume of the solidified product), and the relative cost of binder materials. As a rule, all other factors being equal, the technology that produces the highest packaging efficiency is usually the most cost effective. The primary application of absorption technology is also to reduce free water levels in packages to within established limits.

5.5 DRY SOLID TREATMENT TECHNOLOGIES

Dry solids are treated to reduce their volume rather than to meet transportation and/or disposal requirements as is the case with liquid and semisolid wastes. Volume reduction for dry solids is considered when the high disposal costs resulting from shrinking disposal-site capacity make volume-reduction technologies an economically attractive alternative. The selection of the appropriate volume-reduction technology is initially a function of facility waste generation rate, since treatment is most appropriate where high waste volumes are produced. Other factors to consider in selecting a technology are the radioactive characteristics of the waste (extent of surface contamination and activity levels), shreddability, combustibility, and metal content. It is also important to note that the chosen dry solids treatment technology should not change the disposal classification (A, B, or C) of the waste by increasing the specific activity beyond the class limit, nor should it increase package surface radiation levels to those that require shielded transport or cause handling and/or storage problems. The ranges of treatment and conditioning technologies currently used to process various types of dry solid waste are listed in Table 5-11 and described below.

Compaction. Compaction is a concentration technology in which a press is used to compress the dry solid waste (Figure 5-8) into the final disposal container or into reusable shipping containers. The volume reduction factor achieved during compaction is a function of the void space in the waste, the force applied by the press, the bulk density of the material, and its spring back characteristics. Compactors vary in size, design, and capacity and often are custom designed for the facility's floor space and waste characteristics. A typical compactor system

TABLE 5-11 Technologies Often Used for Processing Various Types of Dry Solid Low-Level Waste

	Trash	Contaminated Equipment	Irradiated Hardware
Transfer Technologies			
Decontamination		×	
Concentration Technologies			
Compaction	×		
Shredding or sectioning	×	×	×
Baling	×		
Transformation Technologies			
Incineration	×		
Conditioning Technologies			
High-integrity containers			×
Solidification (ash)	×		

Source: (DOE 1984).

Figure 5-8 Compactor in use at nuclear power plant. (Reprinted from *Nuclear News*, 1987.)

will contain a power unit, a drive system (hydraulic or mechanical), a platen, a base plate, supporting members, a platform on which the package is positioned, and a control panel. A system to control dispersion of airborne radioactive particulates is also usually incorporated into the compactor system. The control system may include a hood, a shroud placed around the package (i.e., drum) opening, a high-efficiency particulate (HEPA) filter, and an exhaust blower. Volume reduction factors of between 2 and 6 can be achieved for these types of systems when used to compact dry solids such as clothing, laboratory equipment, paper, and plastics. Compaction should not be used with dense or bulky articles where minimal volume reduction would be achieved, with wastes containing free liquids, or with wastes containing explosives.

Baling. Baling is a concentration technology operating on the same principle as compaction, but in which the waste is compressed into generally rectangular bales and secured (banded) to maintain the reduced volume. The bales are then usually placed into disposable containers. Compaction and baling are often accomplished as a sequence of operations. The use of this technique has been pioneered at federal laboratory facilities. Bales come in a range of sizes, processes, and design configuration (hydraulic, electric, and hand-operated platen models). Design variations include continuous extrusion typically for paper and cloth, multiple platen stroke for compressing a variety of wet or dry wastes including scrap metal, and two-stage multiple platen operating at right angles used generally for scrap metal. The units are constructed to operate either in a horizontal or vertical position. In using bales with radioactive waste it is necessary to install contamination control systems to contain radioactive particulates released during the baling process.

Shredding. Shredding is primarily used as a pretreatment operation for dry solid waste prior to incineration or compaction. Shredding can be used on paper, cloth, and plastics achieving a volume reduction factor of about 3. Shredders operate through the intermeshing of a number of motor-driven counterrotating shafts. Upgraded versions will permit reversal of the direction of motion of the motor and shaft to clear jamming, and replacement of cutting teeth to handle different waste forms.

Sectioning. Large metallic and nonmetallic waste objects containing significant void volume are capable of substantial overall volume reduction through sectioning with cutting equipment. Tanks, reactor components, boxes, and contaminated vehicle bodies are among the objects that can be sectioned prior to packaging. The cutting equipment can be operated directly in a hands on fashion when radiation levels are low, or remotely when radiation levels are high.

Combustion. Combustion techniques, which transform the waste to an inert or less reactive form and reduce volume and weight, include commercially available and developmental incineration technology and promising developmental concepts such as acid digestion, molten salt combustion, and pyrolysis. Incineration will be discussed separately (see Section 5.6) because of its history of use in treating radioactive wastes, and because of its future importance as a treatment technique for a range of institutional, governmental, and utility wastes. The concepts discussed below have been developed under federal government sponsorship at DOE laboratory facilities.

The various combustion techniques will achieve volume reduction factors between 20 and 100, when applied to combustible wastes that are essentially organic materials such as paper, cloth, plastic, resins, solvents, and rubber. The residue remaining after the combustion process consists primarily of the inorganic constituent in the waste. These residues are generally of a higher activity level than the original wastes, and must be packaged in the appropriate container for shipment to a disposal site.

Acid digestion is a combustion process in which the organic materials are converted to gaseous end products (CO_2, H_2O, etc.) and insoluble sulfate or oxide residues by digestion in hot concentrated H_2SO_4 in the presence of HN oxidant. The concept was originally conceived at the DOE Hanford Engineering Development Laboratory (HEDL) where work continues at this time. The gaseous stream produced is scrubbed to remove the end products of the reaction to permit release to the environment, and the contaminated liquid scrubber material becomes a secondary waste stream. The major advantages of this system are the wide range of waste that it can process when the waste is sorted and preshredded, the low-temperature, single-stage operation, and the ability to process high levels of radioactivity. Developmental problems with the system include the need to use glass or Teflon-lined containment vessels to achieve corrosion resistance, relatively low processing rates (5 kg/hr maximum), handling and disposal of the acid after depletion, and the need to remove radionuclides from the acid after digestion.

Molten salt combustion is a process by which organic materials are rapidly and completely oxidized by the molten salt medium containing an oxidizing agent to produce CO_2 and H_2O, and the ash and other combustion products are trapped in the molten salt. Thus, the molten salt serves as a heat-transfer agent, as a source of oxygen to accelerate combustion, and as a scrubbing agent to react with the gases generated (except CO_2) and entrap ash in the furnace. This is accomplished by having the waste and air fed below the surface of the salt causing the combustion gases to pass through the melt before release to the environment stripped of everything except CO_2 and H_2O. Periodic removal of the ash and other inorganic materials from the melt is required. Developmental work on this promising concept is continuing.

Pyrolysis is a combustion process in which the organic combustibles are gasified in an oxygen-deficient atmosphere. This developmental concept is being applied to the slagging-pyrolysis incinerator which is discussed further in Section 5.6. While the conditioning process of solidification is primarily required for liquids and semisolids prior to packaging and shipping, the highly dispersible ash produced during combustion may require solidification to meet disposal site constraints.

Decontamination. Decontamination is a transfer technology that involves the removal of surface radioactivity from equipment and structural components using chemical or physical techniques. Decontamination permits reuse of the equipment or buildings and, when used prior to disposal, may enable the equipment to be disposed of as nonradioactive. The decontamination process typically produces waste cleaning fluids containing the radionuclides removed from the cleansed surface, which then must be disposed of as liquid LLW. The decontamination project can involve the application of one or more surface treatment techniques including chemical decontamination, manual decontamination, ultrasonic cleaning, and electropolishing. There are industrial firms that specialize in performing decontamination operations that bring their specialized mobile equipment and trained personnel to a site for the duration of the operations. Although the use of equipment such as high-pressure water and steam cleaning systems, electropolishing systems, wet and dry vacuuming systems, ultrasonic cleaners, and degreasing systems has become more common as the technology has become more sophisticated, much simple surface decontamination is still accomplished by hands-on manual and chemical processes.

5.6   INCINERATION OF RADIOACTIVE WASTES

The combustion process of incineration has been used as a volume-reducing technique for LLW, and transuranic waste since the early stages of the nuclear industry. As noted earlier, increasing pressure to reduce the volume of waste shipped to the disposal sites is resulting in greater use of incineration technology and increasing expenditures on the part of both the DOE and private industry for development of new incineration concepts and improvement of existing technology. Although incineration has been primarily used at DOE facilities in which economics was not an overriding consideration, the emphasis on volume reduction technologies driven by the increased disposal costs resulting from a growing scarcity of existing disposal space, and potentially from the more demanding siting and disposal requirements imposed by 10 CFR Part 61 on the compacts, is making the comparative economics of incineration of commercially generated waste more attractive and spurring the adaptation of incineration technology in the private sector. A further impetus to the use of incineration for commercial wastes results from the increasing volumes of wastes generated and the additional radiological safety personnel required to manage the wastes at these facilities. Incineration converts the combustible waste into radioactive ashes, residues that are chemically inert, nonflammable, and homogeneous, and gaseous effluents containing entrapped radioactive particulates. It is therefore essential to equip the incineration system with highly efficient, multistaged "off-gas" control systems to remove the radioactivity to below environmental release standards before emission of the cleansed gases. Thus, in addition to the ash and other solid residue generated, the off-gas controls will produce a secondary waste stream (e.g., filters) that requires disposal. At least 50% of the solids generated at nuclear fuel cycle facilities are combustible with dry active waste (DAW) being the waste form most suitable for volume reduction by incineration. Certain organic liquids and the semisolid spent ion-exchange resins can also be incinerated. Solid waste forms that are not suitable for incineration, and that need be sorted out prior to treatment, are primarily those that have a radioactivity content sufficiently high as to cause excessive doses to operating personnel or to raise the radioactivity content of the ash above prescribed limits. In addition, wastes with a high rubber or PVC content, containing large metal objects or a high content of other noncombustibles, or having an explosion potential are also excluded from incineration. Since the incinerators are designed to operate with wastes having total heating values between specific limits, wastes with high heating values may have to be excluded whereas those with low heating value will require the addition of supplementary fuel to achieve complete combustion.

An incineration system designed to process radioactive wastes should achieve complete combustion of the wastes while providing radiological protection to operators and the public under normal and accident conditions, and containing the radioactivity within the incineration system and the off-gas treatment system. The generic incineration system for radioactive waste treatment consists of (1) waste feed pretreatment and loading facilities (e.g., shredding), (2) single or multiple combustion chambers, (3) ash collection and unloading equipment, (4) ash transfer and/or immobilization equipment, (5) and off-gas treatment system including particulate removal components, gas scrubber, fans, a stack, and a system for recycling certain secondary waste streams through the combustion chambers, and (6) process equipment instrumentation to monitor critical operating parameters, instrumentation to monitor health and safety-related limits, and controls to enunciate radioactive releases and initiate systems shutdown if required.

Incinerators designed for treatment of radioactive wastes use a variety of combustion approaches either singly or in combination. The principal combustion techniques, categorized by process temperature ranges, are

- Controlled-air incineration, which limits the air supply in the primary combustion phase and requires a secondary combustion phase to achieve complete combustion (800–1100°C).
- Excess-air incineration, in which an excess of oxygen is fed into the primary combustion-phase to permit both the solid and gaseous components to burn directly (800–1100°C).
- Pyrolytic or thermal decomposition, in which the organic waste material is essentially distilled in a highly reducing atmosphere (absence of oxidation) generating combustible liquids and gases. A carbon residue (char) remains after pyrolysis (500–600°C).

- Fludized-bed incineration, in which an inert bed of particles is suspended by air flow through the bed and the preshredded waste is burned upon introduction into the self-sustaining bed (800°C).
- Slagging incineration at high temperature (1400–1600°C) to burn the organic material releasing sufficient heat energy to convert the noncombustibles to a molten slag residue.

TABLE 5.12   Feed Conditions and Operating Variables for the Major Incinerator Types

	Acid-Digestion (HEDL)	Agitated-Hearth (RFP)	Controlled-Air (LANL)	Excess-Air (Mound)
Process Description	Waste is chemically oxidized in H_2SO_4 containing a small percentage of HNO_3. Acid is evaporated to produce residue product	Waste is batch fed into primary chamber. Residue is raked by rotating rabble arm. Final off-gas combustion in separate after-burning chamber	Waste is decomposed in starved-air primary chamber. Volatiles and particulates are final burned in secondary chamber	Waste is burned in a steel storage drum in which air is injected in a spiral pattern to cool drum walls. Continuous-feed option is available
Material of Construction	Digestor and other items—glass, glass lined, or Teflon lined. Off-gas equipment—standard wet system	Incinerator is refractory lined; hearth is refractory lined; off-gas equipment, standard for wet system	Both chambers refractory lined with 5-mm mantic carbon steel outer shell. Hearth is refractory. Off-gas equipment is fiberglass-reinforced plastic, Hastelloy, and mastic	Incinerator is 316 SS; hearth in metal drum, off-gas equipment standard for scrubbing corrosive acids
Solid Feed Capacity (kg/hr)	10	70	45	27
Solid Waste Capabilities	Paper, rags, wood, rubber, PVC, ion-exchange resins, polyurethane; fluorinated wastes are corrosive to glass	Combustible solids with $<3 \times 10^{-3}$ mg Pu per gram; includes paper, polyurethane, PVC, and latex rubber cloth	Cellulosic compounds, polyethylene, PVC, latex rubber, and cartridge filters	Contaminated wastes containing paper, PVC, polyethylene, polypropylene, rubber, cloth, and tramp metal
Feed Pretreatment Noncombustibles	Hand sorted to remove noncombustibles before shredding. Also air classification for metal removal	Tolerates small noncombustibles, but large items must be removed before feeding	Large noncombustibles are hand-sorted out. Small noncombustibles are tolerated	In batch mode, noncombustibles do not interfere with burning. For continuous mode, hand-sorting out of noncombustibles
Acid-Producing Solids—Corrosion	Resistant to acids produced by PVC plastic, etc.; some acid vapors (HNO_3, HCl) escape off-gas scrubbing and are vented to atmosphere	Acid waste or PVC plastic may corrode metal rabble arms	Highly acidic materials may corrode steel incinerator shell and other head-end equipment	Acid gases at 800–1000°C, very corrosive to drum, upper combustion chamber, off-gas header, deluge tank, and transfer pipe between tank and venturi
Volatile Solids	Minimal volatilization at 250°C. Volatiles readily dissolved into acid systems	Some volatilization occurs at 800–1000°C. Lead deposition is possible in incinerator and refractory lining	Some volatilization at 800–1000°C. Lead deposition on refractories occurs if lead-containing materials are burned	Some volatilization at 1100–1300°C. Possible lead deposition
Operating Temperatures	Primary—250°C with H_2SO_4 and HNO_3 oxidation	Primary—600–800°C, slightly reducing atmosphere; secondary—1000°C	Primary—500–800°C, starved air; secondary 1000–1500°C, oxygen enriched	Primary—1100°C, excess air; secondary—none

Source: (DOE 1984).

- Electromelt incineration, in which the waste is burned and sorbed into a glass melt maintained by electromelt joule heating (1200°C).

5.6.1 Types of Incinerators for Combustion of Radioactive Wastes

A variety of incinerator designs has been developed incorporating the combustion techniques described above. These designs, originally developed under DOE sponsorship, have, in a number of instances, been adapted for commercial use.

	Fluidized-Bed (RFP)	Rotary-Kiln (RFP)	Slagging-Pyrolysis (INEL)	Penberthy-Electromelt
Process Description	Waste is fed to fluidized bed of Na_2CO_3 granules for partial combustion. Volatiles are final oxidized in fluidized bed of catalyst granules	Waste is charged into inclined, horizontal kiln that is rotating. Final off-gas combustion occurs in a separate afterburning chamber	Stacked-kiln concept. Waste loaded in top drying zone, descends into pyrolysis zone, and descends into melt zone to be discharged as molten slurry	Waste is burned and sorbed into a glass melt maintained by electromelt joule heating. No second stage. Product perfect for storage with no further treatment
Materials of Construction	Incinerator is made of all 316 SS. Hearth distributor plate of 316 SS. Off-gas equipment is made of non-acid-resistant materials	Incinerator— refractory brick-lined; off-gas equipment—typical of wet scrubbers	Incinerator interior— refractory tamped. Off-gas equipment— standard wet and dry system	Incinerator— refractory lined. Off-gas equipment— standard for wet system sorbing acidic gases
Solid Feed Capacity (kg/hr)	82	41	850	225
Solid Waste Capabilities	Paper, polyethylene, PVC, latex rubber, wood, leaded rubber, organic resins, cartridge filters, HEPA filters, and cloth	Designed for cellulose, polyethylene, PVC, rubber, organic resins, wood, and polyethylene	Designed to handle all kinds of combustibles, most types of noncombustibles, and contaminated soil	Combustible solids such as paper, wood, concrete, rubber, plastics, scrap glass, and metal
Feed Pretreatment Noncombustibles	Large noncombustibles are hand-sorted out before shredding. Further removal is effected by air classification	Tolerates noncom-bustibles within size limitations of ram feeder and ash discharge port	Accepts all types of noncombustibles (soil) within size restraints of waste throat	Tolerates wide range of feeds. No data on noncombustibles. Size limitation on feed
Acid-producing Solids— Corrosion	Acids produced from PVC or other source neutralized by bed of Na_2CO_3	Refractory lining unaffected by acids. Some acid corrosion of steel outer tube or head-end feeding equipment	Acids may corrode metal components. *in situ* neutralization obtained by adding Na_2CO_3	Unknown
Volatile Solids	Low temperature (550°C) minimizes lead deposition. Na_2CO_3 blocks lead, iron, and phosphorus. Chloride remains in off-gas	Refractory can be degraded by low-melting metals such as feed. Some volatiles at 800–1000°C	High temperature (1650°C) volatilizes or melts most metals. Lead in INEL waste could cause deposition and lining problems	Unknown
Operating Temperatures	Primary—525–625°C, 15–25% oxidation, secondary— catalytic at 550°C, oxygen enriched	Primary—600–800°C neutral to slightly reducing, secondary—1000°C oxygen enriched	Primary—1500–1600°C, 30–60% excess air; secondary—1100–1200°C, 40% excess air	1200°C

TABLE 5-13 Advantages and Disadvantages of Major Incinerator Types

Type of Incinerator and Location	Advantages	Disadvantages	Unique Capabilities
Acid-digestion, Hanford Engineering Development Laboratory	Takes wide variety of wastes, but waste must be sorted and shredded; soluble residue for actinide recovery; low-temperature, single-stage operation; and processes high levels of radioactivity	Limited feed rate of 5 kg/hr max, difficult process control for acid feed, VR small without acid recycle, feed requires sorting and shredding, acid gases vented to atmosphere, and useful for only a limited range of organic liquids	Produces soluble inorganic sulfate and oxide residue for Pu recovery and has acid recycle
Agitated-hearth, Rocky Flats Plant or Envirotech	Mechanical agitation of waste during combustion produces efficient oxidation, minimal ash pretreatment, automatic ash removal, and nonrotating refractory has long life	Maintenance of mechanical equipment in the combustion chamber, accommodates only small amounts of activity, possible activity buildup in refractory lining, and short life of seals	Positive agitation yields efficient combustion and automatic ash discharge
Controlled-air, Los Alamos National Laboratory	Limited airflow in primary chamber reduces ash entrainment, built-in TRU assay and X-ray equipment, tolerates small noncombustibles, no shredding of feed, and commercially available	Possible corrosion of off-gas system by HCl, ash removal needed, possible migration of radioactivity in refractory lining, accommodates only low levels of fissile materials	Commercial equipment and minimal ash in off-gas
Excess-air, Mound Laboratory	Low capital cost, no waste pretreatment with batch operation, low waste handling requirements, and adaptable for in-plant operators	Subject to acid corrosion, high particulate loading in off-gas system, relatively high carbon content in ash, accommodates only low levels of fissile material, and reported VR factors do not include produced waste	Allows for incineration of waste in storage drum with sorting or pretreatment
Fluidized-bed, Rocky Flats Plant	*In situ* acid neutralization, low temperature combustion eliminates refractories, agitation	Sorting and shredding required for feed, feed should be free of metals and other combustibles to	Neutralization of acids in fluidized bed of Na_2CO_3, has dry off-gas system,

	by fluidization during combustion, dry off-gas system, continuous ash removal, low-temperature fired ash for actinide recovery, good for high activities, half the size of conventional systems	eliminate unnecessary loading of fluidized bed, expensive catalyst needed for off-gas burning, and some insoluble catalyst in ash	and has no refractories
Rotary-kiln, Rocky Flats Plant	Continuous discharge of ash, minimizing criticality problems, can burn melted or liquid materials, industrially proven success, positive automatic ash removal, minimal waste pretreatment, tumbling action enhances combustion, and processes high levels of fissile material	Rotary seal maintenance, short refractory life possible, incomplete graphite combustion possible, and possible radioactivity migration and buildup in the refractory linings	Positive agitation of wastes for complete combustion and has automatic ash-discharge system
Slagging-pyrolysis, Idaho National Engineering Laboratory	Can process unsegregated wastes with high percentage of noncombustibles, minimal waste sizing or pretreatment, product is a stabilized residue, slag is continuously discharged, in situ neutralization of acids if Na_2CO_3 is used, and is commercially available	Large volume of waste in unit could cause nuclear safety concern, slag residue is unacceptable for recovery of actinides, weight reduction problems because of required additives for slag formation, high capital and operating costs	Slag-type residues are produced requiring no further fixation and system accepts noncombustibles
Penberthy electromelt Savannah River Laboratory	Product has excellent storage properties with no further treatment, tolerates wide range of feeds including liquids, no second stage required, and melt is easily removed	Refractory lining required and off-gases carrying radioactivity will require solidification of scrubber solution resulting in minimal VR advantage for these wastes	Excellent product characteristics and handles wide variety of feeds

Source: (Trigilio 1981; Zeigler 1981).

The primary design concepts for the incineration of radioactive waste and their developmental sources are (DOE 1984)

1. Agitated-hearth incinerator	Rocky Flats Plant (RFP)
2. Controlled-air incinerator	Los Alamos National Laboratory (LANL)
3. Excess-air incinerator	Mound Laboratory (ML)
4. Fluidized-bed incinerator	Rocky Flats Plant (RFP)
5. Rotary-kiln incinerator	Rocky Flats Plant (RFP)
6. Slagging-pyrolysis incinerator	Idaho National Engineering Laboratory (INEL)
7. Penberthy electromelt	Savannah River Laboratory (SRL)

TABLE 5.14   Product Characteristics, Off-Gas Systems, and Secondary Wastes

	Acid-Digestion (HEDL)	Agitated-hearth (RFP)	Controlled-Air (LANL)	Excess-Air HEPA filters
Operation Variables Controlled	Temperature, HNO_3 addition and feed rate	Waste feed, fuel, and air rates, and time and temperature	Feed, fuel, and air rates; temperature is modulated for combustion	Feed, Fuel, and air rates, and temperatures
Product Form	Dry salt cake, rich in sulfates and oxides, thermally stable, inert, and unstable for handling	Inert dry ash, not stable for storage	Dry, thermally, and radioactively stable, inert, and unstable for storage	Dry ash and salt powder for sorption of HCl; non-inert, not stable for storage
Off-Gas System	Dilute acid scrubber, heater, HEPA filter, second scrubber, and final HEPA filter	Potassium hydroxide scrubber, venturi scrubber, gas–liquid separator, and HEPA filter	Water quenching, venturi scrubber, packed bed absorber, condenser, heater, roughing filter, and HEPA filters	Deluge chamber, filter
Secondary Wastes	Scrubbed gases and alkaline scrub solution containing NaCl, $NaNO_3$, $NaNo_2$, and Na_2SO_4	Combustion gases and alkaline scrub solution	Combustion gases and neutralized, spent, NaCl scrubber solution	Combustion gases and spent, neutralized scrub solution
Status	Demonstrated for cold and radioactive wastes at 5 kg/hr. Technology available in the 1980s	Demonstrated for cold wastes at 4 kg/hr. Large-scale unit built for processing 70 kg/hr of LLW	Tested at LANL on cold and radioactive wastes	Demonstrated in cold pilot plant for more than 5 kg/hr. Technology available after 1978
Potential Applications	Processes most solid wastes and high TRU activity wastes, good actinide recovery, best suited for low flow rates, and because of liquid nature, has some advantages for remote operation	With minimal criticality problems, best suited for burning large volume LLW stream. Because of anticipated high maintenance resulting from corrosion of metals and refractories, not a good candidate for remote operation	Good for large volumes of low-level TRU wastes, good nuclear safety features, subject to corrosion from PVC plastics, and needs considerable testing before use in remote applications	The lack of criticality problems makes this unit a good candidate for burning LLW. Especially adaptable to remote operation because of simplicity and low maintenance. Good candidate for incineration of LLW at power plants

Source: (DOE 1984).

Extensive relevant information about these incineration concepts has been summarized by the Department of Energy (DOE 1984) and is presented in the following tabulations:

Table 5-12: Summary of feed conditions and operating variables for the major incinerator types.

Table 5.13: Summary of advantages and disadvantages of the major incinerator types.

Table 5-14: Summary of product characteristics, off-gas systems, and secondary wastes associated with the major incinerator types.

Table 5-15: Capabilities of the major incinerator types for treating liquid radioactive wastes.

Associated with the Major Incinerator Types

	Fluidized-bed (RFP)	Rotary-Kiln (RFP)	Slagging-Pyrolysis (INEL)	Penberthy Electromelt (SRL)
Operation Variables Controlled	Feed, nitrogen, and fuel rates, temperature, and Na_2CO_3 bed makeup	Feed, air, and fuel rates; time, temperature	Waste feed, sand, fuel, and air rates; temperature and pressure	Unknown
Product Form	Inert dry oxide ash and dry salt; inert, nonstable for storage	Inert dry oxide, not stable for storage	Basaltic-type glassy slab, inert, and storage is dependent on leaching requirements	Glass melt containing ash and oxides. Some small tramp metal can be entrapped
Off-Gas System	Dry scrubbing-cyclone filter, sintered metal filter, and HEPA filters	Primary and secondary venturi scrubbers, and HEPA filters	Wet and dry combination, preheater, heat exchanger, sand filter, caustic wet scrubber, preheater, HEPA filter, preheater, and stack	Not complete, but current plans include flue-gas cooler, basic and water scrubbers, demister, reheater, and charcoal and HEPA filters
Secondary Wastes	Bed material (Na_2CO_3) and catalytic material ($Cr_2O_3 + Al_2O_3$) and acid free combustion gas	Combustion gases and acidic and spent alkaline solutions including fly-ash	Wet scrubber liquid and scrubbed gas	Wet scrubber liquids. Sulfate-bearing wastes reduce VR factor because of sorption in liquid scrubber
Status	Demonstrated for cold wastes at 9 kg/hr. A radioactive waste demonstration designed for 80 kg/hr ready for full-scale operation by 1980	Demonstrated at cold pilot plant, 2 kg/hr. Plant being designed for 40 kg/hr of TRU wastes	Mol Belgium (SCK, CEN) plant tested $\beta-\gamma$ burning in 1978	Demonstration unit only
Potential Applications	Good for a wide variety of waste solids and liquids, for acid-producing wastes, and for burning liquids. Can burn ion-exchange resins. Modular design and low maintenance make it a good candidate for remote operation	Good for wastes with recoverable TRU material and wastes with noncombustibles. If seal and bearing maintenance are small, this unit is a good candidate for remote operation	Designed for burning INEL retrievably stored wastes. Good for burning municipal wastes. Remote operation difficult because of equipment size and refractory maintenance	Wet and dry radioactive wastes, particularly glass; small tramp metal not a problem

TABLE 5-15 Capabilities of the Major Incinerator Types for Treating Liquid Radioactive

	Acid-Digestion (HEDL)	Agitated-hearth (RFP)	Controlled-Air (LANL)	Excess-Air (Mound)
Liquid Waste Capabilities	Processing of organic liquids limited because noncombustible and volatile liquids vaporize. Oil and TBP have been digested, paraffins have not	Minimal information. Units theoretically can burn liquids. Organics could be sprayed into the combustion chamber	No demonstration with aqueous and organic liquids. Being tested for liquid and resin combustion	Limited burning of kerosene. Further testing under way
Liquid Waste Capacity Aqueous Liquids	Unknown Tolerates water up to 20 kg/hr	Unknown Aqueous liquids acceptable in absence of fissile material	Unknown Can be modified to accept aqueous	Unknown Capability to handle large quantities of aqueous not demonstrated
Acid-Producing Liquids	Unknown	Chlorinated hydrocarbons may produce acids that react deleteriously with metals in the incinerator	Chlorinated hydrocarbons could present corrosion problems	Should not burn corrosive liquids
Design Purpose	Processing TRU and LLW containing up to kilogram quantities of Pu	Processing nonglove-box waste with background radioactivity	Processing TRU waste, limited to 100 g Pu inventory	Processing TRU wastes, limited to TRU concentration that is safely stored in 210-liter drum
Automation Potential	Housed in glove boxes. Glove-box waste sorting, size reduction, classification, feed conveyor, and extruder are all operated from separate control room	Hand sorting used in feed wastes and for ash removal and replacement. All other procedures automatic	Feed preparation is manual. Feeding, control of incinerator, ash removal, and vacuuming are done automatically	System is hand-operated but batch operation principle eliminates most contact, except barrel preparation and vacuum ash removal

Source: (DOE 1984).

A brief description of the operation of the individual concepts with appropriate references and examples of industrial applications is included in the following:

Agitated-Hearth. The most recent agitated-hearth incinerator developed at the RFP is a 70 kg/hr production scale unit modeled after a small-scale 5 kg/hr unit. It is a stationary, refractory-lined circular steel vessel 2.6 m in diameter by 4.6 m high (Figure 5-9) with rotating agitator arms. The arms move the ram-fed shredded waste through the combustion zone of the hearth. The waste is processed on a batch basis using a semicontinuous feed and the ash is discharged on a batch basis. The off-gas produced in the afterburner is then treated with a series of scrubbers, a gas–liquid separator, and a HEPA filter. At this time, there are no commercial agitated-hearth systems being marketed. Additional information on this system is available in the literature (Trigilio 1981).

Controlled-Air. The controlled-air incinerator, as developed at the LANL, is a multiple-chamber unit (Figure 5-10) in which combustion takes place in both the primary chamber and secondary chamber. Presorted wastes are ram fed into the primary chamber, and the resultant gases and entrained solids are further

Wastes

	Fluidized-Bed (RFP)	Rotary-Kiln (RFP)	Slagging-Pyrolysis (INEL)	Penberthy Electromelt (SRL)
Liquid Waste Capabilities	Burned naphtha solvents, trichloroethylene, TBP, waste oils. Showed 99.9999% burning of polychlorinate biphenyls	Equipped for both organic and aqueous wastes	Should tolerate organic or aqueous liquids. Minimal testing	Good for all kinds of liquid-bearing wastes, organic and inorganic
Liquid Waste Capacity	Estimated 50 liters/hr of combustibles	135 kg/hr of organics	Unknown	Unknown
Aqueous Liquids	Accepts moderate amounts of aqueous waste in presence of auxiliary fuel such as methanol	Accepts aqueous wastes, but glass formers can damage refractories	Tolerates moderate amounts of aqueous along with solids or auxiliary fuels	Unknown, but should tolerate water easily
Acid-Producing Liquids	Acids are neutralized by the Na_2CO_3 bed	Possible acid corrosion of nozzles and other metal components	Combustion acids (from CCl_4, etc.) lead to corrosion without neutralizer such as Na_2CO_3	Acids in off-gas will contribute to secondary wastes in form of spray scrubber liquids
Design Purpose	Processing TRU and LLW wastes, limited to ^{239}Pu up to 10^{-3} g/g of waste	Processing high-level TRU waste for Pu recovery	Processing heterogeneous TRU wastes. Limited to drums with less than 200 g Pu	Burning wide variety of wet and solid wastes
Automation Potential	Housed in a canyon for α containment. Manual barrel opening, but emptying, sorting, and feeding done in glove box. Incineration monitored and controlled from separate room	Contact handling limited to waste sorting of feed. Incinerator operated automatically from a separate room. Ram feeding, feed, and ash transport are automatic	Feed manually sorted for TRU control. Operation and slag removal are automatic	Unknown

oxidized in the secondary unit. The offgas treatment system is, in sequence, a water spray for quenching, a venturi scrubber to remove particulates, a moisture removal system, and a series of roughing and HEPA filters. As a result of the successful performance of the 45 kg/hr unit built at LANL, the private sector has adapted the design for commercial use. Essentially similar controlled-air incinerators for treating solid radioactive wastes are available from Associated Technology, Inc., Koch Process Systems, Westinghouse Electric Company, General Electric, and Waste Technologies, Inc. Additional information on this concept is available in the literature (Kibbey 1980; Trigilio 1981; Bordwin 1980; Stretz 1982).

Excess-Air. An excess-air cyclone incinerator was developed at the Mound Laboratory to process solid TRU wastes. This unit has a combustion chamber consisting of a fixed upper section, and a lower removable combustion vessel (Figure 5-11). Batch quantities of waste are fed into the combustion chamber concurrent with the combustion air that spirals down from the upper section of the chamber. The ignited wastes burn downward and the generated combustion gases move upward inside the spiral. The ash drops through the grate and is conveyed to an interim storage container and the off-

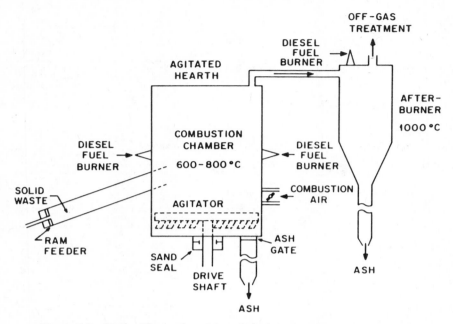

Figure 5-9   Rocky Flats agitated-hearth incinerator. Source: (DOE 1984).

gases pass through scrubbers and filters prior to release. Commercial incinerators, employing the excess air principle, have been developed and are being marketed. These include the Combustion Engineering (CE) system designed to process dry active waste (DAW), resins, and sludges, the Transnuclear incinerator (TN200) designed for dry solids only, and the NGK Insulators LTO incinerator that is essentially the same design as the TN200 unit except for minor variations in the off-gas filtering system. Additional information on this system is available in the literature (Alexander 1980a; Alexander 1980b).

Fluidized-Bed. The RFP fluidized-bed incinerator is a demonstration unit for the treatment of LLW and TRU wastes. This 80 kg/hr production size incinerator is the follow on unit to a successful pilot scale 9 kg/hr unit developed at the RFP. The fluidized-bed incinerator (Figure 5-12) has a primary chamber holding a bed of sodium carbonate granules that is fluidized by an air–nitrogen gas flow. The waste undergoes partial combustion and pyrolysis (flameless) and becomes self-sustaining from the heat generated from the pyrolysis process. The off-gas from the primary chamber passes through a separator to remove entrained solids and into the secondary chamber where complete combustion is achieved using added air in a bed of chromium oxide. The off-gas stream from the secondary chamber containing fly ash, dust, and gases phases through a separator and a series of HEPA filters to remove these constituents. This promising developmental concept has been adapted by Aerojet Energy Conversion Company (AECC) for

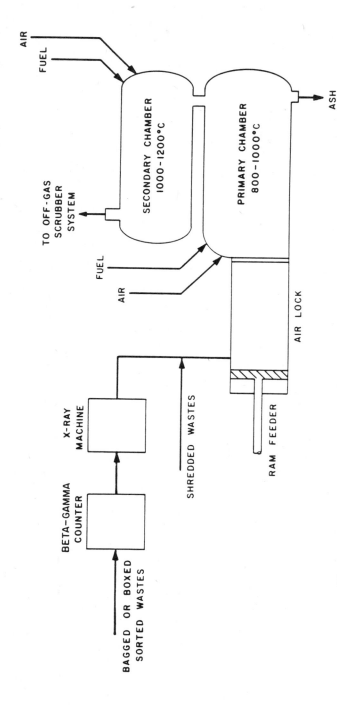

Figure 5-10 Los Alamos National Scientific Laboratory controlled-air incinerator. Source: (DOE 1984).

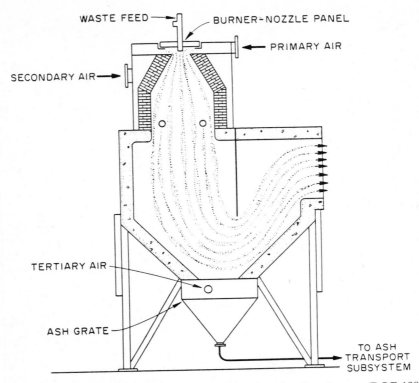

Figure 5-11 Excess-air cyclone incinerator combustion chamber. Source: (DOE 1984).

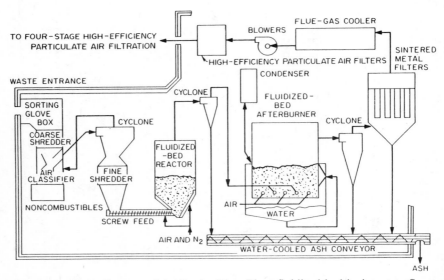

Figure 5-12 Flow diagram of the Rocky Flats Plant fluidized-bed incinerator. Source: (DOE 1984).

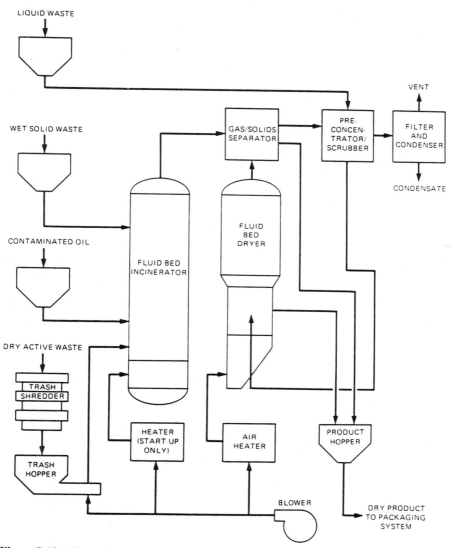

Figure 5-13 Flow diagram of the Aerojet Energy Conversion Company's fluidized-bed dryer/incinerator. Source: (DOE 1984).

commercial use. The AECC system (Figure 5-13) is capable of incinerating solids, resins, heavy sludges, and liquids and has been installed at a number of utilities including the Byron and Braidwood sites of the Commonwealth Edison Co. (CECo) to process DAW, waste oil, and evaporator bottoms. Additional information on fluidized bed incinerators is available in the literature (Ziegler 1982; Kibbey 1980; Trigilio 1981).

Rotary-Kiln. A 40 kg/hr rotary-kiln incinerator is a demonstration unit that has also been developed at the RFP from experience on earlier pilot scale (2 kg/hr) studies with nonradioactive materials. The 40 kg/hr system consists of a horizontally rotating, cyclindrical primary chamber and an afterburner (Figure 5-14). The primary chamber is designed to (1) tumble the wastes to improve combustion efficiency and (2) move the incinerated waste in a continuous, gradual flow toward the ash discharge where it then falls into the collecting container. Off-gases are sequentially treated in an afterburner, wet scrubber, and HEPA filter systems. The RFP system was designed to treat TRU wastes for plutonium recovery, and can also process a variety of solid and liquid wastes including HEPA filters, sludges, and ion-exchange resins. There are no commercially available rotary-kiln incinerators to process radioactive wastes. Additional information on this incinerator type is available in the literature (Kibbey, 1980; Trigilio 1981).

Slagging-Pyrolysis. The slagging-pyrolysis incinerator, a vertical furnace with a primary-stage gas fill and a secondary-stage combustion chamber, was evaluated at INEL for processing soil-contaminated TRU waste. This incinerator type provides a large throughput, has a high volume reduction factor, produces a stable residue suitable for disposal, and is thus suitable for treating contaminated soil. The INEL concept, which represents an adaptation of a commercial

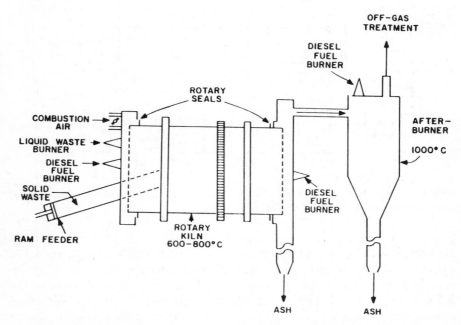

Figure 5-14   Rocky Flats Plant rotary-kiln incinerator. Source: (DOE 1984).

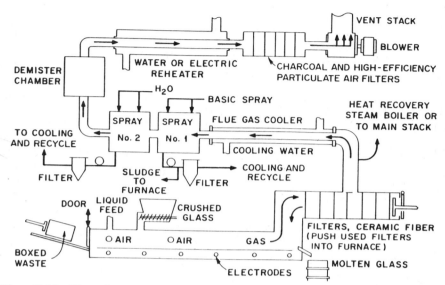

Figure 5-15   Flow diagram of the Penberthy molten-glass incinerator (electromelter) system. Source: (DOE 1984).

slagging-pyrolysis unit developed by ANDCO-TORRAX, exposes the waste to treatment in three stages as it passes down through the gasifier drying, pyrolizing, and combustion/melting ranges. Supplementary heat is added to the gasifier to generate the molten slag that is removed and quenched with water. The gases pass from the gasifier to an afterburner, then a regenerative column, spray dryer, cyclone particulate separator, and sintered metal filter. The concept has not been developed further at INEL to date. Slagging-pyrolysis is discussed further in the literature (FMC 1977; Hedahl 1979).

Penberthy Electromelt (Molten Glass). The Penberthy Electromelt (Figure 3-15) is a commercial incinerator adapted for developmental treatment of LLW at DOE laboratories. The process renders combustible and noncombustible wastes into glass. Waste glass is initially fed into the system, is melted, and becomes electronically conducting as the furnace heats up. An immersed electrode provides the current necessary to maintain self-sustaining joule heating. The liquid and shredded solid wastes are oxidized in the furnace and incorporated into the glass. The final product is removed as a liquid and then solidified. The off-gas system consists of a flue-gas cooler, spray for acid removal, demister, reheater, and charcoal and HEPA filters. The developmental testing of the Penberthy units at DOE laboratories has encompassed a demonstration of vitrification of defense waste (Pacific Northwest Laboratory) and immobilization of combined waste sludges (Savannah River Laboratory). Developmental work continues to establish the parameters for treatment of LLW in an electromelter and permit evaluation of the utility of the process for this purpose. The process is described further in the literature (Kibbey 1980; Chapman 1979).

5.7 STORAGE OF RADIOACTIVE WASTE

The concept of storage of radioactive waste applies to LLW, in terms of interim storage of packaged waste prior to transport, and to both onsite permanent and interim storage of remedial wastes. It also applies to onsite pool and dry storage of spent fuel and away-from-reactor (AFR) storage of spent fuel, and to the storage of HLW and TRU pending the development of a geologic repository for final disposition of these waste forms. HLW and TRU storage is discussed in Chapter 6.

5.7.1 Storage of Low-Level Waste and Remedial Wastes

Interim Storage of Packaged Wastes. Facilities that generate and package LLW for offsite disposal will frequently accumulate packages to obtain sufficient numbers for "exclusive-use" shipments by truck or rail. When packages are accumulated, an onsite storage facility must be provided to store the packages before they are moved to the loading area for loading onto the transport vehicle. The waste package storage area must be kept physically separated from the remainder of the facility. This can be accomplished by using a walled in area or, as a minimum, cordoning off the area with a rope barrier. The storage location must also be posted with "radiation" hazard signs. If the waste package storage area is exposed to the elements, the packages should be covered with tarpaulins that are fixed in position with tie downs. The storage facility must be located under the control of facility security, typically requiring that it be within a security fence and controlled gate that is maintained on a 24-hr basis. The package accountability and inspection records for each package, prepared at the time they are filled and closed, are maintained during this period, showing package locations and other identifying parameters.

Interim Storage of Remedial Waste. High-volume, low-activity waste, consisting of soil and debris (bricks, rubble) from decommissioning of facilities, is often stored in bulk on the site being remediated, or a nearby location, pending the designation of a permanent disposal site. The use of engineered barriers is required to prevent environmental dispersion of this material. Typically, a pad consisting of clay, or an artificial material such as Hypalon, is placed on the ground, and the storage pile constructed on top of the pad. Surface water diversion systems are used to prevent lateral infiltration into the base of the pile, or under the pad. Once the pile is emplaced, it is covered with a tarpaulin or plastic to prevent airborne dispersion of particulates and infiltration of precipitation. If the pile is to remain for a relatively extended period, a more permanent cover is often used such as an asphalt seal or a layer of soil. Monitoring of the air and water pathways is performed to ensure that environmental contamination does not occur.

In some cases the soil and waste from remediated sites has been packaged in drums and stored in a controlled area on an interim basis until a disposal site is designated. Radium-contaminated soil from the remediation of Montclair, West Orange, and Glen Ridge, New Jersey Superfund sites has been drummed and stored while the search was made for a disposal facility that would accept the waste. It is likely that interim storage of remedial soil and waste will continue as long as existing commercial disposal facilities are reluctant to accept this low-activity material, and the public objects to disposal in other facilities created for "sole-source" disposal of remediated wastes. The development of LLW compact sites within the states, whose establishment was in part justified by inclusion of remedial waste, should permit disposal of the stored material. The storage problem created by the inability to dispose of remedial wastes in some offsite facilities is one of the factors moving federal agencies to reconsider the desirability of this remedy instead of use of onsite stabilization or treatment. The recent SARA amendments have, in fact, encouraged onsite remedies for radium-contaminated waste. (See Chapter 7 for further elaboration on this situation.)

Permanent Onsite Storage of Remedial and Other LLW Waste Forms. The concept of permanent onsite storage of remedial waste or other LLW forms, as an alternative to shallow land burial, involves the emplacement of the waste in above- or below-ground engineered vaults or containers generally made as precast concrete structures. This approach, which would provide additional barriers to contaminant dispersion, and potentially permit future retrieval of the waste, has received favorable consideration by the industry and is being incorporated into the design requirements for a number of compact sites. The concept is discussed further in Chapter 6.

5.7.2 Spent Fuel Storage

The storage of spent nuclear fuel rods, after they were removed from the reactor, was not considered a problem in the early days of commercial nuclear power since it was assumed that all spent fuel would be reprocessed in the processing plants then being developed. The government's decision in the 1970s not to permit spent fuel to be reprocessed in this country because of concerns that the recovered plutonium and uranium were subject to pilferage or loss of accountability put the burden of spent fuel storage on the nuclear utilities, which were not prepared for it. The Nuclear Waste Policy Act of 1982 (NWPA 1982) confirmed that utilities owning and generating spent fuel were responsible for storing it until January 31, 1988 when the federal government was scheduled to begin accepting it for disposal.

Onsite water storage in pools originally constructed for short-term storage of irradiated fuel has been utilized for 40 yr at nuclear power plants. However, since the utilities must now store the spent fuel for an extended period, some nuclear utilities are running out of storage space. As one measure of this problem, the DOE estimates that 67 U.S. plants will lose "full core reserve" storage capacity by 2000 (DOE 1986). To alleviate this situation, the utilities have both undertaken to enhance at-reactor storage capacity and are looking into new storage methods. The utilities began pool reracking in the 1970s to accommodate additional fuel rods in existing pools, and have been transshipping spent fuel from one plant to another (usually newer) in the 1980s to make use of unused storage capacity. Among the methods now being considered to enhance at-reactor storage capability are the expansion of existing water storage capacity, more efficient use of available capacity through fuel rod consolidation, the construction of additional at-reactor storage facilities using either wet storage or, alternatively, dry storage of spent fuel in dry storage casks, and the development of away-from-reactor (AFR) storage facilities. These approaches can be grouped, from the standpoint of technologies employed into onsite wet pool storage, dry storage, and AFR storage.

Wet Pool Storage. More than 95% of the existing spent fuel inventory in the United States is stored in water pools at reactor sites, and this storage method will be used for all the LWRs that are currently under construction. A major advantage of wet storage is that it is well suited for removing the radioactive decay heat generated by the spent fuel in the period of several months immediately after its discharge from the reactor, when the decay heat rate is at its highest and dropping off rapidly. A typical spent fuel storage pool is rectangular in shape, constructed of reinforced concrete, and frequently lined with metal plates to which the storage racks are attached. The spent fuel assemblies are placed in the storage racks at the bottom of the pool which are designed both to hold the assemblies securely and maintain the required spacing to prevent criticality. While the pools at reactor facilities vary in length and width, they are generally 12–13 m deep permitting a minimum 3-m layer of water over the fuel assemblies to provide radiation shielding of operating personnel. As described above, the original wet storage pools at many reactor facilities are filling up, and approaches are being sought to enhance existing fuel storage capability.

One attractive approach to provide this enhanced storage is to expand the existing wet storage capacity if space permits. By using the existing storage facility, costs of expansion are minimized. Two ways are available to expand the water volume capacity: reracking of the pool to increase storage density and consolidation of spent fuel rods by dismantling the fuel assembly and rearranging the spent fuel rods into close-packed geometry in a storage canister. Reracking can be accomplished by (1) the replacement of nonpoisoned racks with racks incorporating boron in stainless-steel or aluminum matrices to absorb neutrons and thus permit closer spacing of the fuel assemblies, (2) the replacement of existing non-fuel-storage racks with racks suitable for fuel storage, or (3) the replacement of the original spent fuel storage racks with stainless-steel racks that permit closer spacing of the stored assemblies. Reracking has been used for a number of years and the technology continues to improve. However, storage capacity increases from reracking are limited by the storage pool volume, and constrained by structural considerations for the additional weight. Rod consolidation, which has the potential to substantially increase spent fuel storage capacity at some reactors, is also constrained by structural strength requirements to support the additional weight. Another way some utilities have gained short-term additional storage capacity is to temporarily encroach on the full core reserve (FCR) (space reserved in the spent fuel pool for unplanned discharge of the entire reactor core loading). This occurs when the utility chooses to gamble that the FCR will not have to be used for a period of time and uses this space for spent fuel assemblies. In certain cases it may be cost effective for a utility to construct an additional separate storage pool, particularly if the new pool can be used to serve a number of nuclear power plants.

Dry Storage. It is apparent that none of the wet storage or intrautility transshipment approaches will entirely alleviate the problem of loss of FCR before the government is able to assume responsibility for the spent fuel. Both the DOE and industry have been working to find alternative solutions, with an emphasis on dry storage as a viable alternative. The DOE has used dry storage methods for military reactor fuel for more than 20 yr. Dry storage offers a number of advantages as an interim storage concept. It provides the flexibility to meet a range of utility spent fuel storage needs, permits completely passive cooling, thus precluding safety and performance concerns regarding interruption of coolant flow, permits less extensive cover gas control and monitoring requirements for most dry storage concepts than the comparable requirements for wet storage, permits, in certain cases, the addition of capacity in modules minimizing capital investment, and generates less secondary radioactive wastes as compared to wet storage.

Dry storage modules at reactor sites can be large metal storage casks, concrete storage silos, or drywells. The major developmental dry storage program in the United States is a cooperative effort involving the DOE, the Virginia Power Corporation, and the Electric Power Research Institute (EPRI). Under this program, Virginia Power ships spent fuel to the DOE's Idaho National Engineering Laboratory (INEL) where it is placed in large metal storage casks and monitored. This test program will provide performance data on the storage concept, and promote the development of dry rod consolidation technology and prototypical equipment. The DOE's stated program goal "is to remove conservatism from the licensing of dry storage casks and provide a storage technology that is generically applicable so that the Nuclear Regulatory Commission can license dry storage of spent fuel by rule." There are four dry storage casks to be tested under the program: the Castor V/21 from General Nuclear Systems, Inc. (GNSI) (Figure 5-16), the TN-24P from Transnuclear Inc. (Figure 5-17), the MC-10 from Westinghouse Electric Corporation (Figure 5-18), and the NAC 5100 Nuclear Assurance Corporation (available in 1988). The relevant characteristics of these casks are summarized in Table 5-16. The project is scheduled to be completed in 1988.

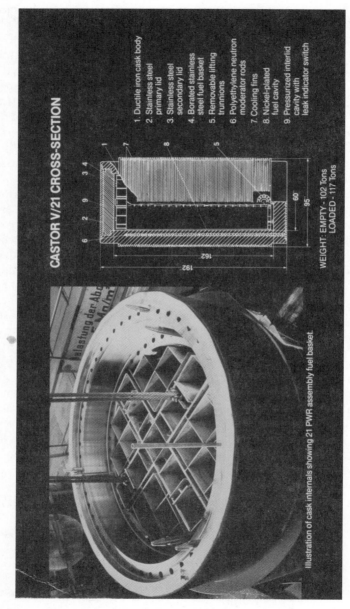

CASTOR V/21 CROSS-SECTION

1. Ductile iron cask body
2. Stainless steel primary lid
3. Stainless steel secondary lid
4. Borated stainless steel fuel basket
5. Removable lifting trunnions
6. Polyethylene neutron moderator rods
7. Cooling fins
8. Nickel-plated fuel cavity
9. Pressurized interlid cavity with leak indicator switch

WEIGHT: EMPTY - 102 Tons
LOADED - 117 Tons

Illustration of cask internals showing 21 PWR assembly fuel basket.

Figure 5-16   General Nuclear Systems castor V/21 cask.

Figure 5-17 Transnuclear TN-24P cask. (Reprinted from *Nuclear News*, 1987.)

The GNSI Castor V/21 is being used in the first dry storage program to be conducted at a commercial nuclear power plant in the United States. The Virginia Power Company's Surry station in Virginia has established an Independent Spent Fuel Storage Installation (ISFSI). The ISFSI, when completed, will consist of three concrete storage pads (230 ft long × 32 ft wide × 3 ft thick), each capable of storing 28 casks. The utility will have five casks loaded by 1987, and plans to load an additional three to four a year at Surry. The utility was licensed in 1986 for 20 yr of dry storage.

Concrete silos make use of the concrete to dissipate the heat generated by the spent fuel, with the spent fuel placed within a sealed chamber in the concrete. There are a number of design concepts for such silos, one of which is planned for a cooperative program between the DOE and Carolina Power & Light (CP&L) Company at their Robinson-2 plant. This plan involves the use of the second currently licensed dry storage system in the United States: the Nutech Horizontal Modular System (NUHOMS) (Figure 5-19), a concrete module and stainless-steel horizontal silo storage system. A 20-yr license for dry storage has been issued by the NRC at Robinson-2. CP&L is currently using three of the modules at the site and may eventually install up to eight of the NUHOMS modules.

Drywells use cylindrical chambers that are sunk into the ground and into

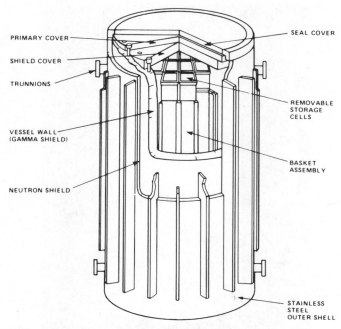

Figure 5-18 Westinghouse MC-10 cask. (Reprinted from *Nuclear News*, 1987.)

TABLE 5-16 **Large Metal Dry Storage Cask Characteristics**

Characteristic	GNSI Castor V/21	Transnuclear TN-24P	Westinghouse MC-10	NAC NAC 5100
NRC license status	Licensed	Unlicensed	Unlicensed	Unlicensed
Capacity	21 PWR	24 PWR	24 PWR	28 PWR
Construction materials	Modular cast iron	Forged steel	Forged steel	Stainless steel/ lead/stainless steel sandwich
Neutron shielding	Polyethylene	Borated plastic	Borated plastic	Stainless steel/ lead/stainless steel sandwich
Maximum weight	115 tons	100 tons	100 tons	Unavailable
Loading of cask at INEL	8/85	12/85	6/86	Not scheduled

Figure 5-19   Concrete modules for the NUHOMS installation at Robinson-2. (Reprinted from *Nuclear News*, 1987.)

which the spent fuel is lowered. The decay heat from the spent fuel is dissipated in the surrounding soil. Drywells have been used by the DOE to store spent fuel at the Nevada Test Site.

Away from Reactor (AFR) Storage.   Away-from-reactor (AFR) facilities, defined as all spent fuel storage facilities not integrated within a reactor plant, include those located at reprocessing plants, disposal sites, or other nuclear fuel cycle facilities. There are currently two AFR wet storage facilities operating in the United States, in storage pools at the Morris, Illinois and West Valley, New York former reprocessing facilities. As dry storage technologies are developed, they will be increasingly used for AFR storage. Although there are no dry AFR storage facilities as yet in the United States, dry AFR facilities have been constructed in Europe, with the Gorleben AFR facility in Germany being a typical example. An AFR design concept being considered in the United States for long-term storage of spent fuel (and possibly high level waste) is the Monitored Retrievable Storage (MRS) facility, which would incorporate such dry storage concepts as sealed concrete storage casks (silos) and drywells.

It is important to note that the wet and dry storage concepts for spent fuel are intended as interim measures pending the operability of a geologic repository. Geologic repository disposal is discussed in Chapter 6.

5.8   PACKAGING AND TRANSPORT OF RADIOACTIVE WASTE

In considering the packaging and transportation of radioactive wastes produced at nuclear fuel cycle facilities, the range of available waste types includes spent fuel, high-level waste (HLW), transuranic waste (TRU), low-level waste (LLW), and uranium mill tailings. However, the types to be considered that undergo transport are limited. Since previously generated HLW and TRU waste is currently being stored pending the development of a geologic repository, and the existing moratorium on reprocessing of spent nuclear fuel from power plants means that no new HLW or TRU waste is generated, there is no transport of HLW or TRU waste being undertaken. In addition mill tailings are generally disposed of onsite and, in the few cases in which they are transported, the radiological concentrations in the tailings are low enough to permit bulk transport with minimal constraints. Thus, we will focus on LLW transport for which there are established transportation systems, and spent fuel transport that has been undertaken in the past and that will be required in the near future. The approach in this chapter will discuss the regulatory organizations and requirements, waste packaging types, shipper requirements and constraints, transport modes and routes, and carrier responsibilities and procedures.

5.8.1 Regulation of Radioactive Waste Packaging and Transport

The U.S. Department of Transportation (DOT) has regulatory responsibility for the transportation of radioactive materials by all modes of transport (rail, road, air, water) in both interstate and foreign commerce and by all means of transport. This statutory authority exists under the Department of Transportation Act of 1966 (DOT 1966) and the subsequent Transportation Safety Act of 1974 (Trans 1974). Postal shipments, however, come under the control of the U.S. Postal Service, not the DOT. The U.S. Nuclear Regulatory Commission (NRC) also has responsibility for safety in the possession and use, including transport, of by-product, source, and special nuclear material as stipulated in the Atomic Energy Act of 1954, as amended. (Atomic Energy 1954). Since both the DOT and NRC have statutory authority to regulate transport of radioactive materials, including wastes, it was necessary for an accommodation to be reached that delineates the responsibilities of each agency to avoid overlap and conflicting standards. This accommodation takes the form of a Memorandum of Understanding (Memo 1979) between the two agencies for regulation of safety in the transportation of radioactive materials. The DOT defines its regulations in Title 49 "Hazardous Materials Regulations," Parts 100–178, with particular emphasis on the requirements of 49 CFR 171, "General Information, Regulations, & Definitions," and 49 CFR 173, "Shippers-General Requirements for Shipments and Packaging." The NRC regulations are defined in 10 CFR 71, "Packaging and Transportation of Radioactive Material."

Taken together 49 CFR 100–178 and 10 CFR 71 provide a body of regulations that is designed to ensure the safety of handling and transport workers and the general public, and to protect the environment against degradation and property against damage. To accomplish this, certain fundamental principles are embodied as objectives in the regulations to be achieved through the incorporation of prescriptive requirements. These include requirements that will prevent accidental criticality when fissile material is transported, ensure containment of the material during handling, packaging, and transport, protect workers and the general public from radiation emitted by the radioactive material, and provide for the safe dissipation of excessive heat generated from thermal decay of the radioactive material (i.e., spent fuel). Thus, the body of regulations imposes primary constraints on the packaging as the principal barrier against release and/or overexposure, and secondarily constrains the transportation mode and means to achieve the safety objectives.

5.8.2 Radioactive Waste Packaging

The fundamental principle governing the extent of packaging is that the constraints in the regulations are intended to be commensurate with the hazards of the type of radioactive material, its quantity, and form. Thus, for certain waste materials that are unregulated for purpose of transportation, no packaging is required; however, at the other extreme, high-integrity containers (HIC), capable of withstanding severe accident conditions without loss of package integrity or spillage of material, are required for other types of radioactive wastes.

Before proceeding, the concept of "packaging" and "package," as defined in the DOT and NRC regulations needs to be clarified. A package consists of the packaging with its content of radioactive material. Packaging is the assembly of components necessary to ensure compliance with the packaging requirements of the regulations. Depending on the type of contents, its activity level, and thermal and radioactive emissions, this assembly may include absorbent materials, radiation shielding, insulation and cooling devices, structural members to provide support and absorb shock, and the tie-down system to the vehicle. There may also be an inner and outer receptacle. Packages are designated as Type A package, Type B package, or a package for low specific activity (LSA) material. A Type A package is designed to maintain its integrity under "normal conditions of transport." The Type A package holds Type A quantities of material. These packages are selected by the shipper without specific regulatory approval. A Type B package is designed to maintain its integrity under both "normal conditions of transport" and hypothetical accident test conditions. The Type B package is able to hold greater quantities of radioactive materials than the Type A package. Some Type B package quantities are referred to as "highway-route controlled." The packages are typically used for spent fuel and high-activity LLW. Type B packages must be certified by the NRC under the provisions of 10 CFR 71 and, in order to be certified, must withstand the aforementioned accident test conditions. These test conditions are, sequentially, (1) a 30-ft free drop onto an unyielding surface, (2) a puncture test requiring a free drop greater than 40 in. onto a 6-in.-diameter steel pin, (3) exposure to a temperature of 1475°F (800°C) for at least 30 min, and (4) immersion in water for an 8-hr period at a depth of 15 ft. If the package is designed to carry fissile materials such as enriched uranium, plutonium, or [233]U, resistance to in-leakage of water at a depth of 3 ft must be demonstrated for a period of 8 hr. Specific packaging designs for LLW materials and spent fuel are discussed below.

The DOT methodology for the classification of radioactive material packages is based on the potential inhalation dose to the public resulting from an accident that assumes a small portion of the package contents, 0.1%, is released. Since the inhalation dose is radionuclide specific, the DOT has assigned dose-related values to some 150 radionuclides. These values (49 CFR 173.433 and 49 CFR 173.435), defined as A_1 and A_2 values, are used not only to limit the activity content in Type A packages, but also to derive specific activity limits for LSA materials, and for exempted limited-quantity packages using fractions of these values. Activities in packages exceeding the A_1 or A_2 values are Type B quantities requiring Type B packaging.

Generally, the regulations require that each package used for shipment of radioactive materials must be designed so that the package can be easily handled and properly secured in a conveyance during transport, has a means for manual handling if the gross package weight is between 22 and 110 lb, and has lifting attachments on the package that do not impose unsafe stresses on the package structure and where failure of the lifting device shall not affect packing integrity. Lifting fixtures must be removed during transit or designed with a strength equivalent to that required for the lifting attachments. In addition, the external surface, as far as practical, must be easily decontaminatable. The outer layer packaging must avoid, as far as practical, pockets or crevices in which water might collect. Each feature that is added to the package at the time of transport, and that is not part of the package, must not reduce the safety of the package. Packages containing LSA radioactive materials and transported as exclusive use are exempted from specification packaging, marking, and labeling.

LLW Packaging. Because of the range of waste types and characteristics that is classified under the generic name LLW, the type of package required for the material can include a wide spectrum of package configurations encompassing "strong tight packages" for LSA waste (the bulk of the LLW shipped), Type A packages, and Type B packages including those containing "highway route controlled quantities." The LSA classification applies to transported materials that are considered to be of low risk or "inherently safe," and thus require less stringent packaging and shipping requirements. LSA materials include solid waste from the fuel cycle, industrial, and institutional sources, tritiated water in concentrations not exceeding 5 mCi/mL, objects with surface contamination not exceeding 0.1 μCi/cm^2 over an area of 1 m^2, and inherently low-activity materials such as uranium and thorium ores. LSA material shipped as "exclusive use" (full load in vehicle) requires only strong tight packages. This provides the shipper the freedom to choose between a variety of different types of packages including plywood boxes, steel boxes, steel drums, concrete containers, and bins. Primarily because of economic considerations, those packages that are costly are frequently designed to be reusable and may contain inner container(s) such as liners or drums that are removed from the outer shipping container at the burial site.

The next step up in container integrity for LLW packaging is the Type A package. Type A packages can contain waste with two upper limits for activity content of radionuclides, "A_1," "A_2." A_1 is the activity limit for "special form" material that would be essentially nondispersible if released from a package. A_2 is the activity limit for "normal form" that is dispersible (i.e., gas, liquid, or powder). Type A packages must achieve specific radiation, containment, and shielding limits under normal conditions of transport. Type A wastes include such diverse types as dewatered filter resins, irradiated hardware, and highly contaminated clothing. Type A packaging includes a range of drum, box, or cask sizes with the DOT Spec. 17C 55-gal steel drum being a versatile "workhorse" for containing this type of material (Figure 5-20). A representative Type A steel box configuration, the UNI-4476 container, is shown in Figure 5-21. Figure 5-22 shows the 10CNS 14-195L cask, a typical transport cask designed to carry 14-55-gal drums. The cask is a steel cylinder with lead γ shielding in the cask wall.

As previously noted, the Type B package can be used to transport high-activity LLW, up to 3000 times A_1 quantities, 3000 times A_2 quantities, or a maximum of 30,000 Ci of activity. Since the Type B casks that are certified for transport of this type of waste (e.g., nonfissile, irradiated, and contaminated hardware and neutron source components) are used primarily for transport of spent fuel, they will be discussed in the following section under Spent Fuel Packaging. Other Type B packages designed specifically to transport radioactive waste include steel boxes with lead shielding such as the CNS 15-160B transport

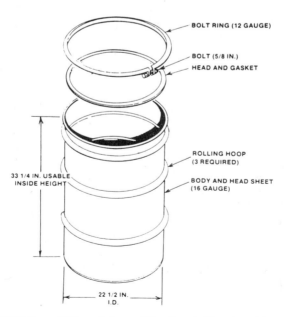

Figure 5-20   DOT Spec. 17C steel drum (55 gallons). (Reprinted from ASME 1986.)

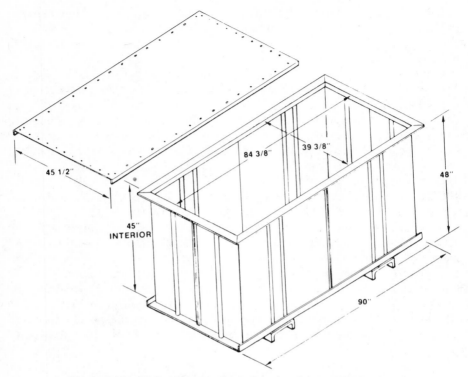

Figure 5-21   UNI-4476 container. (Reprinted from ASME 1986.)

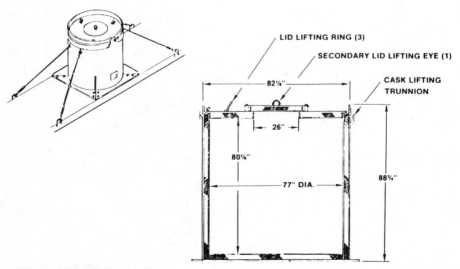

Figure 5-22   CNS 14-195L Type A transport cask. (Reprinted from ASME 1986.)

cask (Figure 5-23) which holds 15-55-gal drums or 2–80 ft³ disposable liners. A popular design cylindrical cask is the Vandenberg cask (Figure 5.24) which is capable of holding 3-55-gal drums of 1–60 ft³ disposable liner.

In addition to transport containers, a class of onsite storage container has been developed for use at nuclear facilities. The on-site storage container (OSSC) manufactured by ATCOR Engineering Systems, Inc. (Figure 5-25) is used for a storage shield in which HICs or carbon steel solidification liners are placed. The shielding provided by an OSSC can allow the processing and temporary storage of higher activity Class B and C resins, sludges, or filters. The use of the OSSC avoids the cost of shipping cask demurrage and allows a package waste volume to be maximized over an extended period.

Spent Fuel Packaging. Type B packaging, commonly referred to as a shipping cask, is used to transport the limited amount of spent fuel shipped in the United States. Because of the low utilization factor, and high expense in developing and certifying these casks, there are not many of these casks available. Table 5-17 provides relevant parameters for the casks currently being used, or with near-term availability. These casks generally have common features. They are classified by size as for either truck or rail, are reusable, have a cylindrical vessel, are equipped with impact limiters to alleviate impacts from collisions, contain two layers of radiation shielding in the walls (one for γ radiation and one for neutron radiation), contain features to dissipate decay heat from the fuel assemblies, and are designed to contain the fuel under any accident scenario. Truck casks fall into two sizes: legal-weight truck (LWT) casks weighing about 25 tons when loaded with either one PWR or two BWR fuel assemblies, and overweight (OWT) casks that weigh about 40 tons when loaded. Rail casks can also be subclassified by weight, with the lighter rail casks weighing about 75 tons loaded with 7 PWR or 18 BWR fuel assemblies, and the heavier cask of 100 tons holding about 12 PWR or 32 BWR assemblies. As representative of the types of spent fuel shipping casks described in Table 5-17, Figure 5-26 shows an LWT cask, the NLI-1/2, Figure 5-27 shows an OWT cask, the TN-9, and Figure 5-28 shows a rail cask, the IF-300.

As a result of the growing shortage of spent fuel storage space at many nuclear power plants, a concept that is gaining increasing attention is the use of a cask that can provide dual short-term storage and eventual transport capability. With this in mind, there are several casks under development for spent fuel storage with features that permit their use for transport at a later date. These casks generally provide large capacities, with enhanced shielding and heat rejection capabilities. (See Section 5.7 for discussion of dry storage casks.)

Other Types of Radioactive Waste Packaging. The high-integrity container (HIC), as its name would imply, is a high-strength package that is designed to resist crushing from static loads and corrosion from the contained waste and the soil media. The HIC, whose concept grows out of 10 CFR 61 requirements for land burial facilities, supports the disposal unit cover and thus minimizes creation

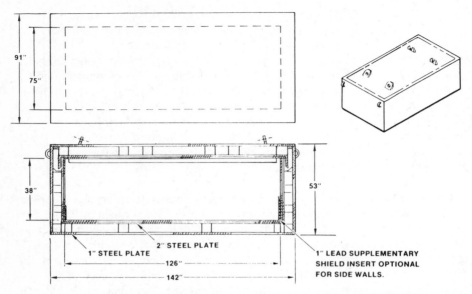

Figure 5-23   CNS 15-160B transport cask. (Reprinted from ASME 1986.)

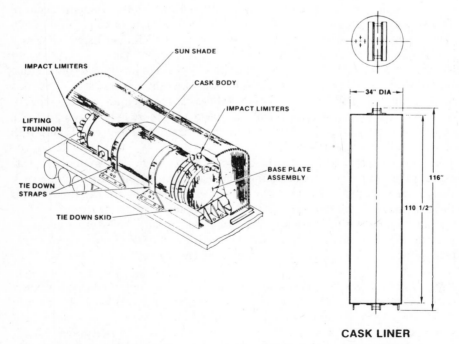

Figure 5-24   CNS 3-55 Vandenberg cask. (Reprinted from ASME 1986.)

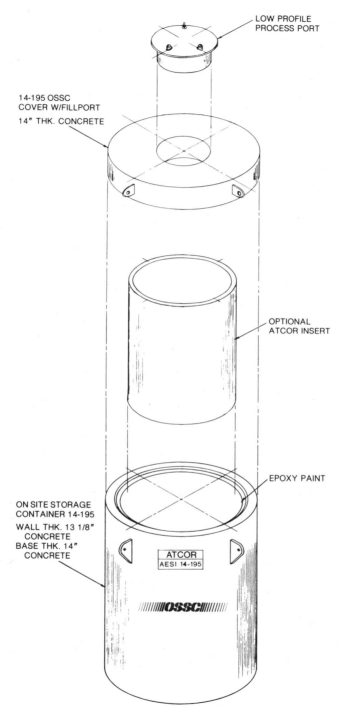

LOW PROFILE
PROCESS PORT

14-195 OSSC
COVER W/FILLPORT
14″ THK. CONCRETE

OPTIONAL
ATCOR INSERT

EPOXY PAINT

ON SITE STORAGE
CONTAINER 14-195
WALL THK. 13 1/8″
CONCRETE
BASE THK. 14″
CONCRETE

ATCOR
AESI 14-195

OSSC

Figure 5-25 ACTOR on-site storage container (OSSC) used as a process shield. (Photo Courtesy of the Manufacturer.)

TABLE 5-17 U.S. Spent Fuel Shipping Casks

Shipping Cask	Number Available	Weight (tons)		Capacity (Intact Assemblies)		Licensed Use	Transport Mode	Owner/ Operator
		Empty	Loaded	PWR	BWR			
Current Truck Casks								
NLI-1/2	5	22	23	1	2	LWR	LWT	Nuclear Assurance Corp.
TN-8/TN-9	2/2	37	39	3(TN-8)	7(TN-9)	LWR	OWT	Transnuclear
NAC-1	6	24	25	1	2	LWR	LWT	Nuclear Assurance Corp.
FSV-1	3	22–23	23.5–25.5	1	3	HTGR	LWT	General Atomics
Current Rail Casks								
IF-300	4	63–65	68–70	7	18	LWR	Rail /OWT	General Electric
NLI-10/24	2	90	97.5	10	24	LWR	Rail	Nuclear Assurance Corp.
Future Spent Fuel Cask Concepts (Developmental)								
TN-12	0[a]	87	97	12	32	—	Rail	Transnuclear
LWT cask	—	24	25	2	5	—	Truck	—
OWT cask	—	34	37	4	10	—	Truck	—
Light rail cask	—	63	71	10	24	—	Rail	—
Heavy rail cask	—	81	93	14	36	—	Rail	—

[a] Not yet licensed in the United States. Used in Europe.

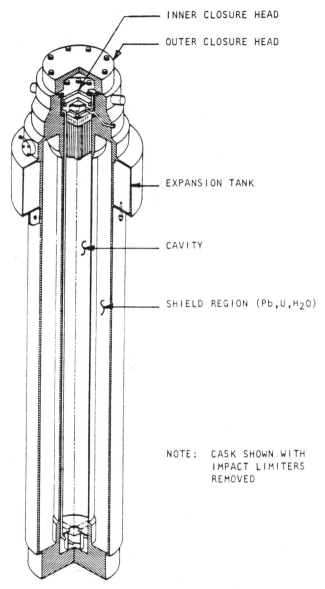

INNER CLOSURE HEAD

OUTER CLOSURE HEAD

EXPANSION TANK

CAVITY

SHIELD REGION (Pb,U,H$_2$O)

NOTE: CASK SHOWN WITH
IMPACT LIMITERS
REMOVED

Figure 5-26 NLI-1/2 legal weight truck cask. (Reprinted from ASME 1986.)

RADIOACTIVE WASTE TECHNOLOGY

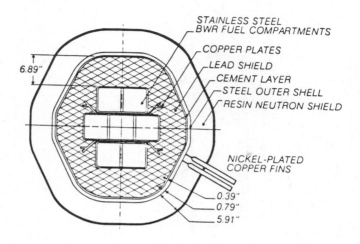

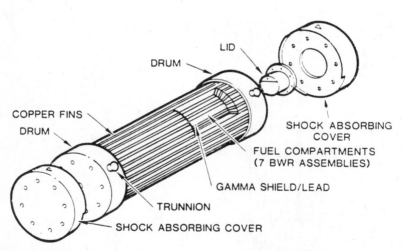

Figure 5-27 TN-9 overweight truck spent fuel cask. (Reprinted from ASME 1986.)

of voids in the cell with resultant subsidence and infiltration. The use of the HIC eliminates the need for addition of solidification agents or adsorbents to immobilize the waste in the container. The use of HICs is growing, particularly for ion-exchange resins and filter media, and as liners used in conjunction with an outer container. Table 5-18 lists the performance criteria developed by the State of South Carolina for use of HICs at the Barnwell disposal site.

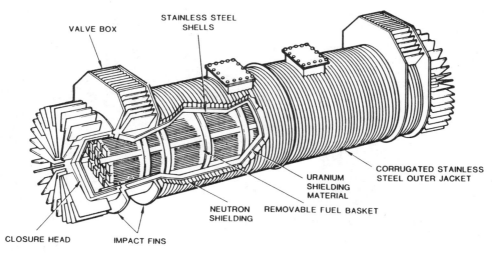

Figure 5-28 IF-300 spent fuel shipping cask. (Reprinted from ASME 1986.)

TABLE 5-18   State of South Carolina Criteria for High-Integrity Containers

The general criteria for high-integrity containers to be used for high concentration waste forms are as follows:

1. The container must be capable of maintaining its contents until the radionuclides have decayed, approximately 300 yr, since two of the major isotopes of concern in this respect are strontium-90 and cesium-137 with half-lives of 28 and 30 yr, respectively.

2. The structural characteristics of the container with its contents must be adequate to withstand all the pressure and stresses it will encounter during all handling, lifting, loading, offloading, backfilling, and burial.

3. The container must not be susceptible to chemical, galvanic, or other reactions from its contents or from the burial environment.

4. The container must not deteriorate when subjected to the elevated temperatures of the waste streams themselves, from processing materials inside the container, or during storage, transportation, and burial.

5. The container must not be degraded or its characteristics diminished by radiation emitted from its contents, the burial trench, or the sun during storage.

6. All lids, caps, fittings, and closures must be of equivalent materials and constructed to meet all of the above requirements and must be completely sealed to prevent any loss of the container contents.

Source: (DOE 1984).

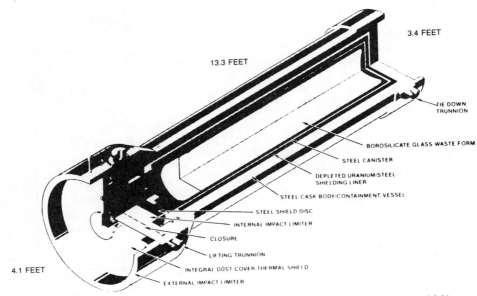

3.4 FEET

13.3 FEET

TIE DOWN
TRUNNION

BOROSILICATE GLASS WASTE FORM

STEEL CANISTER

DEPLETED URANIUM/STEEL
SHIELDING LINER

STEEL CASK BODY/CONTAINMENT VESSEL

STEEL SHIELD DISC

INTERNAL IMPACT LIMITER

CLOSURE

LIFTING TRUNNION

INTEGRAL DUST COVER/THERMAL SHIELD

EXTERNAL IMPACT LIMITER

4.1 FEET

Figure 5-29 Defense high-level waste truck cask. (Reprinted from ASME 1986.)

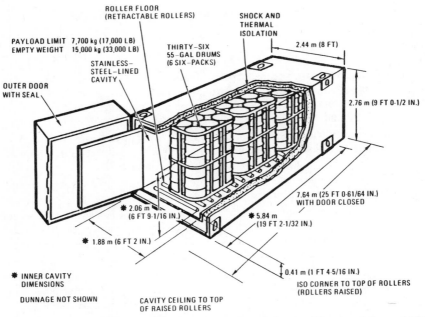

ROLLER FLOOR
(RETRACTABLE ROLLERS)

SHOCK AND
THERMAL
ISOLATION

PAYLOAD LIMIT 7,700 kg (17,000 LB)
EMPTY WEIGHT 15,000 kg (33,000 LB)

THIRTY–SIX
55–GAL DRUMS
(6 SIX–PACKS)

2.44 m (8 FT)

STAINLESS–
STEEL–LINED
CAVITY

OUTER DOOR
WITH SEAL

2.76 m (9 FT 0-1/2 IN.)

7.64 m (25 FT 0-61/64 IN.)
WITH DOOR CLOSED

✱ 2.06 m
(6 FT 9-1/16 IN.)

✱ 5.84 m
(19 FT 2-1/32 IN.)

✱ 1.88 m (6 FT 2 IN.)

0.41 m (1 FT 4-5/16 IN.)

✱ INNER CAVITY
DIMENSIONS

DUNNAGE NOT SHOWN

CAVITY CEILING TO TOP
OF RAISED ROLLERS

ISO CORNER TO TOP OF ROLLERS
(ROLLERS RAISED)

Figure 5-30 TRUPACT-1 loaded with six-packs of drums. (Reprinted from ASME 1986.)

In anticipation of future shipments of HLW and TRU waste, transportation systems for these waste forms are under development, with initial use anticipated for shipment to the Waste Isolation Pilot Plant in the 1989–1990 time frame. Two concepts, the Defense High-Level Waste (DHLW) cask and the Transuranic Package Transporter (TRUPACT), are in the design stage. The DHLW (Figure 5-29) is designed to carry 5000 lb by truck and incorporate the necessary shielding, impact limiters, and thermal shielding to handle normal transport or accident conditions without loss of integrity. The TRUPACT (Figure 5-30) is a bimodal system that can hold 36-55-gallon drums of TRU waste.

5.8.3 Shipper Requirements and Constraints

The shipper bears the major responsibility for ensuring the adequacy of the waste packaging, that the carrier is licensed and follows proper procedures, and that the burial site operator is certified to receive the packages. The mechanism for confirming that these responsibilities are being met is the shipping paper or "Radioactive Waste Manifest" that accompanies each shipment of waste. The manifest contains the following information as a minimum:

1. DOT proper shipping name and identification number for the material (from 49 CFR 172.101); NRC Certificate Identification, if relevant;
2. Shippers certification that the waste package has been properly prepared for transport;
3. Description of the chemical and/or physical form of the waste;
4. Name of each radionuclide and the activity of the radioactive material in each package(s);
5. Type of packaging and transport index assigned to each package;
6. Category of label applied to each package; and
7. Waste form and classification (from 10 CFR 61).

Prior to loading the packages on the transport vehicle, the shipper must affix warning labels on two opposite sides of the package. There are three different warning labels, distinguished as "Radioactive-White I, Radioactive-Yellow II, and Radioactive-Yellow III," that indicate the external radiation level and imply specific handling procedures. The maximum permissible dose rate at any point on the external surface of any package of radioactive materials may not exceed 200 mrem/hr (by regulation) or 10 mrem/hr at 1 m from the surface. If the waste packages are to be shipped by "exclusive use" vehicle, higher radiation levels are permitted: 1 rem/hr at the package surface for a closed transport vehicle, 200 mrem/hr at the external surface of the vehicle, 10 mrem/hr at 2 m from any lateral external surface of the vehicle, and 2 mrem/hr in any position of the vehicle that is occupied by a person (transport worker).

In addition to the warning label, the shipper must mark each package with the applicable DOT package specification, proper shipping name (from

49 CFR 172.101 list of hazardous materials), and identification number, the appropriate package Type (A or B), and the gross weight (if greater than 110 lb).

In order to ensure compliance with the waste classification requirements of 10 CFR 61, the NRC has developed a position (NRC 1983) that requires that its licensee, prior to packaging of LLW for shipment, must make a reasonable effort to ensure a realistic representation of the distribution of radionuclides within the waste, must classify the waste in a consistent manner, establish an onsite compliance program specific to that facility that considers the different radiological and other characteristics of the different waste streams generated by the facility, and adhere to an objective of achieving measured or inferred radionuclide concentrations accurate to within a factor of 10. Licensees may use at least four basic methods either individually or in combination to characterize packaged wastes; materials accountability, classification by source, the conversion factor technique using a calibration curve, or a direct measurement of individual radionuclides. For the conversion factor technique, which is used by many industrial and fuel cycle facilities, external gross γ radiation readings are converted to an estimated curie content of the package using a calibration curve specific to the type of material in the package.

The typical procedure at the shipper's facility for loading of the waste in drums (the most common form of packaging for LLW) is to

1. Inspect each drum prior to use for obvious defects (e.g., hole, split seam).
2. Place drums on pallets at the drumming station and load with waste material. All liquid wastes are immobilized prior to closure of the drum.
3. Place the lid with the gasket on the drum. Ensure that there is a good fit between drum and lid.
4. Attach the bolt ring to the drum. The bolt ring must engage both the lid and rim of the drums. Screw in bolts, attach nuts, and screw on tight.
5. Reinspect the drum for any defects.
6. Affix appropriate marking/label to the drum.
7. Weigh the drum and record the weight on the label affixed to the drum. Also record assigned "container number" on drum.
8. Perform a radiation and contamination survey of the external surface of the drum to validate that the radiation level on the external surface does not exceed 200 mrem/hr and that the level of removable radioactive contamination is kept ALARA. The results of the surveys are to be recorded and kept on file. If ALARA levels are not achieved, decontaminate the external surface and repeat the survey.
9. Determine the activity level of the drum (presumes use of the conversion factor technique).
10. Using a fork lift, move the pallet with the drum on it to the temporary storage area and record the location of the drum.

5.8.4 Transport Modes and Routes

Radioactive waste packages can be, by law, transported by air, truck, rail, and water (barge). Neither air nor water transport is used with any frequency in the United States; the large majority of the waste is shipped by either truck or rail, with truck transport being the predominant mode. Truck transport of radioactive waste is a specialized business conducted by a few companies.

For gross vehicle weights of less than 80,000 lb, truck shipments are categorized as legal weight trucks (LWT), and the vehicle is permitted to use major highways. When the gross vehicle weight exceeds 80,000 lb, the trucks operate in an overweight (OWT) mode that requires a special permit in each state that is traversed, and generally involves restrictions on hours of travel. OWT loads exceeding 105,000 lb are not widely used because of additional restrictions imposed over this weight and the need for specialized equipment.

Truck transport of radioactive waste has been a controversial issue in a number of localities during the 1980s. State and local governments have attempted to prevent transport of waste from power plants or other nuclear facilities because the transport would be through populated areas. One prominent example of this is the attempt on the part of the City of New York to prevent waste shipment from the Brookhaven National Laboratory on Long Island from passing through the Borough of Queens enroute to a disposal site. A recent judicial decision (1986) has confirmed the authority of the federal government to regulate radioactive waste shipments. In part as a result of this controversy, and consistent with the desire to minimize the possibility of accident or sabotage, and the potential population exposure under either normal or accident conditions, the NRC and DOT have established selected preferred routes for waste transport that, for the most part, avoid urban areas and use the federal interstate highway system. These routes are preinspected to ensure that they are safe. It is also required (10 CFR 73) that advance notification be given to the NRC and the states traversed prior to the proposed shipment of radioactive waste. This notification must include the type and quantity of waste to be shipped, the name of the shipper and carrier, the points of origin and destination, the proposed route, and the anticipated shipping schedule. The NRC will then assess the suitability of the proposed shipping route and possibly assign compliance personnel to monitor the shipment.

In the case of rail shipments, when the weight of the car and contents is below 263,000 lb and certain size constraints are met, the shipment can be made in an "unrestricted interchange mode" in which standard rolling stock may be employed and there are no special operating limitations. When this weight or the size envelope is exceeded, specialized rail cars are required, and the available track may be restricted. During the major part of the last 40 yr, the railroads in the United States did not encourage routine shipments of radioactive waste. Requirements such as limited speeds that increased trip times, and booking of cars adjacent to the ones carrying the waste made an already costly transport mode prohibitive in many cases. It is only in recent years that the railroads have begun to relax some of their restrictive requirements in an effort to encourage more use of the rails by waste shippers. Another constraint that limits rail use is that not all waste-generating facilities have a rail spur on or adjacent to their property. If intermodal truck and rail shipment is then required, the economics of the situation would tend to favor using just truck shipment.

5.8.5 Transport Procedures and Responsibilities

The process of transporting waste packages by truck to a disposal facility, which is the most common mode of waste shipment, begins with an inspection of the truck onsite after arrival to ensure that it is roadworthy and is capable of transporting the consigned load safely. In addition, the empty truck is surveyed for residual contamination prior to loading and the levels recorded. All radwaste package loading operations are performed under the direction of the Radiation Safety Officer (or equivalent), and are monitored to ensure that the prescribed radiation limits are not exceeded and that radiation protection measures are observed when loading the truck. The packages are loaded onto the truck with the heavier ones (typically drums) placed on the bottom, plywood placed over the bottom layer, and the additional packages placed on the plywood. Once the loading is complete, the packages are blocked and braced to prevent the load from shifting during transport. A photographic record is then generally made of the truck interior showing location and bracing of the packages. The truck door is then closed, latched, and the shipper's seal is applied. The seal number is recorded. Although not quite identical, a similar approach is used to load waste packages onto rail cars.

The truck driver is provided specific instructions to cover the transport of the packages including the maintenance of exclusive use shipment controls (if relevant) and emergency procedures and instructions defining response to accidents while carrying the waste packages. It should be noted that the primary responsibilities for ensuring that radioactive waste is safely transported belong to the shipper and carrier. However, if an accident occurs, the state or local governments in whose jurisdiction the accident occurred assume the lead responsibility for protecting people in the vicinity of the accident from overexposure, appointing an emergency response team, and coordinating communication between the various agencies responding to the accident. The federal agencies will provide support as requested to state and local agencies. The driver, after receiving the transport instructions and emergency procedures, confirms in writing his (her) understanding and acceptance of the procedures.

The vehicle is placarded with the diamond shaped "Radioactive" placard for exclusive use shipments of LSA material, or for any shipment containing packages bearing the "Radioactive-Yellow III labels (200 mrem/hr dose limit on package surface). The placards are attached to the front, back, and both sides of the vehicle. The specific hazardous material identification number must also be displayed in proximity to each of the placards. After loading of the vehicle is complete, it is surveyed on all four sides and the truck cab for radiation levels. The radiation levels are recorded on a plot sketch of the truck.

Individual commercial burial facilities and the host states may also impose their own additional requirements on the shipment. For example, the State of Nevada requires that shippers of waste to the Beatty disposal facility complete a "Low-Level Radioactive Waste Shipment Compliance Certification" warranting that the waste shipment was inspected within at least 48 hr prior to shipment and conforms in all respects to federal and state requirements for shipment, transportation, and disposal. In addition, the state requires certification by the shipper that the sections of the U.S. Ecology Site Operations Manual dealing with radioactive material possession limits and receipt requirements have been read and adhered to. These certifications are provided by the shipper in conjunction with a valid users permit for shipments to the state, an overweight permit (if required), an inspection report showing compliance with all regulatory requirements, the bill of lading, and the manifest.

For shipments of certain types of radioactive waste such as spent fuel, federal regulations (10 CFR 73) require that the shipment be provided with additional security protection against the possibilities of theft or sabotage. This security protection includes the use of drivers and guards with special training, control of the transport route, establishment and use of a rapid communication network, and transport equipment that can be immobilized and that also provides protection against attack.

REFERENCES

(Alexander 1980a) Alexander, B. M. "Incineration, of LWR-Type Waste in the Mound Cyclone Incinerator: A Feasibility Study." MLM 2792, Miamisburg, Ohio, 1980.

(Alexander 1980b) Alexander, B. M., and J. W. Doty, Jr. "Incineration of LWR Type Waste at Mound Facility." Proceedings of the 18th DOE Air Cleaning Conference, Vol. 2, 939, CONF801038, San Diego, California, 1980.

(ASME 1986) American Society of Mechanical Engineers, American Nuclear Society. "Radioactive Waste Technology," A. Moghissi, H. Godbee, and S. Hobart, eds., 1986.

(Atomic Energy 1954) "The Atomic Energy Act of 1954," as amended, Public Law 83–703, 68 stat. 919, 42-USC, Chapter 23.

(Bordwin 1980) Bordwin, L. and A. Taboas, "U.S. DOE Radioactive Waste Incineration Technology: Status Review." LA-UR-80-692, Proceedings of the 2nd U.S. DOE Environmental Control Symposium, CONF-800334, Reston, Virginia, 1980.

(Chapman 1979) Chapman, C. C., et al. "Utilization of Hanford Wastes in a Joule Heated Ceramic Melter and Evaluation of Resultant Canisterized Product." PNL 2904, Pacific Northwest Laboratory, Richland, Washington, 1979.

(Clark 1978) Clark, W. E. "The Use of Ion Exchange to Treat Radioactive Liquids in Light Water Cooled Nuclear Reactor Power Plants." NUREG/CR 0143, ORNL/NUREG/TM 204, 1978.

(DOE 1984) U.S. Department of Energy, Oak Ridge National Laboratory, for EG&G,

Idaho, "Low-Level Waste Treatment Technology Handbook." DOE/LLW-13TC, 1984.

(DOE 1986) U.S. Department of Energy, "Spent Fuel Storage Requirements." DOE/RL-86-5, October 1986.

(DOT 1966) "The Department of Transportation Act of October 15, 1966." Public Law-89.670, 80 Stat 937, 49USC 1651.

(FMC 1977) FMC Corporation, "Selection of Waste Treatment Process for Retrieved TRU Waste at Idaho National Engineering Laboratory." Final Report R-3689, Santa Clara, California, 1977.

(Furhmann 1981) Furhmann, M., R. M. Nielson, Jr., and P. Columbo, "A Survey of Agents and Techniques Applicable to the Solidification of Low-Level Radioactive Wastes." BNL-51521, 1981.

(Godbee 1978) Godbee, H. W., and A. H. Kibbey, "The Use of Evaporators to Treat Radioactive Liquids in Light-Water-Cooled Nuclear Reactor Power Plants." NUREG/CR-0142, ORNL/NUREG-42, 1978.

(Hedahl 1979) Hedahl, T. G., and M. D. McCormack, "Research and Development Plan for the Slagging Pyrolysis Incinerator." TREE 1309, Idaho Falls, Idaho, EG&G, Idaho, June 1979.

(Kibbey 1978) Kibbey, A. H., and H. W. Godbee, Oak Ridge National Laboratory, "The Use of Filtration to Treat Radioactive Liquids in Light-Water-Cooled Nuclear Reactor Power Plants." NUREG/CR 0141, ORNL/NUREG 41, 1978.

(Kibbey 1980) Kibbey, A. H., and H. W. Godbee, "A State of the Art Report on Low-Level Radioactive Waste Treatment." ORNL/TN 7427, 1980.

(Liu, 1973) Liu, K. H. "Use of Ion Exchange for the Treatment of Liquids in Nuclear Power Plants." ORNL 4792, 1973.

(McCabe 1976) McCabe, W. C., and J. C. Smith, *Unit Operations of Chemical Engineering*, 3rd ed. McGraw-Hill, New York, 1976.

(Memo 1979) "Transportation of Radioactive Materials; Memorandum of Understanding." *Federal Register* (44 F.R. 38690), July 2, 1979.

(Nuclear News 1987) American Nuclear Society, "Waste Management Today." *Nuclear News*, 30(3), March 1987.

(NWPA 1982) "Nuclear Waste Policy Act of 1982." Public Law 97-425, Signed into law by President Reagan on February 7, 1983.

(Perry 1973) Perry, R. H., and C. H. Chilton, eds. "Evaporators." In *Chemical Engineer's Handbook*, 5th ed., pp. 11-27–11-38. McGraw-Hill, New York, 1973.

(Stretz 1982) Stretz, L. "Controlled-Air Incineration of Hazardous Chemical Waste at the Los Alamos National Laboratory." Proceedings of Waste Management 1982. Vol. 1, Tucson, Arizona, 1982.

(Trans 1974) "Transportation Safety Act of 1974," as amended, Public Law 93-633, 88 Stat. 2156, 49USC 1808, January 3, 1975.

(Trigilio 1981) Trigilio, G. "Volume Reduction Techniques in Low-Level Radioactive Waste Management." NUREG/CR-2206, 1981.

(Yamamoto 1968) Yamamoto, Y., N. Mitsuishi, and S. Kadoya, "Design and Operation of Evaporators for Radioactive Wastes." Technical Report Series, No. 87, STI/DOC/10/81, IAEA, 1968.

(Ziegler 1981) Ziegler, D. L., G. D. Lehmkuhl, and L. T. Melle, "Nuclear Waste Incineration Technology Study, Transuranic Waste Management Program." RFP-3250, July 1981.

(Ziegler 1982) Ziegler, D. L., et al. "Rocky Flats Plant Fluidized Bed Incinerator." RFP-3372 of December 1982, and RFP-3249 of March 1982, Golden, Colorado.

CHAPTER 6

RADIOACTIVE WASTE DISPOSAL TECHNOLOGY

The viable concepts for the permanent disposal of high-level waste (HLW), spent fuel, transuranic wastes (TRU), low-level waste (LLW), and tailings are discussed in this chapter. The parameters of technologies currently in use, and those being developed for near-term application, are described.

6.1 OVERVIEW OF VIABLE WASTE DISPOSAL TECHNOLOGIES

The designated concept for the disposal (permanent isolation) of HLW, TRU, and spent fuel is to emplace the appropriate canisters of these materials in a deep geologic repository. Construction on the first geologic repository is currently scheduled to begin in 1998 and Phase I operation, involving the receipt of spent fuel, in 2003. The development of this facility is the major component of the DOE's HLW program.

Alternative concepts for the disposal/storage of HLW forms include

1. Ocean disposal of HLW canisters (a) in the region between the plate forming the continental shelf and the adjoining plate in which the gradual movement of the plates would forever seal the canisters in the earth's interior or (b) in drilled holes in the bottom of deep valleys in the ocean containing no known life forms and where the potential dispersion of the waste is extremely remote.

2. Disposal in the thick antarctic ice sheet by placing the canisters in a drilled hole. The decay heat generated in the canisters of HLW or spent fuel would cause the canister to sink through the ice sheet, with the ice sealing the hole above the canister, until bedrock is reached.

3. Disposal in deep space by sending a space vehicle loaded with HLW canisters into the sun to be destroyed (transmuted) by the heat of the sun or into a selected orbit from which return is impossible. Space disposal of HLW has been studied by NASA and the DOE and remains an unlikely prospect because of the dangers associated with a launch or orbital accident requiring destruction of the vehicle over populated areas.

4. Transmutation of the radionuclides in the waste to a less dangerous (active) form through an as yet undeveloped technique such as nuclear fusion. Although promising in concept, the approach requires continued storage of HLW forms for an undetermined period until the technology is developed. Transmutation is at best a long term alternative.

Geologic disposal is discussed in Section 6.10 and the other alternative HLW disposal technologies are discussed in Sections 6.11–14.

The technology currently in use for the disposal of LLW at the three commercial burial sites and the DOE facilities is shallow land burial (SLB) in trenches, pits, or other near surface disposal units. As a result of the adoption of the LLW regulation, 10 CFR 61 (NRC 1982), the technology for near-surface disposal has been upgraded to incorporate selected minimum site selection, facility design, and operational requirements to meet defined performance objectives. Thus, any new LLW disposal facilities will, as a minimum, be required to use this improved shallow land burial (ISLB) technology to achieve 10 CFR 61 standards. However, as a result of past inadequacies associated with SLB and concern about long-term containment capability, this disposal concept, even as enhanced by 10 CFR 61 standards, is not at present an acceptable option to some states and interstate compacts. As a result of public or local governmental pressure, they are either mandating or seriously considering using alternatives to SLB even though federal governmental and private authorities in the field consider SLB adequate to contain LLW over the long term with no degradation of groundwater or other damage to the biosphere. The alternatives to SLB involve combinations of variations in location (above ground, near surface, far underground) and use of engineered enclosures such as concrete liners, vaults, or bunkers. They include

1. Above-ground vault (AGV)—An engineered structure at grade level designed to withstand long-term weathering and to deter intruders.
2. Below-ground vault (BGV)—An engineered structure below grade level designed to be compatible with the soil in which it is buried.
3. Earth-mounded concrete bunker (EMCB)—A hybrid approach in which higher activity waste could be encased in concrete below ground, and lower activity waste placed in earthen mounds above the grade level.
4. Mined cavity disposal (MCD)—Hollowed out regions in mineral deposits such as salt mines would be used for LLW package disposal (a similar approach to HLW disposal).

5. Intermediate depth disposal (IDD)—Emplacement of LLW at a depth comparable to MCD but using an engineered structure for primary containment of the waste.

6. Shaft disposal (SD)—Use of a lined vertical hole (shaft) instead of the conventional SLB horizontal trench. The hole would be capped after filling. The concept would place the waste at greater depth and pose a decreased likelihood of need for retrieval than SLB.

7. Modular concrete canister (MCC)—Use of an overpack or "mini-vault" to encase a number of drums or crates and thus make the packaging provide a barrier to waste migration.

Shallow land burial (SLB) and the various alternative technologies being evaluated (by the DOE, NRC, and states) are discussed in Sections 6.8.1 and 6.8.2, respectively.

High-volume semisolid uranium and phosphate tailings or waste sludges are disposed of in surface impoundments. As in the case of shallow land burial, concern about past practices failing to provide adequate protection against dispersion has led to regulation-driven revisions in design concepts incorporating the use of engineered barriers in the impoundments. These barriers include natural and artificial liners and caps, control of surface water flow, and use of clay cores in retaining dams. Variations to the traditional approach of damming a "box canyon" or a natural depression, even with the incorporating of barriers, continue to be investigated to obtain designs that offer better potential for long-term isolation from the biosphere. These alternative disposal concepts include

1. Above-grade disposal in which the tailings are sealed in a pocket created by earthen embankments, compacted subsoil, or other impermeable material at the bottom, and clay or other dense material as a cap.

2. Disposal of tailings slurry in an available open pit mine that uses a clay liner and clay caps to encapsulate the material. Compacted fill is placed at the bottom of the mine cavity to the level of the water table, a layer of clay emplaced over the fill, and tailings above the clay.

3. Disposal of dewatered tailings in an available open pit mine. Again, fill is placed to the level of the water table, but since the tailings are dewatered no side or underlying liner is needed. A clay cap is used on top of the tailings.

4. Disposal of tailings in a below-grade pit excavated from natural impermeable subsoil (shale or clay) to avoid the need for a liner. A clay cap is also used to prevent water infiltration.

5. Disposal in an excavated below-grade trench that is buried and capped to encapsulate the tailings. The trench is excavated, lined, filled, and reclaimed progressively; tailings are not exposed for an extended period.

6. Disposal in a specially designed above-grade, lined, and capped impoundment with a very flat embankment slope and topography to provide good wind protection.

In each of the above-described design concepts overburden and topsoil are placed over the clay caps and a vegetative cover developed. Surface impoundments and the viable alternatives are discussed further in Section 6.9.

High-volume, low-activity remedial wastes and contaminated soil are disposed of in a variety of ways, including use of commercial LLW burial facilities, in-place stabilization, incorporation into another site undergoing remediation (e.g., UMTRA tailings pile), or surface impoundments such as are described for tailings (see Section 6.9 and Chapter 7 for elaboration). Ocean burial, which was used in the 1950s for disposal of LLW packages and stopped because of migration of the waste from the burial site, is once again being considered for disposal of radioactive waste, namely the bulk burial of low-activity soil and rubble from remediation of contaminated sites. Ocean disposal is discussed further in Section 6.12.

6.2 DISPOSAL SITE SELECTION

The criteria and methodology for the selection of radioactive waste disposal facilities is specified by the NRC in the Code of Federal Regulations and in supporting regulatory guides and position papers. Potential license applicants, including industrial facility owners and operators, the states, and the DOE, must demonstrate in the license application adherence to the criteria and overall conformance with the selection process. The variations in standards and approach between the classes of radioactive waste disposal facilities will be noted in this section; however, the emphasis is on the highly structured and documented process for the selection of sites for the land disposal of LLW.

6.2.1 Selection of Sites for Land Disposal of Low-Level Waste

Section 61.50, "Disposal Site Suitability Requirements for Land Disposal," of 10 CFR 61 establishes technical requirements for the site, and Subpart C of Part 61 lists performance objectives that must be met by the disposal facility. The technical requirements generally address specific conditions of the site or locale that could affect the long-term stability of the site and thus the ability to ensure waste isolation. The natural conditions of the site are considered to be the primary line of defense in achieving waste isolation and site stability and thus protecting the population over the long term against releases of radioactive constituents from the waste. The site suitability requirements are intended to create this situation by eliminating from consideration land that has certain unfavorable hydrogeologic, geologic, land use, and demographic conditions that could adversely affect the site and its environs. While meeting the performance objectives of 10 CFR 61 and protection of the public's health and safety is a function of the site characteristics, facility design, operation, and closure requirements as a total process, the careful selection of the site is expected to "limit the potential for radionuclide leaching, provide long pathways to minimize potential radionuclide releases, prevent erosion and inundation of the disposal site to minimize active maintenance, and avoid areas in which detrimental human activities are occurring" (NRC 1987). The following minimum site suitability requirements of paragraph 61.50 are designed to eliminate areas having unfavorable site characteristics.

1. Capable of being characterized [61.50 (a) (2)]—The disposal site should be capable of being characterized, modeled, analyzed, and monitored. To achieve this, the site components, their unique physical characteristics, and a general representation of each site component sufficient to enable predictions of site performance should be provided. In turn, these parameters can be best provided if the sites are hydrologically simple and contain processes that occur at consistent and definable rates, and if input assumptions are representative for all site conditions.

2. Population distribution and land use [61.50(a) (3)]—The site should be located in an area of low population density where the potential for future population growth and development is limited and not likely to affect the ability of the disposal facility to meet the performance objectives. It should be a minimum of 2 km from the residential property limits of the nearest urban community.

3. Natural resources [61.50(a) (4)]—Areas should be avoided if they contain natural resources in quantities or of such quality that future exploitation could affect the ability to meet the performance objectives or achieve waste isolation.

4. Site must be generally well drained [61.50(a) (5) (6)]—The disposal site must be generally well drained and free of areas of flooding and frequent ponding. Thus, a 100 yr floodplain, coastal high-hazard area, wetland, or areas in which natural ground slope is steep and flood velocities could cause damage to the disposal facility are not suitable for waste disposal. Furthermore, upstream drainage areas must also be minimized to decrease runoff and resultant erosion.

5. Depth to the water table [61.50(a) (7)]—The disposal site must be sufficiently above the water table that groundwater intrusion (perennial or otherwise) into the waste will not occur. Areas with a known or suspected high water table should therefore be avoided. In no case will waste disposal be permitted in the zone of fluctuation of the water table.

6. Groundwater discharge [61.50(a) (8)]—The hydrogeologic unit used for disposal should not discharge groundwater to the surface to form surface features such as springs, seeps, swamps, or bogs within the disposal site. The NRC "prefers long flow paths from the disposal site to the point of groundwater discharge in order to increase the amount of time for decay of the radionuclides, increase the hydrodynamic dispersion within the aquifer, and increase the likelihood of retardation of reactive radionuclides in the aquifer" (NRC 1987).

7. Tectonic and geomorphic processes [61.50(a) (10)]—Areas of known significant tectonic activity (e.g., faulting, folding, seismic activity, or vulcanism) must be avoided if they affect the ability of the site to meet the performance objectives or preclude defensible modeling or prediction of long-term impact. In addition, a site should not have the potential for significant mass wasting, erosion, slumping, landsliding, or weathering.

8. Adverse impacts from nearby facilities [61.50(a) (11)]—The site must not be located near any facilities or activities that could adversely affect the ability of the site to meet the performance objectives or significantly mask the site monitoring program.

The performance objectives of Part C of 10 CFR 61 are designed to provide reasonable assurance that radiation exposure to humans is within established limits during operation, closure, and postclosure periods. They are included in Table 6-1. Together with the minimum site suitability requirements, they provide the fundamental prescriptive requirements for LLW disposal sites, and establish a benchmark for screening candidate sites.

The site selection process has evolved from a somewhat unstructured and inconsistent approach used in the selection of the original LLW disposal sites in the 1960s to a highly sophisticated, structured methodology now required to meet licensing requirements. The generic site selection process, which is subject to some variation in emphasis and content between the states (or compacts), proceeds iteratively from a broad general screening of regions of interest to the successive identification of candidate areas within the regions, candidate sites within the areas, and then the detailed assessment of proposed sites involving site characterization. This four-step process is summarized in Table 6-2, and consists of the following:

Region of Interest Screening. Defining the region of interest within the state (or compact) allows for the elimination of unfavorable areas and identification of candidate areas for further consideration. At this stage, relevant data are obtained from published and open file documents on land use, transportation, and geophysical information, aerial photographs, and other literature sources. In addition to applying the exclusionary minimum site suitability requirements, other land use factors are considered to narrow the region of interest. These include economic viability of land areas, and parcels of land with institutional commitments (e.g., national or state parks, military reservations). Transportation factors to be evaluated include potential site access, distance from major waste generators, and population impact along projected transportation routes.

Candidate Area Evaluation. Candidate areas are now further evaluated to identify candidate sites within the areas. Typical information sources for this evaluation include land-use plans, U.S. Geological Survey (USGS) and State geological survey reports, aerial photographs, and open file data. The 10 CFR 61 screening requirements also are used in the candidate area evaluation, as are the

TABLE 6-1   Land Disposal Facilities Performance Objectives

§61.40   General requirement.

Land disposal facilities must be sited, designed, operated, closed, and controlled after closure so that reasonable assurance exists that exposures to humans are within the limits established in the performance objectives in §§61.41 through 61.44.

§61.41   Protection of the general population from releases of radioactivity.

Concentrations of radioactive material that may be released to the general environment in groundwater, surface water, air, soil, plants, or animals must not result in an annual dose exceeding an equivalent of 25 mrem to the whole body, 75 mrem to the thyroid, and 25 mrem to any other organ of any member of the public. Reasonable effort should be made to maintain releases of radioactivity in effluents to the general environment as low as is reasonably achievable.

§61.42   Protection of individuals from inadvertent intrusion.

Design, operation, and closure of the land disposal facility must ensure protection of any individual inadvertently intruding into the disposal site and occupying the site or contacting the waste at any time after active institutional controls over the disposal site are removed.

§61.43   Protection of individuals during operations.

Operations at the land disposal facility must be conducted in compliance with the standards for radiation protection set out in Part 20 of this chapter, except for releases of radioactivity in effluents from the land disposal facility, which shall be governed by §61.41 of this part. Every reasonable effort shall be made to maintain radiation exposures as low as is reasonably achievable.

§61.44   Stability of the disposal site after closure.

The disposal facility must be sited, designed, used, operated, and closed to achieve long-term stability of the disposal site and to eliminate to the extent practical the need for ongoing active maintenance of the disposal site following closure so that only surveillance, monitoring, or minor custodial care is required.

more specific requirements defined in U.S. NRC Regulatory Guide 4.18, "Standard Format and Content of Environmental Reports for Near-Surface Disposal of Radioactive Waste."

Candidate Site Evaluation. The purpose of the candidate site evaluation is to identify the proposed site. At this stage of the site selection process, review of available data is generally supplemented by onsite visits, reconnaissance, and physical inspections, and a limited amount of sampling (e.g., soils, surface water). Previously evaluated sources are reexamined, local utility officials are contacted to establish the locations of electric and water distribution systems, and title searches of the candidate sites are conducted. It is essential, at this stage, to be aware of the site suitability requirements and parameters elaborated on in Regulatory Guide 4.18 to ensure that the sites being considered can meet licensing standards.

A number of Geographic Information Systems (GIS) have been developed to assist in the organizing and analysis of the substantial data base compiled on the candidate sites and aid in the selection of the proposed site. The GIS technique has evolved from the earlier use of overlays representing individual exclusionary and comparative selection parameters that permitted a visual determination of suitable sites. The more sophisticated GIS approach uses computer mapping of encoded geophysical, land-use, and demographic data to create the data base. The encoded data are referenced to coordinate systems on separate parameter maps generated from the same base map, and the individual parameter maps are composited to produce a single derivative map. The compositing is done on a cell-by-cell basis where the parameters within each cell are summed and a total numerical "score" developed based on a comparative numerical value or "weight" assigned by the user to each of the mapped parameters. Each score level is distinguished by a user-supplied symbol or color and a composite map can be generated. Obviously, the assignment of weights to the individual parameters (e.g., existence of surface water, agricultural use of land, and extent of faulting) controls the final composite score requiring that these weights be carefully assigned. The resultant computer-generated composite map, that replaces the earlier-used overlays, indicates areas most suited for siting LLW disposal facilities (typically as uncolored or unpatterned) based on the assigned weights to the parameters.

Evaluation of the Proposed (Prime) Site. Once a proposed site has been identified, it is necessary to continue the site data compilation and evaluation process until the potential applicant (owner) is satisfied that the site is licensable. To be licensable, the proposed site must meet the minimum technical requirements of 10 CFR 61.50 and the performance objectives of Subpart C of 10 CFR 61, satisfy NEPA requirements as defined in NRC Regulatory Guide 4.18 (see discussion of NEPA requirements in Chapter 4), and the health and safety, organization, and procedural requirements of NUREG-1199, "Standard Format and Content of a License Application for a Low-Level Radioactive Waste Disposal Facility." The continuing evaluation of the data base for the proposed site, if determined to be licensable, evolves directly into the site characterization (Section 6.3).

6.2.2   Selection of Sites for Disposal of Tailings and Other Semisolid Wastes

The criteria and requirements for the selection and licensing of a facility for the impoundment of high-volume, semisolid radioactive waste typically classified as by-product material (i.e., uranium mill tailings) are incorporated in Appendix A to 10 CFR 40 (NRC 1985a). This appendix establishes technical, financial, ownership, and long-term site surveillance criteria relating to the siting, operation, decontamination, decommissioning, and reclamation of mills and tailings. An applicant for a by-product materials license must clearly demonstrate how the requirements and objectives set forth in Appendix A have been addressed. The criteria of Appendix A are for the most part performance standards, with certain minimal prescriptive requirements, allowing flexibility to achieve an optimum tailings management program on a site-specific basis. The following summarizes the site selection criteria of Appendix A.

1. *Features of Selected Site.* The features to be optimized in the site selection process and thus to ensure that the site meets the objective of isolating the tailings for thousands of years without ongoing active maintenance are (1) remoteness from populated areas, (2) hydrologic and other natural conditions as they contribute to continued immobilization and isolation of contaminants from usable groundwater sources, and (3) potential for minimizing erosion, disturbance, and dispersion by natural forces over the long term. Overriding consideration is to be given to site features (rather than engineering design) to achieve long-term isolation of the tailings.

2. *"Prime Option" for Disposal of Tailings.* Below-grade disposal, either in mines or specially excavated pits (where the need for any specially constructed retention structure is eliminated), is to be considered the "prime option" for disposal of tailings. "Where full below grade disposal is not practicable, the size of retention structures and size and steepness of slopes of associated exposed embankments shall be minimized by excavation to the maximum extent reasonably achievable or appropriate given the geologic and hydrologic conditions at a site. In these cases it must be demonstrated that an above grade disposal program will provide reasonably equivalent isolation of the tailings from natural erosional forces."

3. *Site Criteria.* Site criteria that must be adhered to for either above- or below-grade disposal are (1) upstream rainfall catchment areas must be minimized to decrease erosion potential and the size of the maximum possible flood that could erode or wash out sections of the tailings disposal areas, (2) topographic features should provide good wind protection, and (3) the impoundment shall not be located near a capable fault that could cause a maximum credible earthquake larger than that which the impoundment could reasonably be expected to withstand.

The site selection process, as in the case of LLW disposal sites, proceeds iteratively from a broad general screening of regions of interest through successive identifications of candidate areas and sites and then the detailed characterization of proposed sites. In evaluating the proposed site, the NEPA requirements defined in the NRC Regulatory Guide 3.8, "Standard Format and Content of Environmental Reports for Uranium Mills," and the health and safety, organization, and procedural requirements of NRC Regulatory Guide 3.3, "Standard Format and Content of a License Application for a Uranium Mill" are applicable.

TABLE 6-2 Site Selection Process

Category	Step 1: Region of Interest	Step 2: Candidate Areas	Step 3: Candidate Sites	Step 4[a]: Proposed Sites
Study area	Compact, state, or geographic region	A homogeneous area. Sites within an area will contain same general environmental characteristics	Sites that are potentially licensable	The site for which the applicant is seeking a license
Criteria to be reviewed	General exclusionary data pertaining to health and safety, areas protected by law	General compact or state criteria, general screening requirements from §61.50, and Regulatory Guide 4.18	General review of compact or state criteria, §61.50, and information in Regulatory Guide 4.18	Evaluate compact or state criteria, §61.50, Regulatory Guide 4.18
Data to be reviewed	USGS and state geologic maps, federal and state regulations, aerial photographs	USGS and state geologic maps, topographic maps, university research, local government ordinances and surveys, aerial photographs	USGS and state geologic maps, topographic maps, university research, local government ordinances and surveys, and local utility maps. Actual field observation	Data collected to this point and collect original data
Level of analysis	Reconnaissance-level map reviews, literature and regulation reviews	Reconnaissance review of local maps, high-level aerial photographs, literature, and regulations	Reconnaissance information and site visits (surface-water samples, low-level aerial photos, onsite photos, air analysis, windshield surveys, etc.)	Demonstrate fulfillment of site characterization requirements. Prepare environmental report as necessary
Purpose	Identify candidate areas	Identify candidate sites	Identify proposed site for characterization	Meet site licensing requirements

Reprinted from NRC 1987.

[a]Step 4 involves site characterization. Steps go from most general (Step 1) to most detailed (Step 4).

6.2.3 Selection of Sites for Disposal of High-Level Radioactive Wastes in Geologic Repositories

The criteria and requirements for the siting and licensing of a geologic repository for the disposal of high-level waste (HLW) and spent fuel are defined in 10 CFR 60 (NRC 1983). Part 60 is unique in that it prescribes rules governing the licensing by the NRC of a federal agency, the Department of Energy (DOE), rather than a commercial applicant, to receive and possess source, special nuclear, and by-product material at a geologic repository as stipulated in the Nuclear Waste Policy Act of 1982. Subpart E of 10 CFR 60 establishes the technical criteria for the geologic repository, Sections 60.111–113 define the repository's performance objective, and Section 60.122 lists the siting criteria. The siting criteria are classified as "favorable conditions" and "potentially adverse conditions." As in the case of land disposal sites for LLW, the natural conditions of the site are the primary factors in achieving long-term waste isolation and stability and protecting the public against radioactive releases. If potentially adverse human activity or natural conditions exist that cannot be remedied, it must be shown by analysis that these conditions do not affect significantly the ability of the geologic repository to meet the performance objectives relating to isolation of the waste. The performance objectives for the geologic repository are described in Table 6-3. The favorable siting conditions for a geologic repository are

1. *Geologic conditions*—The nature and rates of tectonic, hydrogeologic, geochemical, and geomorphic processes operating within the geologic setting during the Quaternary Period, when projected, would not affect or would favorably affect the ability of the geologic repository to isolate the waste.

2. *Hydrogeologic conditions in the saturated zone*—For disposal in the saturated zone, hydrogeologic conditions that provide (1) a host rock with low horizontal and vertical permeability, (2) downward or dominantly horizontal hydraulic gradient in the host rock and immediately surrounding hydrogeologic units, and (3) low vertical permeability and low hydraulic gradient between the host rock and the surrounding hydrogeologic units.

3. *Geochemical conditions*—Geochemical conditions that (1) promote precipitation or sorption of radionuclides, (2) inhibit the formation of particulates, colloids, and inorganic and organic complexes that increase the mobility of radionuclides, or (3) inhibit the transport of radionuclides by particulates, colloids, and complexes.

4. *Thermal loading of minerals*—Mineral assemblages that, when subjected to anticipated thermal loading, will remain unaltered, or alter to mineral assemblages having equal or increased capacity to inhibit radionuclide migration.

5. *Emplacement depth for waste*—Conditions that permit the emplacement of waste at a minimum depth of 300 m from the ground surface.

TABLE 6-3   Geologic Repository Performance Objectives

§60.111   *Performance of the geologic repository operations area through permanent closure.*

(a) *Protection against radiation exposures and releases of radioactive material.* The geologic repository operations area shall be designed so that until permanent closure has been completed, radiation exposures and radiation levels, and releases of radioactive materials to unrestricted areas, will at all times be maintained within the limits specified in Part 20 of this chapter and such generally applicable environmental standards for radioactivity as may have been established by the Environmental Protection Agency.

(b) *Retrievability of waste.* (1) The geologic repository operations area shall be designed to preserve the option of waste retrieval throughout the period during which wastes are being emplaced and, thereafter, until the completion of a performance confirmation program and Commission review of the information obtained from such a program. To satisfy this objective, the geologic repository operations area shall be designed so that any or all of the emplaced waste could be retrieved on a reasonable schedule starting at any time up to 50 yr after waste emplacement operations are initiated, unless a different time period is approved or specified by the Commission. This different time period may be established on a case-by-case basis consistent with the emplacement schedule and the planned performance confirmation program.

(2) This requirement shall not preclude decisions by the Commission to allow backfillng part or all of, or permanent closure of, the geologic repository operations area prior to the end of the period of design for retrievability.

(3) For purposes of this paragraph, a reasonable schedule for retrieval is one that would permit retrieval in about the same time as that devoted to construction of the geologic repository operations area and the emplacement of wastes.

§60.112   *Overall system performance objective for the geologic repository after permanent closure.*

The geologic setting shall be selected and the engineered barrier system and the shafts, boreholes, and their seals shall be designed to ensure that releases of radioactive materials to the accessible environment following permanent closure conform to such generally applicable environmental standards for radioactivity as may have been established by the Environmental Protection Agency with respect to both anticipated processes and events and unanticipated processes and events.

§60.113   *Performance of particular barriers after permanent closure.*

(a) *General provisions.* (1) *Engineered barrier system.* (i) The engineered barrier system shall be designed so that assuming anticipated processes and events: (A) Containment of HLW will be substantially complete during the period when radiation and thermal conditions in the engineered barrier system are dominated by fission product decay; and (B) any release of radionuclides from the engineered barrier system shall be a gradual process that results in small fractional releases to the geologic setting over long times. For disposal in the saturated zone, both the partial and complete filling with groundwater of available void spaces in the underground facility

TABLE 6-3   (*Continued*)

shall be appropriately considered and analyzed among the anticipated processes and events in designing the engineered barrier system.

(ii) In satisfying the preceding requirement, the engineered barrier system shall be designed, assuming anticipated processes and events, so that

(A) Containment of HLW within the waste packages will be substantially complete for a period to be determined by the Commission taking into account the factors specified in §60.113(b), provided, that such period shall be not less than 300 yr nor more than 1000 yr after permanent closure of the geologic repository, and

(B) The release rate of any radionuclide from the engineered barrier system following the containment period shall not exceed one part in 100,000 per year of the inventory of that radionuclide calculated to be present at 1000 yr following permanent closure, or such other fraction of the inventory as may be approved or specified by the Commission, provided that this requirement does not apply to any radionuclide that is released at a rate less than 0.1% of the calculated total release rate limit. The calculated total release rate limit shall be taken to be one part in 100,000 per year of the inventory of radioactive waste, originally emplaced in the underground facility, that remains after 1000 yr of radioactive decay.

(2) *Geologic setting.* The geological repository shall be located so that prewaste-emplacement groundwater travel time along the fastest path of likely radionuclide travel from the disturbed zone to the accessible environment shall be at least 1000 yr or such other travel time as may be approved or specified by the Commission.

(b) On a case-by-case basis, the Commission may approve or specify some other radionuclide release rate, designed containment period, or prewaste-emplacement groundwater travel time, provided that the overall system performance objective, as it relates to anticipated processes and events, is satisfied. Among the factors that the Commission may take into account are

(1) Any generally applicable environmental standard for radioactivity established by the Environmental Protection Agency;

(2) The age and nature of the waste, and the design of the underground facility, particularly as these factors bear upon the time during which the thermal pulse is dominated by the decay heat from the fission products;

(3) The geochemical characteristics of the host rock, surrounding strata, and groundwater; and

(4) Particular sources of uncertainty in predicting the performance of the geologic repository.

(c) Additional requirements may be found to be necessary to satisfy the overall system performance objective as it relates to unanticipated processes and events.

6. *Population density*—A low population density within the geologic setting and a controlled area that is remote from population centers.

7. *Groundwater travel time*—Prewaste emplacement groundwater travel time along the fastest path of likely radionuclide travel from the disturbed zone to the accessible environment that substantially exceeds 1000 yr.

8. *Hydrogeologic conditions in the unsaturated zone*—For disposal in the unsaturated zone, hydrogeologic conditions that provide (1) low moisture

flux in the host rock and in the overlying and underlying hydrogeologic units, (2) a water table sufficiently below the underground facility such that fully saturated voids contiguous with the water table do not encounter the underground facility, (3) a laterally extensive low permeability hydrogeologic unit above the host rock that would inhibit the downward movement of water or divert downward moving water to a location beyond the limits of the underground facility, (4) a host rock that provides for free drainage, or (5) a climatic regime in which the average annual historic precipitation is a small percentage of the average annual potential evapotranspiration.

The potentially adverse conditions of primary significance for the siting of a geologic repository are

1. *Flooding potential*—Potential for flooding of the underground facility from either floodplains or from failure of man-made surface water impoundments.

2. *Adverse human activity*—Potential for foreseeable human activity to adversely affect the groundwater flow system such as groundwater withdrawal, extensive irrigation, subsurface irrigation of fluids, underground pumped storage, military activity, or construction of large-scale surface water impoundments.

3. *Effect of natural phenomena*—Potential for natural phenomena such as landslides, subsidence, or volcanic activity of such magnitude that large-scale surface water impoundments could be created that could change the regional groundwater flow system and thereby adversely affect the performance of the geologic repository.

4. *Structural deformation*—Structural deformation, such as uplift, subsidence, folding, or faulting that may adversely affect the regional groundwater flow system, or that occurred during the Quaternary Period.

5. *Changes in hydrologic conditions*—Potential for changes in hydrologic conditions that would affect the migration of radionuclides to the accessible environment, such as changes in hydraulic gradient, average interstitial velocity, storage coefficient, hydraulic conductivity, natural recharge, potentiometric levels, and discharge points. In addition, potential for changes in hydrologic conditions resulting from reasonably foreseeable climatic changes.

6. *Groundwater conditions in host rock*—Conditions, including chemical composition, high ionic strengths, or ranges of eH-pH, that could increase the solubility or chemical reactivity of the engineered barrier system, and conditions that are not reducing.

7. *Geochemical processes*—That would reduce sorption of radionuclides, result in degradation of the rock strength, or adversely affect the performance of the engineered barrier system.

8. *Evidence of dissolutioning*—Such as breccia pipes, dissolution cavities, or burial pockets.

9. *Earthquakes*—Which have occurred historically that if they were to be repeated could affect the site significantly, indications that either the frequency of occurrence or magnitude of earthquakes may increase, and/or more frequent occurrence of earthquakes or earthquakes of higher magnitude than is typical of the area in which the geologic setting is located.

10. *Quaternary Period activity*—Evidence of igneous activity since the start of the Quaternary Period or of extreme erosion during this period.

11. *Naturally occurring materials*—Presence of naturally occurring materials within the site in such form that economic extraction is currently or potentially feasible. In addition, evidence of subsurface mining for resources within the site, or of drilling for any purpose within the site.

12. *Need for complex engineering measures*—Rock or groundwater conditions that would require complex engineering measures in the design and construction of the underground facility or in the sealing of boreholes and shafts.

13. *Design of underground opening*—Geomechanical properties that do not permit design of an underground opening that will remain stable through permanent closure.

14. *Rising of the water table*—Potential for the water table to rise sufficiently so as to cause saturation of an underground facility located in the unsaturated zone.

15. *Perched water bodies*—Potential for existing or future perched water bodies that may saturate portions of the underground facility or provide a faster flow path from an underground facility located in the unsaturated zone to the accessible environment.

16. *Gaseous radionuclide movement*—Potential for the movement of radionuclides in a gaseous state through air-filled pore spaces of an unsaturated geologic medium to the accessible environment.

The net result of applying the criteria for favorable and potentially adverse conditions to the siting of a geologic repository is to severely limit the land areas available for consideration. This is essential to ensure that the site(s) finally selected will provide the maximal capability of isolating the waste and assuring long-term protection of the public health and the environment as is demanded by the highly active radiological and thermal characteristics of the HLW, and spent fuel to be emplaced in the repository.

6.3 DISPOSAL SITE CHARACTERIZATION AND LICENSING

A body of detailed information about the characteristics of the proposed site and the region surrounding it is required as part of the data requirements for licensing and design of a radioactive waste disposal facility. To a limited extent, selected characterization data are also required for the backup alternative sites. This information is required to assess the general suitability of the site for the proposed waste disposal operations and to provide the parameters on which to base the technical analyses to demonstrate achievement of performance objectives and minimum site suitability requirements. Thus, the data base aids in the assessment of the site's ability to maintain long-term isolation of the waste, in identification of the interactions between the site characteristics, the waste, and the packaging, and establishment of the data collection points and a baseline of data for the site monitoring program. Although there are some variations in emphasis in the characterization data collected for each type of disposal facility, in general the following is representative of the information that is required in the facility design and licensing process.

1. *Geography (location) and demography*—The population distribution, the distances and directions of population centers within a specified distance of the site, and description of the site and environs including disposal area and buffer zone, significant surface and subsurface features, and major transportation routes.

2. *Land use*—Current and projected land use within a specified distance of the proposed disposal site including an inventory of natural resources, and current zoning and project trends.

3. *Ecology*—Terrestrial and aquatic ecology of the site and vicinity, including an inventory of flora and fauna, and identification of unique ecological characteristics, potential for endangered species, and/or flora and fauna of "importance."

4. *Meteorology and air quality*—Local meteorology and climatology data including temperature, precipitation, snow cover, evaporation rate, wind speed and direction, freeze-thaw cycles, etc., frequency and probability of severe meteorological phenomena, and onsite air quality including radioactive particulate and gas concentration and distribution.

5. *Geohydrology and water quality*—Surface water flow in the vicinity of the site, the movement of water through the unsaturated zone, the movement of groundwater (direction, velocity, rate of flow, etc.) in the region, and the uses of groundwater in the vicinity of the site, and chemical characteristics of surface and groundwater.

6. *Geology and seismology*—The geology of the region (stratigraphy, lithology, soils, and structure), geotechnical characteristics including the physical characteristics of surface soils, topography of the site and adjacent land area, and the history of seismic events in the vicinity of the site and distance to the nearest active fault.

7. *Radiological characteristics*—Existing radiation levels in the media at the site comprising the radiological baseline.

8. *Regional historic, archaeologic, architectural, scenic, cultural, and natural*

landmarks—Listings in National Register of Historic Places, National Register, or National Park Service Register.

9. *Socioeconomics*—Historic, current, and projected trends of population, transportation, political structure, labor availability and skills, historic commuting patterns, schools, health systems, tax base, economic structures, housing stock, and sociocultural parameters in terms of quality-of-life indicators.

The data compiled during the site characterization phase, along with other more general information on the applicant, its operating and safety procedures, and financial status is submitted to the Nuclear Regulatory Commission in the form of an application for a license for disposal. The commission reviews the application in accordance with administrative procedures established by rule.

The application submitted to the Commission consists of general information, specific technical information, institutional information, and financial information and must be accompanied by an Environmental Report. The general information required includes the identity of the applicant, qualifications of the applicant, and a description of the disposal site location, proposed activities, types and quantities of waste, plans for uses of the disposal facility other than disposal of radioactive wastes, proposed facilities, and equipment and proposed schedules.

The specific technical information includes information needed to demonstrate that the performance objectives will be met. This includes a description of the site, the design of the facility, the construction and operation of the facility, the site closure plan, the radioactive materials to be received, possessed, and disposed of at the facility, the quality control program, the radiation safety program, the environmental monitoring program, and the administrative procedures that the applicant will apply to control activities at the land disposal facility. Known natural resources at the disposal site must also be identified. Several technical analyses must also be submitted to demonstrate that the performance objectives will be met. These include pathway analyses to demonstrate protection of the general population from releases of radioactivity, protection of individuals from inadvertent intrusion, protection of individuals during operations of the facility, and analyses of the long-term stability of the disposal site and the need for ongoing active maintenance after closure.

The institutional information required includes a certification by the federal or state government that owns the disposal site that they will assume responsibility for custodial care after site closure and postclosure observation and maintenance. Where the proposed disposal site is on land not owned by the federal or state government, the applicant must submit evidence that arrangements have been made for assumption of ownership in fee by the federal or a state government. The financial information submitted must be sufficient to demonstrate that the financial qualifications of the applicant are adequate to carry out the activities for which the license is sought and to meet other financial assurance requirements.

The Environmental Report submitted must discuss the environmental impact of the proposed action, any adverse effects that cannot be avoided if the proposal were implemented, alternatives to the proposed action, the relationship between local short-term uses of man's environment and the maintanance and enhancement of long-term productivity and any irreversible and irretrievable commitments of resources that would be involved in the proposed action if it were implemented. The environmental report should include a benefit–cost analysis that considers and balances the environmental effects of the facility and the alternatives available for reducing or avoiding adverse environmental effects, as well as the environmental, economic, technical, and other benefits of the facility. It should also include a discussion of the status of compliance of the facility with applicable environmental quality standards and requirements that have been imposed by federal, state, and regional agencies having responsibility for environmental protection. The content and format of the Environmental Report are discussed in detail in Chapter 4.

After receipt of the application, the NRC has 60 days to determine the acceptability of the application from the standpoint of completeness and adherence to regulatory requirements. If the application is accepted for action, the NRC conducts an in-house review, culminating in the preparation of an Environmental Impact Statement (EIS) by the NRC staff or a contractor. The review process will involve

- Distribution of the Environmental Report and supporting material for review to other involved government agencies (e.g., EPA), and private parties (e.g., public interest groups) with an interest in the proceedings. Comments are obtained and weighed.
- Conducting of public hearings to obtain input from interested parties on the proposed project.
- Independent analysis of the viability of the proposed site and alternatives, and verification of the applicants studies, conducted by NRC staff or consultants.
- Preparation of a Draft Environmental Impact Statement (DEIS) summarizing the parameters of the proposed action and the NRC's position on the application. This may include stipulation as to additional actions or constraints.
- Circulation of the DEIS to the involved parties for comment, resolution of the comments, and revision of the DEIS in accordance with the resolution.
- Publication of a final Environmental Impact Statement (FEIS).

The review process will be accompanied by the preparation of documentation reflecting the analyses and decisions made, and culminating in the development of a "Record of Decision."

Upon completion of the above-defined process, authorization would be issued by the NRC to develop the site in accordance with the parameters specified in the application, as modified by the conditions established (if any) in the FEIS. The operating license would be issued upon verification that, in fact, the facility had been developed in accordance with the approved conditions.

During the operational phase, the licensee carries out the disposal activities in accordance with the requirements of the governing regulations (e.g., 10 CFR 61 for an LLW disposal facility) and any conditions on the license. Periodically, during the operational phase, at least 30 days prior to the license expiration date, the licensee must file an application to renew the license to conduct the above-ground operations and dispose of waste. This application is accompanied by a report, generally following the format of the original Environmental Report, that summarizes the operational history, and describes changes in facility character-istics and operating conditions. At this time, the Commission will review the operating history and make a decision to permit or deny continued operation. When disposal operations are to cease, the licensee applies for an amendment to the license to permit site closure. After final review of the licensee's final site closure and stabilization plan, the Commission may approve the final activities necessary to prepare the disposal site so that ongoing active maintenance of the site is not required during the period of institutional control.

The disposal site closure phase is the period in which the final site closure and stabilization activities are being carried out. The licensee must then remain at the disposal site for a period of 5 yr for postclosure observation and maintenance to assure that the disposal site is stable and ready for institutional control. The Commission may alter the duration of this period if conditions warrant. At the end of this period, the licensee applies for a license transfer to the disposal site owner. After finding of satisfactory disposal site closure, the Commission will transfer the license to the state or federal government that owns the disposal site.

During the institutional control phase and under the conditions of the transferred license, the owner will carry out a program of monitoring to ensure continued satisfactory disposal site performance, physical surveillance to restrict access to the site, and minor custodial activities. At the end of the prescribed period of institutional control, the license will be terminated by the Commission.

For the siting of a geologic repository, the DOE is required to conduct a program of site characterization that includes *in situ* exploration and testing at the depths that the wastes would be emplaced in addition to the surface characterization program previously described. The subsurface drilling program must be carefully conducted to limit adverse effects on the long-term performance of the repository. This is accomplished by limiting the number of exploratory boreholes to the extent practical and locating them where construction shafts or large pillars are planned in the repository. The DOE is also required to develop a site characterization plan prior to initiating the characterization program that includes (1) a general plan for site characterization activities, (2) a description of the waste form package(s) to be emplaced in the repository, and (3) a conceptual design for the repository operations area.

The licensing procedure for the construction and operation of the repository by the DOE, while paralleling in most respects the procedure for licensing a land burial site, differs in the following significant respects:

- The Safety Analysis Report (SAR) submitted with the license application, in addition to the Environmental Report, must focus on analyses to determine the extent that each of the favorable and potentially adverse conditions present contributes or detracts from isolation, an evaluation of the performance of the proposed geologic repository for the period after permanent closure, the effects of engineered and natural barriers against the release of radioactive materials, and an analysis of the performance of the major design structures, systems, and components.
- The general information provided with the license application must include (1) a certification that DOE will provide at the repository operations area such safeguards as it requires at comparable surface facilities, (2) a description of the physical security plan for protection against radiological sabotage, and (3) a description of site characterization work actually conducted by DOE at all sites considered in the application.

The license, when issued for the repository, must include specific conditions restricting the form and radioisotope content of the waste, the geometry and materials of the waste packaging, and the amount of waste permitted per unit volume of storage space.

6.4   DISPOSAL FACILITY DESIGN AND CONSTRUCTION

6.4.1   Low-Level Waste Near-Surface Disposal Facility Design

A near-surface LLW disposal facility includes all of the land and buildings necessary for the operations to dispose of the radioactive waste. The actual disposal site, however, is that portion of the disposal facility that is used for disposal of waste. It consists of the disposal units and the buffer zone. Although other activities such as waste receipt and inspection, storage, and treatment can be conducted within the confines of the disposal site, the disposal of waste is the primary activity of the disposal site.

There are two primary objectives guiding the development of the design for an LLW disposal facility as presented in 10 CFR Part 61: that the minimum technical requirements for the disposal systems are met and that the performance objectives are achieved. The design approach must continue to enhance the sites ability to achieve this objective which also guided the initial selection of the site. Flexibility is allowed in achieving these objectives, however; specific design concepts and parameters are not prescribed, but are left to the license applicant to develop based on individual site conditions. Preliminary disposal site design activities begin during site selection as part of the basis for identifying a preferred site. As information is compiled and evaluated during the site characterization phase, a preliminary site design is developed and iteratively refined. The facility designers must consider both the need for achieving waste isolation during site operation and over the long term after the facility is closed, and the avoidance of the need for continuing active maintenance after closure.

10 CFR Part 61 establishes the minimum technical requirements for near-surface disposal site design as follows:

1. Site (facility) design features must be directed toward long-term isolation and avoidance of the need for continuing active maintenance after site closure.
2. The disposal site design and operation must be compatible with the disposal site closure and stabilization plan and lead to disposal site closure that provides reasonable assurance that the performance objectives will be met.
3. The disposal site must be designed to complement and improve, where appropriate, the ability of the disposal site's natural characteristics to ensure that the performance objectives will be met.
4. Covers must be designed to minimize water infiltration to the extent practical, to direct percolating or surface water away from the disposed waste, and to resist degradation by surface geologic processes and biotic activity.
5. Surface features must direct surface water drainage away from disposal units at velocities and gradients that will not result in erosion that will require ongoing active maintenance in the future.
6. The disposal site must be designed to minimize to the extent practical the contact of water with waste during storage, the contact of standing water with waste during disposal, and the contact of percolating or standing water with wastes after disposal.

In addition to these minimum technical requirements, the design basis for the disposal facility includes other NRC regulations, federal, state, and local codes and standards, and quality assurance requirements. NRC regulations include the requirements in 10 CFR 20, "Standards for Protection Against Radiation," which provide the exposure limits for the performance objective of protection of individuals during operation. Other pertinent federal design codes and standards for disposal facilities are 29 CFR 1926, OSHA "Safety and Health Regulations for Construction," the American Concrete Institute "Code Requirements for Nuclear Safety Related Concrete Structures (ACI 349–80)," and relevant parts of the Uniform Building Code, the Uniform Mechanical Code, and the National Electric Code.

The quality assurance program, as defined in the facility license application, must provide for a quality control program for the determination of natural disposal site characteristics and for quality control during the design, construction, operation, and closure of the facility and for the receipt, handling, and emplacement of the waste. For quality assurance purposes, structures and components at an LLW disposal facility are classified in terms of two quality levels: the more stringent quality level "Q" and the "non-Q" level. Q level systems, structures, and components are those that are essential for safe operational control or whose failure could result in undue radiological risk to employees or to public health and safety. Thus, all engineered safeguards and other aspects of the facility design that directly prevent or mitigate the consequences of potential releases of radioactivity that might cause undue radiological risks to employees or the public are covered by Q level standards, as are items whose failure would cause degradation of performance or reliability relative to facility operations.

Whereas the six 10 CFR Part 61.51 minimum technical requirements establish broad objective design requirements to achieve performance objectives, the requirements relating to facility design features derive from other provisions of 10 CFR Part 61, and draw extensively upon prior LLW facility design and operating experience. Among the more significant requirements relating to disposal facility design are the utilization of available space at the facility, the contouring of the surface, and the configuration of the disposal unit.

The objective in allocating available space is to bound the disposal site to permit efficient land use and maximum waste volume allocation while, at the same time, maximizing the achievement of long-term stability and isolation. A first consideration is to establish a buffer zone between the disposal units and the site boundary to provide an area in which monitoring stations can be set up to detect any migration from the disposal units and thus permit remedial actions to be taken to intercept this migration. Although no waste disposal is allowed in the buffer zone, other surface activities (e.g., storage, office space) that do not interfere with monitoring can be conducted. Among factors that need be considered by the facility designers in establishing the dimensions of the buffer zone are surface topography, underlying soil characteristics, the direction and velocity of groundwater flow, and the location of offsite receptors (wells and other water users). Once the disposal area is defined, the layout of disposal units within this area should allow sufficient space between units to ensure long-term wall integrity, to provide for surface water drainage, and to provide sufficient space to emplace the disposal unit covers. In this regard, the disposal facility is to be designed so that closure and stabilization are ongoing processes, and are not performed in total at the cessation of disposal operations. Thus, the sequence of use for disposal units over the facility lifetime, location of roads, management of surface water, and access to borrow areas should reflect the need to perform closure operations on each unit as it is filled.

In contouring the surface, final slopes are to be designed to minimize erosion and failure of the slopes. This can be accomplished by controlling the slope angle, particle size of the soil, degree of compaction or cementation, and type and management of the vegetative cover.

The disposal unit is the primary focus of the facility design to achieve long-term waste isolation because it provides the initial barrier the site offers against radionuclide migration. It must be designed to reflect the characteristics of the site, as well as those of the waste form to be buried therein. To achieve the necessary long-term waste isolation, the critical design features intended to prevent migration and/or intrusion such as barriers, covers, and contours must be designed to last several hundred years. Flexibility exists in selecting the type of disposal unit to be used, depending on the type (classification) of waste to be received; the classical trench design is not required. The most stringent design requirements are applied to the disposal units for Class C waste that require intruder protection. This may involve the use of separate units for Class C waste. Class B waste may be buried in trenches, similar to those currently in use, cells, boreholes, or other type disposal unit as long as the design selected meets the minimum technical requirements of 10 CFR Part 61.51. Class B waste must be structurally stable, thus permitting disposal unit designs different from those used for Class A waste, which is not required to achieve similar stability requirements. If Class A waste is not stable, it must be buried in separate units from Class B or C wastes. The design of the segregated disposal units for Class A waste would range from conventional trenches to cells or other types typically used for sanitary landfills.

Disposal unit size is primarily dependent on the dimension and topography of the disposal site, the type and volume of waste, and the dimensions of the waste packages. Disposal unit depth is also site specific, constrained by the requirements that the waste not be in contact with standing water and be well above maximum seasonal fluctuations in the water table and the zone of capillary rise. In order to minimize the probability that erosion will expose waste and/or water will collect in one end of the disposal unit and flow out the top, site surface slopes are not to be so steep as to result in significant elevation differences between sidewalls of a disposal unit. The elevation differences longitudinally between the ends of the disposal unit should also be less than the combined thickness of the backfill over the waste and the cover. Ground surface spacing between disposal units must be sufficient to ensure disposal unit integrity, permit controlled surface water drainage, and allow for safe use and movement of equipment.

The design of the disposal unit cover is of particular importance in ensuring long-term isolation of the waste by minimizing the infiltration of water into the disposal unit and resultant mobilization of the waste. The cover should also direct percolating or surface water away from the buried waste and resist degradation through erosion, geologic processes, or biotic activity. To accomplish this, both natural earthen and artificial covers are employed, either individually or in combination. Clay is generally preferred as a component of a natural cover because of its relatively low permeability, but it must have surface protection against erosion and other external degrading influences. Preferred man-made cover materials include concrete, geotextiles and geomembranes, and soil-cement asphalt. A vegetative layer or a layer of gravel (in arid regions) may be used as the surface of the cover to stabilize the system. It is important that the cover extends beyond the disposal unit sidewalls and be sealed to the sidewall to create a continuous barrier against water infiltration. In addition, the cover is an integral component of the surface drainage system and acts to direct surface runoff away from the disposal unit. The sidewalls may also be of natural materials or incorporate artificial barriers. They are generally constructed of the same materials as the cover to ensure compatibility at junction points. The sidewall slope angle is dictated by the strength of the soil. The disposal unit floor is generally covered with a material such as sand to permit ready water infiltration and collection and is sloped to ensure water removal through a french drain to a collection sump.

In addition to the packages of waste, material will be emplaced in the disposal unit to fill the void spaces between packages and to backfill over the packages. A freely draining, noncohesive material, such as clean sand or gravel, is best for a fill material because it readily fills the void spaces and promotes rapid water movement through the disposal unit and thus minimizes contact time with the waste packages. Low permeability materials should be used for backfill to provide a barrier against water infiltration. The potential for waste and water contact is also minimized through the development of a surface water management system that collects surface flow and transports it away from the disposal units to a discharge point. A combination of drainage ditches, surface contouring, and control of surface roughness parameters generally comprises the surface water management program.

6.4.2 Tailings Impoundment Design

The underlying technical criteria governing the design of uranium tailings surface impoundments are defined in Appendix A to 10 CFR Part 40. These criteria, as in the case of the impoundment site selection criteria, provide a foundation of minimal prescriptive requirements, but permit the facility developer substantial flexibility in selecting design configuration and parameters. The relevant impoundment design criteria in Appendix A to 10 CFR Part 40 are directed toward the protection of groundwater. In summary, they require that the seepage of toxic materials into the underlying groundwater is to be reduced to the maximum extent reasonably achievable. Should any seepage occur, it cannot result in deterioration of existing groundwater supplies from their current or potential uses. The techniques to be considered to minimize seepage are (1) installation of low permeability bottom liners, (2) adoption of well-proven designs that provide the maximum practical recycle of solutions and conservation of water to reduce the net input of liquid to the tailings impoundment, (3) dewatering of tailings by process devices and/or *in situ* drainage systems (required at new impoundments), and (4) neutralization to promote immobilization of toxic substances. It is emphasized that the primary method of protecting groundwater is isolation of the tailings and tailings solutions. However, disposal involving contact with groundwater will be considered if it can be demonstrated that the proposed disposal and treatment methods will not degrade groundwater from current or potential uses.

Other tailings impoundment design criteria relate to stabilization measures to be undertaken after facility closure (see Section 6.6), and are intended to prevent surface erosion over the long term.

6.4.3 Geologic Repository Design

A geologic repository encompasses the operative area including both surface and subsurface areas, where waste handling activities are conducted, and the portion of the geologic setting that provides isolation of the wastes. As in the case of the LLW disposal facility, the primary design objectives for a geologic repository for the disposal of HLW (as presented in 10 CFR Part 60) are to achieve specified minimum technical criteria and meet the defined performance objectives. Although the DOE retains some flexibility in defining design concepts and parameters to achieve these objectives, the highly radioactive nature of the wastes and the need to ensure their long-term isolation require that the minimum technical criteria be rigorous and extensive. The design process will be iterative, evolving from a preliminary design developed during the site selection and characterization phase to a refined design provided with the license application.

Following is a summary of the significant general design criteria for the geologic repository operations area as presented in 10 CFR Part 60:

1. The geologic repository operations area shall be designed to maintain radiation doses, levels, and concentrations of radioactive materials in air in restricted areas within 10 CFR Part 20 limits.

2. Structures, systems, and components important to safety shall be designed so that natural phenomena and environmental conditions anticipated at the operations area will not interfere with necessary safety functions, and to withstand dynamic effects such as missile impacts that could result from equipment failure, or other conditions that could lead to loss of their safety precautions.

3. Structures, systems, and components important to safety shall be designed to perform their safety functions during and after credible fires or explosions in the operations area, to maintain control of radioactive wastes and effluents, and permit prompt termination of operations and evacuation of personnel during an emergency, and to include onsite facilities and services that ensure a safe and timely response to emergency conditions and facilitate the use of available offsite services that may aid in recovery from emergencies.

4. Each utility service system that is important to safety shall be designed so that essential safety functions can be performed under both normal and accident conditions.

5. All systems for processing, transporting, handling, storage, retrieval, emplacement, and isolation of radioactive waste shall be designed to ensure that a nuclear criticality accident is not possible unless at least two unlikely, independent, and concurrent or sequential changes have occurred in the conditions essential to nuclear criticality safety.

6. The design shall include provisions for instrumentation and control systems to monitor and control the behavior of systems important to safety over anticipated ranges for normal operations and accident conditions.

7. The design of the geologic repository shall include such provisions for worker protection as may be necessary to provide reasonable assurance that all structures, systems, and components important to safety can perform their intended function.

8. Hoists important to safety shall be designed to preclude cage free fall, with a reliable cage location system, and with two independent indicators to indicate when waste packages are in place and ready for transfer.

In addition to the general design criteria for the repository, criteria are also specified for surface facilities, the underground facility, the seals for shafts and boreholes, and for the waste package and its components. The surface facility criteria establish requirements for facilities for receipt and retrieval of waste, surface facility ventilation, effluent control and monitoring, waste treatment, and decommissioning. The underground facility criteria establish general requirements and specific requirements related to design flexibility, waste retrieval, control of water and gas, underground openings, rock excavation, underground facility ventilation, and thermal loads. The seal and shaft criteria require that their designs do not permit them to become pathways that compromise the repository's ability to meet the performance objectives. The waste package design criteria establish requirements for the HLW package design, forbid the inclusion of explosive, pyrophoric, and chemically reactive materials or free liquids, require unique identification of the package, and define waste form criteria dealing with solidifiication, consolidation, and combustibles.

The geologic repository operations area must also be designed so as to permit implementation of a performance confirmation program to provide data that indicate whether subsurface conditions are within limits specified during the licensing action, and whether operational systems and barrier systems are functioning as intended.

Since the design of the first geologic repository is currently in progress, the feasibility and compatibility of the full scope of design criteria are yet to be validated. It is likely that further elaboration and/or modification to the technical requirements will be necessary as feedback is obtained from the site selection and facility design process.

6.5 DISPOSAL FACILITY OPERATION

6.5.1 Low-Level Waste Near-Surface Disposal Facility Operation

The regulatory approach to operation of an LLW disposal facility is similar to that applied to site selection and facility design. 10 CFR Part 61 prescribes minimum technical requirements for facility operation, but permits the site operator flexibility in establishing operating procedures and parameters consistent with achieving the performance objectives. The minimum technical requirements that apply to LLW disposal facility operations are

1. Wastes designated as Class A pursuant to 10 CFR 61.55 must be segregated from other wastes by placing the waste in disposal units that are sufficiently separated from disposal units for the other waste classes so that any interaction between Class A wastes and other wastes will not result in the failure to meet the performance objectives. This segregation is not necessary for Class A wastes if they meet the stability requirements in 10 CFR 61.

2. Wastes designated as Class C pursuant to 10 CFR 61.55 must be disposed of so that the top of the waste is a minimum of 5 m below the top surface of the cover or must be disposed of with intruder barriers that are designed to protect against an inadvertent intrusion for at least 50 yr.

3. Wastes must be emplaced in a manner that maintains the package integrity during emplacement, minimizes the void spaces between packages, and permits the void spaces to be filled.

4. Void spaces between waste packages must be filled with earth or other material to reduce future subsidence within the fill.

5. Waste must be placed and covered in a manner that limits the radiation dose rate at the surface of the cover to levels that at a minimum will permit the licensee to comply with all provisions of 10 CFR 20 ("Standards for Protection Against Radiation") at the time the license is transferred.

6. The boundaries and locations of each disposal unit (e.g., trenches) must be accurately located and mapped by means of a land survey. Near-surface disposal units must be marked in such a way that the boundaries of each unit can be easily defined. Three permanent survey marker control points, referenced to United States Geological Survey (USGS) or National Geodetic Survey (NGS) survey control stations, must be established on the site to facilitate surveys. The USGS or NGS control stations must provide horizontal and vertical controls as checked against USGS or NGS record files.

7. A buffer zone of land must be maintained between any buried waste and the disposal site boundary and beneath the disposed waste. The buffer zone shall be of adequate dimensions to carry out environmental monitoring activities specified in paragraph 61.53(d) of 10 CFR 61 and take mitigative measures if needed.

8. Closure and stabilization measures as set forth in the approved site closure plan must be carried out as each disposal unit (e.g., each trench) is filled and covered.

9. Active waste disposal operations must not have an adverse effect on completed closure and stabilization measures.

10. Only wastes containing, or contaminated with, radioactive materials shall be disposed of at the disposal site.

The rationale for the waste segregation requirements described in (1) and (2) above derives from the waste classification system defined in 10 CFR Part 61.

Class A segregated waste, containing radionuclides of low activity and short half-life, is expected to become nonhazardous prior to the end of the postclosure active maintenance period and is thus considered to be the least hazardous waste category. Therefore, Class A segregated waste need meet only minimum waste form and packaging requirements. If this waste form were not segregated, its inherent instability could lead to differential settlement (subsidence) of the various waste forms in the disposal unit and the disposal unit cover. Concentrations of Class B waste are higher than those for Class A segregated waste but are limited by ceilings on individual radionuclides. Class B and C wastes also are required to achieve greater stability than Class A waste to ensure that the waste form does not degrade and reduce the overall stability of the disposal unit. Class C represents the greatest potential radiological hazard of waste accepted for near-surface disposal, and must be disposed of in units having additional barriers against inadvertent intrusion. Among the intruder barriers that may be considered are use of Class B waste as overburden on top of Class C waste, concrete wall and/or cap disposal units, layered earth materials using materials with obvious physical differences in each layer, and other engineered structures and liners.

During facility operations, an environmental monitoring program is conducted typically at the same stations established in the preoperational monitoring program and at stations in the buffer zone. The program provides data to evaluate the potential health and environmental impact during facility operation and to enable the evaluation of long-term effects and the need for mitigation measures. The monitoring system has to provide an early warning of radionuclide releases from the disposal units before the releases leave the site boundary. If the monitoring data indicate that the boundary will be breached or the performance objectives otherwise violated, corrective measures can then be taken to stop the migration. The monitoring program is continued after site closure to validate the adequacy of closure measures.

Although not specifically defined as an operational criterion, inherent in the operational controls and procedures at the facility is the achievement of ALARA release levels and occupational and population exposures. Various emission controls, administrative constraints, and institutional controls are performed to ensure that ALARA levels are maintained during operation.

6.5.2 Tailings Impoundment Operation

In many respects the operation of a tailings impoundment is constrained by concepts and objectives similar to an LLW disposal facility, particularly by the achievement of ALARA releases and exposures. Strict control of emissions from the impoundment is required to ensure that occupational and population exposures are reduced to the maximum extent reasonably achievable and to avoid site contamination. The greatest potential source of offsite radiation exposure is dusting and radon release from dry surfaces of the tailings disposal area not covered by tailings solution. To control these releases, it is required that the uncovered tailings be wetted or chemically stabilized during operation to prevent or minimize blowing and dusting to the maximum extent reasonably achievable. Institutional controls, such as extending the site boundary and exclusion area, may also be employed to minimize population exposure to airborne releases. An extensive environmental and effluent monitoring program is conducted to ensure continued isolation of the tailings from the biosphere and provide an early warning if a release should occur. A comparable cycle of preoperational, operational, and postclosure monitoring is conducted as is used at an LLW disposal facility. Frequent inspections are conducted of the tailings retention system to ensure that no failures resulting in a release have occurred.

6.5.3 Geologic Repository Operation

The primary objective of the operational phase will be to maintain the necessary controls to achieve the performance objectives, and to ensure that the facility is left in condition to achieve the necessary isolation after closure. Since the geologic repository is yet to be built, operational constraints and procedures have not been developed. The minimum technical criteria that have been established in 10 CFR Part 60 specify design testing of critical safety features (e.g., boreholes, shaft seals, backfill, and thermal interaction effects) to be initiated during construction and before full-scale operation proceeds to seal boreholes and shafts. A program to monitor the condition of the waste packages upon receipt is also specified. This program will continue up to the time of permanent closure. Although not yet specified, a significant program to monitor all potential release pathways both at the source and before the site boundary is reached will be required. The focus of this monitoring program will be the detection of releases that could migrate to the groundwater, and of groundwater in-flow into the repository chambers.

6.6 DISPOSAL FACILITY CLOSURE

The period in the lifetime of a radioactive waste burial site after closure when the waste is to be contained because of its danger to the public health and the environment is, by far, the longest phase in the lifetime cycle, dwarfing the 20–40 yr of active operation of the facility. To ensure that the waste remains isolated from the biosphere during this postclosure period requires that closure performance be a major consideration during each of the prior phases in the facility's life cycle. This concept is reflected, for an LLW disposal facility, in 10 CFR 61.44, which states "The disposal facility must be sited, designed, used, operated, and closed to achieve long-term stability of the disposal site and to eliminate to the extent practicable, the need for ongoing active maintenance of the disposal site following closure so that only surveillance, monitoring, or minor custodial care are required." It is further emphasized in 10 CFR 61.51(a)(2), which states "the disposal site (facility) design and operation must be compatible with the disposal site (facility) closure and stabilization plan and lead to disposal site (facility) closure that provides reasonable assurance that the performance objectives of 10 CFR 61, Subpart C will be met."

Thus, to isolate the radioactive wastes from the biosphere and to keep them isolated after closure until the radioactivity has decayed to levels that are no longer harmful, it is necessary to select a site with minimal potential transport pathways, and design and construct the facility to prevent the known transport pathways from being operative during site operation and after site closure. During the operational phase, activities are directed toward ensuring that the facility meets the objectives of the design at the time of closure.

During the actual closure phase, on-site contaminated buildings and equipment are decommissioned, disassembled, and disposed of. The administration building and other uncontaminated supporting facilities would be retained. Since closure and stabilization measures must be carried out as each disposal unit is filled and covered [10 CFR 61.52(a)(2)], extensive stabilization activities should not be required at closure. However, compacting, grading, and capping of the final disposal units (e.g., trenches) will be performed after all accumulated water is pumped from the open trenches. Some recontouring, filling, and soil removal may be required to establish final site contours and provide additional erosion and runoff control. At this time the vegetative cover on the entire site should be completed and permanent identification markers established for locating disposal excavations and monitoring wells after closure. The components of the site environmental monitoring program and the site security plan that will be continued after closure are developed and put into practice. Technical analyses are performed of the long-term stability of the disposal site to validate that the performance objectives will be met with a reasonable assurance that there will not be a need for ongoing active maintenance of the disposal site following closure. The activities to be performed in connection with the site closure, and the documentation to be prepared, are developed in a Site Closure Plan, the contents of which are summarized in Table 6-4.

While the procedure for closure of a tailings impoundment containing radioactive wastes is essentially comparable to that for an LLW disposal facility, Appendix A technical criteria to 10 CFR 40 defines prescriptive standards of the final earth cover as

> sufficient earth cover, but not less than three meters, shall be placed over tailings or wastes at the end of milling operations to result in a calculated reduction in surface exhalation (flux) or radon emanating from the tailings or wastes to less than two picocuries per square meter per second ... Direct gamma exposure from the tailings or wastes should be reduced to background levels.

The technical criteria also encourage phased covering and reclamation of tailings impoundments as sections of the impoundment are filled to help in controlling particulate and radon emissions during operations. The analytic model correlating surface radon exhalation with thickness of cover material is described in Section 7.11.3. This model is used by licensees to establish stabilization cover thicknesses in closure plans for uranium tailings impoundments.

TABLE 6-4   Site Closure Plan

The site closure plan includes the following elements:

1. *Site and facility description*—Site location map showing disposal area and adjacent properties, contour map of site and adjacent watershed, diagram and description of the site facilities, including waste storage areas, surface water management facilities, and on-site storage.

2. *Waste inventory*—Final inventory of wastes and their locations, physical, chemical, and radiological description of wastes (as obtained from internal waste receipt records).

3. *Closure schedules*—Last dates of waste receipt and burial, date final disposal unit closed, date final cover in place (soil, clay, and vegetation), date closure is expected to be completed, schedule of closure monitoring activities.

4. *Plan for decommissioning and disposal of contaminated buildings and equipment*—Personnel and equipment requirements, schedule, occupational dose assessment (collective and individual), environmental impact of decommissioning and disposal.

5. *Plan for stabilization of the site*—Compacting, grading, and capping of final trenches (or other excavations), maintenance of grounds, structures, and equipment until closure completed, recontouring, filling, soil removal (as required) to establish final site contours, and emplace final cover material, additional erosion and runoff control (as required), water management to eliminate liquids that have accumulated in trenches, site revegetation cover emplacement and maintenance, permanent identification markers for locating disposal excavations and monitoring wells after closure, environmental impact of site stabilization.

6. *Plan for final radiological survey of site*—Radiological survey procedures and protocol to be conducted after closure of the final trench and site stabilization performed, measurement plan for residual radioactivity levels for surface soils to ensure compliance with site threshold quantity requirements.

7. *Monitoring programs*—Measurement of contaminant releases during site closure activities, components of site environmental monitoring program to be continued after closure (station locations, collection and measuring procedures, number and types of samples and analyses to be performed), soil sampling program (surface, waste incorporation zone, below waste incorporation zone) to validate concentration limits.

8. *Emergency response plans*—Maintenance of emergency response plans, facilities, and equipment until closure is completed.

9. *Surveillance and security plans*—Security systems to prevent unauthorized entry or removal of equipment or material during closure, passive security system for postclosure activities.

10. *Technical analyses*—Validation that long-term performance objectives will be met through analyses of long-term stability of disposal site, that provides reasonable assurance that there will not be a need for ongoing active maintenance of the disposal site following closure. Analyses will consider active natural processes (erosion, mass wasting, slope failure, settlement of wastes and backfill, infiltration through covers over disposal areas and adjacent soils, and surface drainage of the disposal sites).

11. *Cost estimate*—Cost components for each closure-related activity, including decommissioning and disposal of contaminated buildings and equipment, site stabilization, radiological surveys, monitoring programs, and security programs.

A primary objective of the final stabilization process is to contour the final slopes, on and in the vicinity of the impoundment, to grades that are as close as possible to what would be provided if tailings were disposed of below grade, and thus minimize long-term erosion potential. After the slope contouring is performed, a self-sustaining vegetative cover is to be established or a rock cover employed to reduce wind and water erosion to negligible levels. Furthermore, all impoundment surfaces are to be contoured to avoid areas of concentrated runoff or abrupt or sharp changes in slope gradient.

The closure plan incorporated in the DOE application to amend the geologic repository license to permit closure will emphasize the achievement of long-term isolation of the emplaced HLW and spent fuel through (1) use of measures such as land use controls, construction of monuments, and preservation of records, (2) compilation of relevant geologic, geochemical, hydrologic, and other site data, and (3) tests and analyses relating to backfill of excavated areas, shaft sealing, and waste interaction with the host rock.

6.7 POSTCLOSURE CARE AND INSTITUTIONAL CONTROL PHASES

After site closure activities are completed at a land disposal facility, the licensee carries, out a transitional period during which no waste management operations are conducted. During this period the licensee observes and monitors the site to validate that the site is stable, waste isolation is being maintained, and closure measures are satisfactory. In addition a custodial and remedial maintenance program is conducted consisting of routine site maintenance, preventative maintenance, water management, and remedial measures to ensure integrity of the stabilization cover and adherence to performance objectives. While the licensee must maintain responsibility for the disposal site for 5 yr, the period for post closure observation and maintenance can vary from this interval depending on site-specific conditions. Physical surveillance is conducted during the postclosure period to restrict access to the site and to ensure the integrity of the waste disposal site.

The elements of the postclosure period are defined by a postclosure plan that is an extension of the site closure plan. A preliminary postclosure plan is developed in conjunction with the closure plan prior to initiation of operations at new sites, and is incorporated in summary form in the environmental report. The final postclosure plan, revised to reflect operational and closure experience, new data, and updated plans, is issued prior to performance of the postclosure activities.

Once the period of postclosure observation and maintenance is satisfactorily completed, the license will be amended to permit transfer to the federal or state government agency that is the disposal site owner. Satisfactory completion of the postclosure phase occurs when the NRC finds that closure of the site is in conformance with the site closure plan, that reasonable assurance has been provided by the licensee that the performance objectives are met, that the required funds and records are available for transfer to the site owner, that the postclosure monitoring program remains operational for implementation by the site owner, and that the site owner is prepared to assume responsibility for the site. The assumption of licensee responsibility by the disposal site owner initiates the institutional control phase. During the institutional control phase, the site owner performs routine custodial care and continues the postclosure monitoring programs. Site access remains restricted, pertinent waste inventory records are maintained, and waste movements are controlled. At the completion of the period of institutional control, validated primarily by demonstrated stability and evidence that ongoing active maintenance is not necessary to preserve isolation, the NRC will terminate the license.

The above-described postclosure and institutional control phases generally apply to both LLW disposal sites and tailings impoundments. In the case of the geologic repository the DOE, as the site owner and operator, follows a conceptually similar approach of extensive postclosure monitoring to demonstrate stability and long-term isolation of the wastes, provide maintenance, and ensure adequate record keeping. These activities, which would be expected to extend for a longer period than for a near-surface or surface facility because of the more hazardous nature of the waste, are performed in accordance with an NRC-approved postclosure plan. The NRC will terminate the repository's license when it determines that (1) the final disposition of wastes has been made in conformance with the DOE's plan, (2) the final state of the repository operations conforms to the DOE's plans for permanent closure and decontamination and/or decommissioning of surface facilities, and (3) the termination of the license is authorized by law. After the license is terminated, the DOE will continue to own the facility and ensure that it is restricted from public access.

6.8 LAND BURIAL TECHNOLOGY

Burial in the soil was the earliest method used for disposal of low-level radioactive materials (LLW). Experience with this technology at both government and commercial sites is summarized in Chapter 3 (Sections 3.4 and 3.5). That discussion also describes the use of this technology for waste that is now considered TRU and that will be disposed in a geologic repository. Depending on site conditions related to factors such as distance to groundwater and soil sorptive properties, both liquid and solid waste has been disposed by this method.

There are several variants of land burial technology that have been applied. The choice among them is related to different site characteristics, waste characteristics, and costs. This section will discuss several alternatives currently being evaluated for near-surface disposal at new facilities to be developed in accordance with the requirements of the Low-Level Radioactive Waste Policy Amendments Act of 1985 and for intermediate-depth disposal at government and/or commercial sites.

6.8.1 Near-Surface Disposal

Near-surface disposal is generally defined as emplacement of the waste within about the top 100 ft (30 m) of the soil surface. Any such disposal is regulated in accordance with the provisions of 10 CFR Part 61 as described in Chapter 3 (Section 3.5). Each of the three operating commercial LLW disposal facilities uses this technology. The regulatory framework under which any new near-surface disposal facility will be sited, designed, operated, closed, and monitored was summarized as follows by the Nuclear Regulatory Commission in the Final Environmental Impact Statement on 10 CFR Part 61 (NRC 1982):

Principle	Contribution to Performance Objectives
Long-term stability	Reduces water infiltration and potential for *migration*
	Reduces need for long-term *maintenance*
	Reduces likelihood and impact of inadvertent *intrusion*
	Reduces *occupational hazards* and *offsite releases* in accidents
Reduce contact of water with waste	Reduces potential for *migration*
	Reduces need for active *maintenance*
	Reduces *intruder impact* due to waste degradation
	Reduces *occupational hazards* and *offsite releases*
Institutional care and other intruder controls	Custodial care reduces potential for water *infiltration and migration*
	Ensures proper *maintenance*
	Reduces likelihood and impact of *inadvertent intrusion*
	Reduces *occupational hazards*

There are several main alternatives that have been proposed for achieving the principles outlined above. They have been evaluated for conformity to the regulations by the U.S. Army Corps of Engineers for the NRC (NRC 1984, 1985b). Conceptual designs for alternative technologies were developed and compared for the DOE by Rogers and Associates Engineering Corporation (DOE 1987) as part of the department's program of technical assistance to the states and interstate compacts that is managed by EG&G Idaho, Inc. Several potential disposal facility developer/operators have also proposed specific disposal unit designs incorporating alternative technologies (CN 1988; *W* 1986). The major differences among the alternatives are the materials used as barriers between the waste and the biosphere, methods of waste emplacement and disposal unit closure, and location of waste with respect to the soil surface. The alternatives receiving most attention, and described in the following paragraphs, are shallow land disposal, modular concrete canister disposal, below ground vaults, earth-mounded concrete bunkers, and above-ground vaults. It is important to remember that site characteristics are required under 10 CFR Part 61 to be the primary barrier to waste migration. Engineered barriers may only augment a site's ability to meet the performance objectives of 10 CFR Part 61. They may not be necessary for the site to meet the performance objectives.

Shallow-Land Disposal (SLD). SLD is accomplished by constructing trenches that are approximately 30 ft deep. Figure 6-1 is a schematic diagram of an SLD disposal facility. It illustrates the ordered stacking of waste packages to minimize voids within the trench that may lead to postclosure trench cover failure. Also illustrated in this figure is the multilayer construction of the trench cover. The trench cover is the primary barrier preventing (or retarding) infiltration of surface water into the disposal unit. It also reduces surface radiation levels to regulatory limits or less and reduces the likelihood of inadvertent intrusion into the waste after the end of the institutional control period.

Current conceptual designs provide for segregated disposal of LLW by classification as defined in 10 CFR Part 61.55. Class A waste trenches would be larger (270 × 54 ft at base) than those for Class B and C waste (66 × 22 ft at base). The smaller B–C waste trenches provide for a thinner layer of waste (12 ft) and greater thickness of earthen cover than do the Class A trenches in which waste is stacked to a height of 21 ft. These differences provide additional shielding for the B–C waste and minimize radiation levels at the surface of a closed trench. They also enhance the disposal unit's stability and reduce the likelihood of someone's inadvertently coming in contact with the waste because of the increased depth of soil cover. Figure 6-2 illustrates the differences in dimensions of the two waste classification trenches. As indicated in this figure, trench floors are constructed to direct any infiltrating water to one side of the trench where a water collection (French drain) system is provided. This design minimizes the length of time water

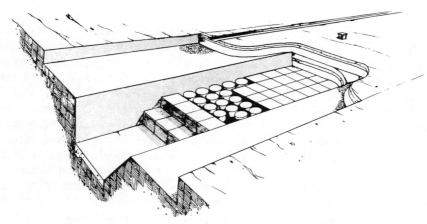

Figure 6-1 Schematic of SLD facility. Source: (DOE 1987).

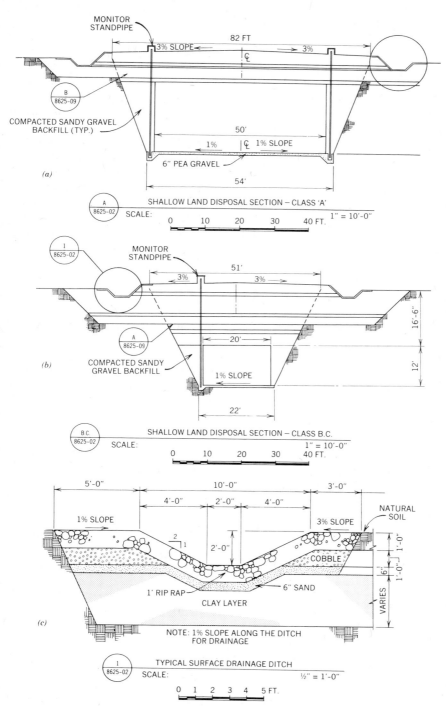

Figure 6-2 Cross sections of SLD trenches. (a) Class A waste trench; (b) segregated class B–C waste trench; (c) surface drainage ditch details. Source: (DOE 1987).

is in contact with the waste and provides a method for measuring the rate of water infiltration and collecting the water for monitoring, treatment if required because of levels of contained radionuclides, and eventual disposal. The measurements of the amount and radionuclide content of the water (during operation and after closure) are prime inputs into decisions on the need for any remedial action (such as trench cover repair) during the institutional control period.

Waste is emplaced in the trenches using cranes. The space between the waste packages, and between the stacked waste and the trench walls, is backfilled with sandy soil to improve short-term stability of the disposal unit. Stability requirements for Class B–C waste (pursuant to 10 CFR Part 61.56) are met by the waste form and package rather than by the trench. Trench cover design provides for several layers of different materials. These include sandy materials near the waste to minimize water retention and the possibility of container corrosion and/or radionuclide leaching, clay to minimize water infiltration, larger stones or cobbles to deter intruders and in some cases retard wind erosion, and topsoil to support growth of natural vegetation that also reduces erosion. The specific combination and depth of the individual materials will be based on site conditions and the waste emplaced in a given trench. Class A waste trenches generally have shallower covers with fewer different layers than do Class B–C waste trench covers.

Modular Concrete Canister Disposal (MCCD). MCCD provides an additional engineered barrier between the emplaced waste and the biosphere compared to SLD. This is achieved by placing the waste into steel-reinforced concrete canisters that are then placed in trenches similar to those used for SLD. Figure 6-3

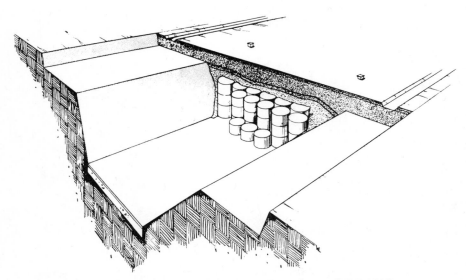

Figure 6-3   Schematic of MCCD facility. Source: (DOE 1987).

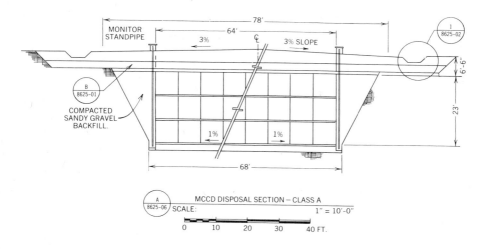

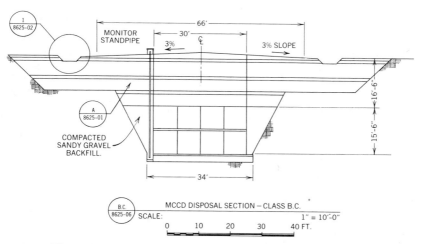

Figure 6-4 Cross sections of MCCD trenches. Source: (DOE 1987).

illustrates a typical MCCD facility. The trench cross-sectional view is shown in Figure 6-4.

The space between individual waste packages within the canister is filled with cement grout. Therefore, each separate canister provides structural stability. This is in addition to the stability provided for Class B–C wastes by the waste form itself. The primary objectives of this extra barrier and stability are to delay the time at which trench cap failure might occur and reduce the rate of water coming into contact with the waste. The concrete canister provides additional shielding for high-activity waste and reduces the radiation levels at the surface of a closed trench. It also reduces the likelihood of human, plant, or animal intrusion into the waste. Spaces between canisters, and between the stacked canisters and the trench wall, are filled with sandy soil. Segregated disposal of Class A and Class B–C waste is also planned for MCCD. Trench dimensions would provide for cover thicknesses similar to those for SLD. Other dimensions would likely be somewhat greater than for SLD because of the additional non-waste-containing volume from the concrete canisters and the structural stability they provide.

Waste handling at the disposal facility would be increased compared to SLD because of the additional steps to emplace the waste in the concrete canisters. This additional handling increases occupational exposure by more than a factor of two compared to SLD. Once in the canisters the waste would be emplaced in the trench by crane in a manner similar to SLD. Several different canister shapes (including hexagons and interlocking modules) have been proposed to reduce space between canisters and to simplify evaluation of the response of the emplaced waste under potential seismic forces.

One of the technical issues still being evaluated is the anticipated service life of different types of concrete. This will affect how well the disposal unit meets the long-term stability criteria and how quickly radionuclides might be available for migration. Analyses performed for the conceptual design comparison (DOE 1987) indicate an anticipated dose reduction of 10% to an intruder (who farms the site after institutional control ceases) and 15% to a member of the general public (at an adjacent farm) compared to SLD.

Below-Ground Vaults (BGV).   BGV disposal, as illustrated in Figure 6-5, provides an engineered barrier between the waste packages and the biosphere as well as structural stability by use of an engineered concrete structure below natural grade. The structure consists of reinforced concrete walls and roof. Some designs use a concrete floor whereas others employ a floor using natural materials but with engineered features such as sloping and water collection and monitoring systems. The BGV vault is backfilled with sandy gravel after waste emplacement

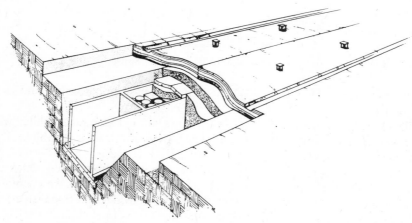

Figure 6-5   Schematic of BGV facility. Source: (DOE 1987).

in much the same way as the SLD trench and MCCD excavation. Waste segregation and differences in the Class A and Class B–C structures for this technology are shown in Figure 6-6. As indicated in this figure, Class B–C waste is stacked to allow a 6 ft 6 in. layer of sandy gravel backfill to be placed over the waste before the vault roof is emplaced. The current conceptual design provides for emplacement of the concrete vault roof after the disposal unit is full and the backfill layer over the Class B–C waste provides shielding to reduce occupational exposure while the roof is being put in place. There is in addition a 10-ft earthen cover over the Class B–C vault roof compared to a 6 ft 5 in. cover over the Class A waste vault roof. The vaults are actually composed of a number of individual cells (36 for Class A waste vaults and 3 for Class B–C waste vaults) that are closed individually as the cell is filled. This reduces the amount of water that may get into the vault from precipitation and provides shielding for workers emplacing

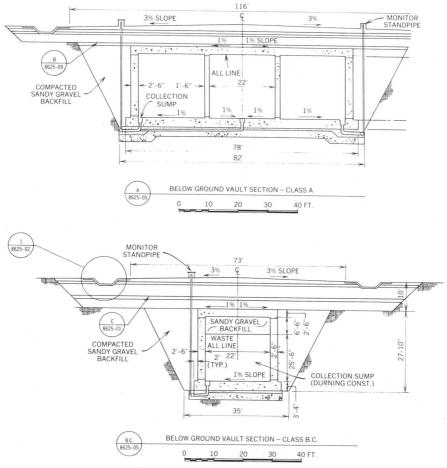

Figure 6-6   Cross sections of BGV structures. Source: (DOE 1987).

subsequent waste packages. As with the other concepts, each cell is discharged to a French drain system that runs the entire length of the vault and from which water may be collected and analyzed.

Analyses performed for the conceptual design study indicate that occupational exposures for a BGV facility will be about 60% greater than those for SLD, primarily due to placing the roof over individual cells. The effect of the structure on delay of the time at which groundwater comes in contact with the waste results in a 45% reduction in the whole body dose to a member of the public (15% reduction for thyroid dose), compared to SLD. Doses to an intruder are projected to be 50–90% lower for BGV than SLD for the case in which contact is due to farming and drinking water from a well onsite or construction and direct contact with the waste.

Earth-Mounded Concrete Bunkers (EMCB). EMCB technology, which was originally employed at the waste facility at Centre de la Manche, France, disposes of the Class A waste above grade with an earthen cover. The Class B–C waste is disposed below grade in a concrete bunker. The above-grade structure is called a tumulus. The below-grade structure is called a monolith. Figure 6-7 illustrates the EMCB concept and shows a tumulus located over a monolith. As shown in Figure 6-8, the waste in the tumulus is placed in concrete containers similar to those described for MCCD. These containers provide structural stability for the tumulus. The earthen cover placed over the stacked waste utilizes multiple layers of different soil types to direct infiltrating water away from the waste and deter human, plant, and animal intrusion into the waste. Tumulus sides are graded to enhance long-term stability and rip-rap (large stones or boulders) and vegetation are included to minimize erosion by wind and water.

Class B–C waste packages are grouted in place within the below-grade bunker. The product is a concrete monolith that provides structural stability for the waste. The tumulus above the monolith provides the required intruder barrier for the Class C waste. If the tumulus and monolith are located at different parts of

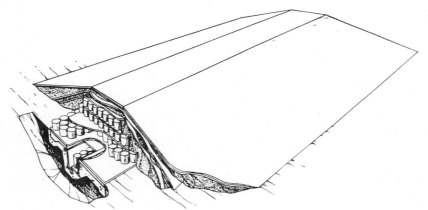

Figure 6-7   Schematic of EMCB facility. Source: (DOE 1987).

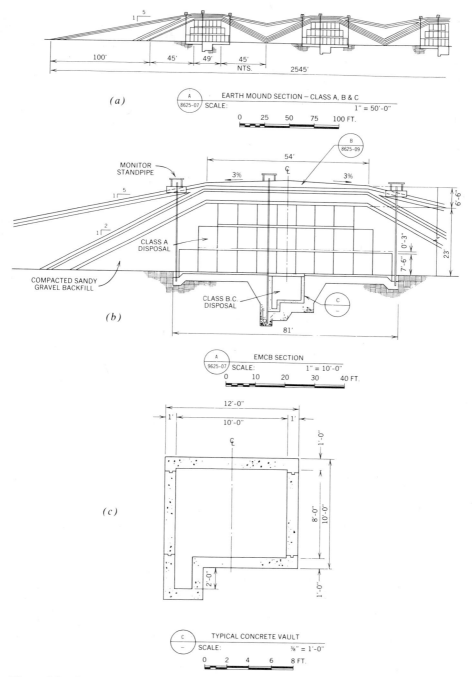

Figure 6-8 Cross section of colocated EMCB tumulus and monolith. (a) Overall layout; (b) tumulus detail; (c) concrete detail. Source: (DOE 1987).

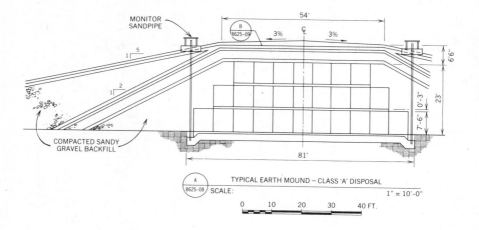

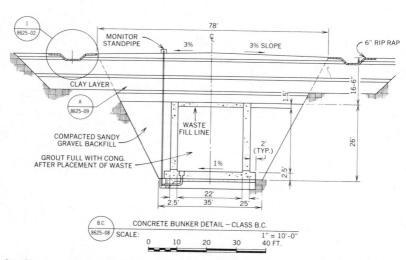

Figure 6-9 Cross sections of separate EMCB tumulus and monolith. Source: (DOE 1987).

the site as illustrated in Figure 6-9, additional earthen cover may be placed over the waste in the monolith to supply the necessary intruder barrier. Whether colocated or separate, water collection and recovery systems are provided for the tumulus and monolith.

Above-Ground Vaults (AGV). AGV disposal provides for emplacement of waste in a reinforced concrete structure above the natural grade. Like BGVs, AGVs are composed of multiple individual cells that are filled with waste, backfilled

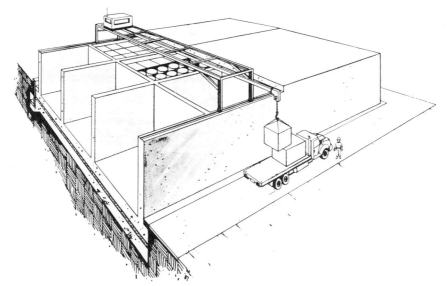

Figure 6-10   Schematic of AGV facility. Source: (DOE 1987).

with sandy soil to enhance removal of any water that may enter the vault, and covered with a concrete roof. As illustrated in Figure 6-10, access to individual cells is expected to be via overhead crane. Vault cross-sections, again with segregated Class A and Class B–C disposal, are shown in Figure 6-11. Note that provision is made to collect and monitor water releases from individual disposal cells rather than from the entire vault. Also, Class B–C waste is covered with a 6 ft 6 in. backfill layer to provide shielding prior to emplacement of the vault roof.

A major technical uncertainty with AGV technology is the performance of the structural concrete for the hundreds to thousands of years for which the waste is to be isolated. This uncertainty, common to all the alternatives that use structural concrete (that is, except SLD), is crucial for AGVs because there is no earthen cover to provide containment of the waste subsequent to failure of the engineered structure. Location of the AGV at the soil surface results in at least some part of the release resulting in exposure via the surface water pathway rather than only through groundwater as analyzed for below-ground facilities. Analyses for the conceptual design report indicate that whole body equivalent dose to a member of the general public is four times higher for AGV than for SLD. Thyroid doses exceed those from SLD by a factor of seven. These doses exceed the regulatory limit in 10 CFR Part 20 as described in Chapter 3. The uncertainty concerning concrete performance also affects the confidence with which doses to inadvertent intruders can be projected. Assuming that the concrete remains intact for 500 yr and then fails, the Conceptual Design Report analyses project doses to intruders that exceed those from SLD by a factor of four if someone builds on the waste site. As indicated in Figure 6-12, the dose rate estimates are very sensitive to the

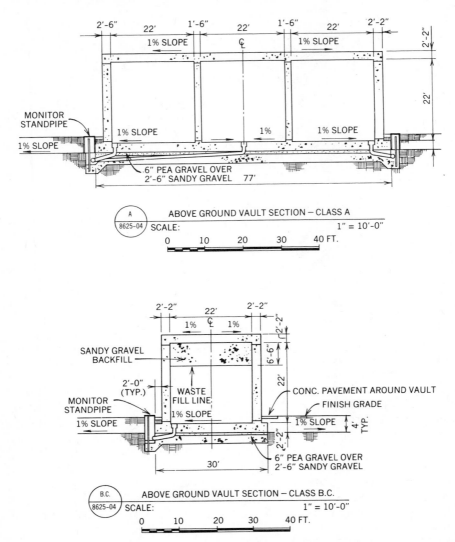

Figure 6-11   Cross sections of AGV structures. Source: (DOE 1987).

assumed concrete service life. An AGV facility would need extensive monitoring during the institutional control period to verify that the concrete was successfully resisting environmental attack such as acid rain, freeze–thaw cycles, airborne pollutants, and earthquakes. This technology is expected to require the greatest amount of active maintenance during and beyond the institutional control period.

To facilitate upcoming choices among disposal technologies by compacts or states, the Conceptual Design Report Study modeled reference facilities using

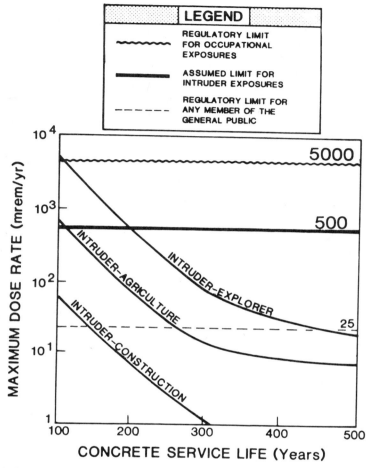

Figure 6-12 Sensitivity of dose rate estimates to variation in concrete service life. Source: (DOE 1987).

consistent assumptions for site conditions, waste characteristics, facility lifetime, and financing. The results of this study are summarized in Table 6-5. This table shows that, for the assumptions used to describe a representative or generic facility using the several alternatives, each is capable of meeting the performance objectives of 10 CFR Part 61. (In addition to the specific provisions of Part 61, a 500 mrem/yr limit on intruder exposure has been assumed.) This analysis assumes that the concrete performs as designed for at least 500 yr. The AGV structure is the one exception to this successful compliance with the regulations even under this assumption. If the AGV structure is breached at an earlier time, the corresponding doses increase as indicated in Figure 6-12.

TABLE 6-5   Summary Comparison of Alternative Near-Surface Low-Level Waste Disposal Technologies

Technology	Maximum Public Whole Body Dose (mrem/yr)	Maximum Public Thyroid Dose (mrem/yr)	maximum Intruder Whole Body Dose (mrem/yr)	Cumulative Worker Dose (man-rem/yr)	Probable Worker Lost-Time Injuries	Total Development Time (months)	Total Life Cycle Cost ($ millions)	Disposal Charge Private ($/ft³)	Disposal Charge Public ($/ft³)
SLD	7.5	46	7.5	10	7	60	196	39	33
BGV	4.1	39	4.1	16	14	70	294	56	50
AGV	34	330	20	15	15	84	395	65	61
MCCD	6.5	40	6.7	16	14	72	300	57	50
EMCB	3.4	33	3.5	14	18	72	434	76	69

Source: (DOE 1987).

6.8.2 Intermediate Depth Disposal

Intermediate depth disposal of LLW using shafts has been practiced at several U.S. DOE facilities (Nevada Test Site, Oak Ridge National Laboratory, Savannah River Plant, and Los Alamos National Laboratory) as well as in Canada (at the Bruce Nuclear Generating Plant and the Chalk River National Laboratory). It has been used for selected disposal of relatively high-activity low-volume waste. The shafts are generally drilled into approximately the top 300 ft of soil using augers or other drilling methods and the technology is sometimes referred to as augered hole disposal. Shafts may be lined or unlined, and of varying depth and diameter depending on soil properties affecting shaft stability and interaction with the waste package as well as the depth to the water table and projected travel time between the emplaced waste and the water table.

Disposal at depths greater than for SLD reduces the radiation level at the ground surface and the potential for inadvertent intrusion by humans, plants, or animals because of the greater cover thickness. Shaft disposal is also described as greater confinement disposal (GCD) in some reports for this reason. Other advantages of the method are that shaft cover integrity should be more easily maintained than would be the case for a SLD trench because of the smaller diameter of disturbed soil. Greater disposal depth reduces the probability of surface water infiltration into the waste and should result in lower radionuclide concentrations in groundwater compared to disposing of the same waste by SLD.

As implied in the discussion above, the method may be limited in applicability to sites having substantial depth to groundwater. Further, drilling techniques impose an effective limit on the dimensions of material that can be emplaced in the shafts. Therefore, it is expected that shaft disposal, at appropriate sites, would likely be used as it has been at the sites mentioned above for portions of the waste stream with high activity and surface exposure levels.

6.8.3 Mined Cavities

Enclosed cavities developed in the removal of natural resources have also been proposed as useful disposal units for LLW although none has been adopted to date. The major difference between such cavities and the geologic repositories discussed in Section 6.10 is that for LLW the cavity may be located closer to the ground surface. To minimize interaction between the waste packages and the surrounding soil, geologic formations of limestone or bedded salt are considered to be most attractive. Both previously mined and newly constructed cavities may be used. Preliminary evaluations of mined cavities indicate that an appropriately sited facility, operated and closed in accordance with regulatory requirements (both those in 10 CFR Part 61 and supplemental criteria specific to this technology), would result in minimal public radiation exposure because there is no contact between groundwater and the emplaced waste. However, because of the need to handle the waste both upon receipt at the facility and subsequently at the disposal location below ground it is expected that occupational exposure would be substantially higher for this alternative than for others available for LLW disposal. NRC estimated that the number of people involved in direct handling of the waste would be about a factor of two higher than for SLD. The necessity for close proximity to the waste results in a total occupational exposure of approximately four times that experienced in a near-surface disposal facility (NRC 1981).

6.9 SURFACE IMPOUNDMENTS

Surface impoundments are employed for the permanent disposal of uranium mill and phosphate tailings. They may also be used for the disposal of other semisolid or liquid radioactive wastes from processing operations. Tailings impoundment design has undergone significant revision in the last 15 yr to improve the isolation and stability characteristics of the impoundment. The primary objectives of the impoundment function are to (1) eliminate or minimize radiochemical leachate migration and resultant impacts on groundwater, (2) eliminate or minimize airborne radioactive emissions, including radon gas and particulates, and resultant environmental dispersion, and (3) ensure long-term stability and isolation of the tailings without the need for continued active maintenance.

These objectives are achieved by a combination of tailings treatment steps to remove radioactive constituents prior to emplacement in the impoundment (e.g., barium chloride precipitation of radium-226) and the careful selection of disposal site locations and impoundment design parameters incorporating the use of natural and artificial migration barriers. In general, the location of tailings disposal has been, and continues to be, above grade in traditional or upgraded surface impoundments. While the NRC has expressed a preference in 10 CFR Part 40 Appendix A for consideration of below-grade, near-surface, disposal, this approach has not been used in any significant manner primarily because very few new uranium projects have been undertaken in the 1980s since adoption of the Appendix A criteria, and because the NRC is willing to consider above-grade options if properly designed. Below-grade disposal, in the context of use of open mine pits, will be discussed in this section as an alternative to use of surface disposal concepts.

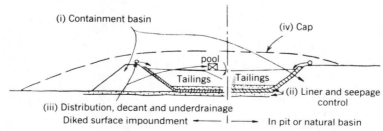

Figure 6-13   Cross section through tailings impoundment. Source: (NRC 1979).

A tailings impoundment can be divided into four major components (Figure 6-13): (1) a physical containment basin and structures (i.e., dams or dikes), (2) natural or synthetic liner, singly or in combination, and seepage control measures, (3) tailings management system including tailings distribution and water decant facilities, and (4) natural or artificial cover, singly or in combination, and other stabilization and reclamation features including vegetative or overburden layers and sealants. The extent of incorporation of specific aspects of these four components is a function of site-specific characteristics such as topography, hydrogeology, climate, and soils parameters. The four components function in concert to produce a desired tailings management program, and therefore a disposal facility design must combine them to produce the desired tailings management program. For example, the seepage potential of the underlying soil in the selected containment basin establishes whether a liner is needed or whether an impermeable bottom of the impoundment can be created from the natural soils.

The containment basin and associated structures comprise the design features that provide the physical volume of the impoundment to hold both the permanent solids fraction of the semisolid discharge and the temporary volume of fluids in the discharge. The surface impoundments can be constructed as ring-dike impoundments that are four-sided structures in relatively flat areas. They can also be formed as valley dam impoundments by constructing a dam in an existing natural drainage area such as a valley or canyon. In the latter case, which has been the predominant practice to date, an impoundment basin is formed by placing a dam wall across a valley and using the natural basin sides to provide containment. Below-grade (in-pit) disposal can be achieved by using existing open pit mines, or excavations to function as the basin.

In preparing the bottom and sides of the basin prior to introduction of tailings, a number of options are available to ensure that leachate migration does not occur and the tailings remain isolated from the groundwater. If the underlying soil or rock is generally impermeable, or if there is no danger of affecting the groundwater, the preparation could be limited to soil compaction to increase soil density and thus reduce permeability. However, in most cases the use of natural clay liners or synthetic liners is required to obtain the necessary isolation characteristics. Clay liners, generally between 1 and 3 ft in thickness, can be used as a sealant over compacted soil to inhibit seepage, and are a superior ion-exchange medium. Permeabilities of 10^{-7} cm/sec and lower can be attained. Although a number of different kinds of clay exist, bentonite clay is a favored type because of its availability in the Western states. The bentonite clay is attractive because it has a high content of montmorillonite, an expanding-lattice-clay mineral that swells when wet. Synthetic liners, in the form of sheeting or membranes, are being increasingly used, typically in combination with clay or other natural materials. Although synthetic liners are capable of obtaining permeabilities of from 10^{-9} to 10^{-10} cm/sec they have historically posed problems in this type of application. To avoid mechanical failures, they require careful preparation of the base to eliminate rocks and sudden slope changes; they tend to lose their flexibility when exposed to sunlight or chemicals; and it is questionable whether they will retain their integrity over the long term when subjected to the chemical and physical environment in the impoundment. Failure would tend to be catastrophic, resulting in a sudden release of contaminants. However, recent advances in the development and application of reinforced liners give promise of overcoming these shortcomings. Hypalon, a nylon-reinforced elastomer, is considered to be preferable for use with uranium tailings. Other materials under consideration include polyvinyl chloride (PVC), Neoprene, gunite, cement grout, and asphalt or asphaltic concrete.

Tailings management encompasses, in addition to the storage and isolation capability provided by the basin and liner, the system for distributing the tailings within the impoundment, the liquid removal system(s) by means of decantation or drainage under the tailings, and the dam operating procedures and constraints. A tailings management program should successfully control (1) the volume and locations of the fines and slimes through distribution and segregation of the tailings solids within the impoundment, (2) the amount of water stored in the impoundment to minimize the seepage head, and (3) the amount of tailings under water or kept damp to minimize airborne releases.

Stabilization of the tailings with a cover, after removal of the water, is accomplished either as an ongoing measure or after the impoundment is full. A variety of stabilization techniques is employed typically involving multiple layers of material as is done with covers at LLW disposal facilities. A layer of clay, 1–3 ft in thickness, is generally used to reduce water infiltration into the tailings and to reduce airborne radon emissions. The clay, however, should be covered to keep it damp and prevent wind or water erosion of the fine clay particles. A natural soil cover may be used over the clay, or an artificial cover or sealant used as an additional barrier. Where rainfall is sufficient, native flora can be planted on top of the soil cover. In semiarid regions, a layer of coarse gravel or crushed rock (rip-rap) can be used to stabilize the surface and prevent soil erosion.

A number of alternative tailings disposal siting and design concepts have emerged from the variety of possible combinations of locations, impoundment types, and engineered barriers that can be used on tailings impoundments. The following represent a range of tailings management alternatives (NRC 1979):

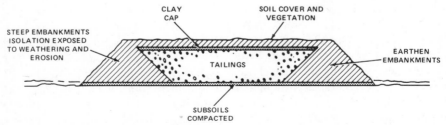

Figure 6-14   Above-grade disposal of uranium tailings. Source: (NRC 1979).

- *Alternative 1—Above-Grade Disposal with Continued Active Care* (Figure 6-14):
  An earthen berm (dike) is constructed on the sides of the surface impoundment. Tailings are conveyed to the impoundment by slurry pipeline and the water recycled to the mill. As the beach areas dry out, they would be covered with compacted clay and a native soil cover. A vegetative cover would then be developed in a layer of top soil. Active care and maintenance would be required indefinitely to prevent erosion of the cover.

- *Alternative 2—Disposal of Tailings in Below-Grade Mines or Pits*:
  Tailings are emplaced in an open pit mine or excavated pit and then covered with a clay cap, natural soil (overburden), and a vegetative cover developed. Figure 6-15 shows one variation of this concept in which the tailings slurry is deposited in a mine pit lined with clay and partially backfilled to the level of the water table. Other variations include dewatering the tailings prior to emplacement eliminating the need for extensive liners on the sidewall, and creation of an impoundment by digging a specially excavated pit in an isolated area with relatively impermeable soils, eliminating the need for a liner. In each case the tailings would be permitted to dry before the cover materials were emplaced. For these below-grade, near-surface disposal alternatives, the tailings are isolated from erosional forces thus eliminating the need for ongoing care. These alternatives are representative of the tailing disposal concepts being encouraged by the NRC.

- *Alternative 3—Disposal in Specially Excavated Below-Grade Trench:*
  For this below-grade, near-surface disposal concept (Figure 6-16), a pit is excavated in the form of a trench in sections, with construction, filling with tailings, drying, sealing, backfilling, and restoration moving sequentially along the length of the trench. This approach permits the reclamation of the

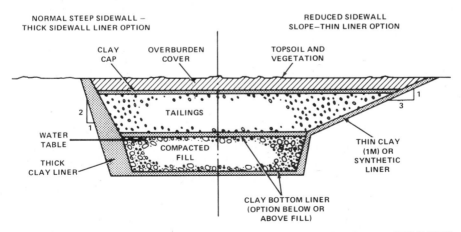

Figure 6-15 Disposal of tailings slurry in available open-pit mine. Source: (NRC 1979).

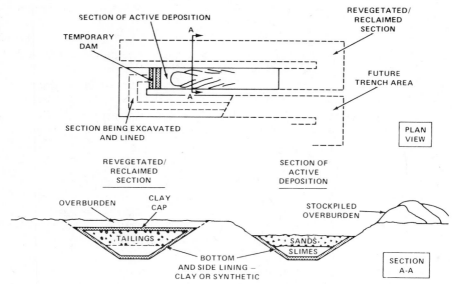

Figure 6-16   Disposal in specially excavated below-grade trench. Source: (NRC 1979).

tailings to be phased, reduces the area of exposed tailings prior to reclamation, and allows for segregation of the slimes from the sands with the slimes being covered by the sands during the deposition stage. This design approach is considered to be quite promising for future tailings disposal concepts.

- *Alternative 4—Disposal in Upgraded Conventional Above-Grade Impoundment:*
  In this case, the intent is to incorporate design and siting features that make the upgraded impoundment alternative essentially equivalent to below-grade burial from the standpoint of resisting erosion and achieving long-term isolation. Among the siting and design features (Figure 6-17) that would, taken in combination, help achieve this objective are (1) siting of the impoundment where the upstream drainage area is small and where the topography shelters the face of the tailings dam from the wind, (2) constructing the dam incorporating features accepted as standard geotechnical engineering practice (e.g., earthen dam with clay core, appropriate slope), (3) incorporating features to cause deposition of sediment on the tailings area from any runoff, and (4) creating gradually sloped embankments during final reclamation, using appropriately thick cover layers, and stabilizing the surface with a continuous vegetative cover or rip-rap as appropriate to minimize erosion potential. This approach incorporates those features that would make above-grade disposal reasonably equivalent to that provided by the below grade concepts (alternatives 2 and 3).

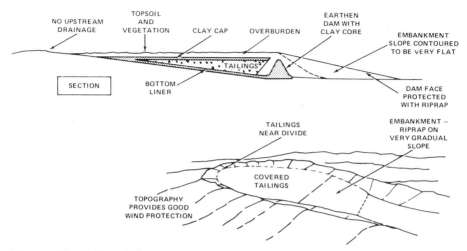

Figure 6-17 Tailings Disposed above grade with special siting and design features. Source: (NRC 1979).

It is apparent that the design of tailings impoundments will continue to evolve to achieve the objectives of the regulations and ensure long-term isolation of the waste from the biosphere.

6.10 GEOLOGIC REPOSITORIES

Conventionally mined repositories in deep underground geologic formations have been identified as the preferred disposal method for spent fuel, high-level waste, and transuranic elements. This decision followed investigation in a variety of disciplines and the preparation and publication of an environmental impact statement discussing the anticipated effects on public health and safety and the environment of managing this waste (both commercial and defense waste). There are several main reasons why this technology was chosen (rather than one of the alternatives discussed in succeeding sections):

1. The long distance from the surface provides shielding from direct exposure to the highly radioactive elements contained therein.
2. The long distance and the methods used to seal any excavated areas reduce the probability of inadvertent contact with the waste in the future and make intentional removal (for whatever reason) much more difficult.
3. Appropriately sited repositories will have little, if any, water in contact with the waste. Therefore, the potential exposure via the groundwater pathway is minimized.
4. Geologic phenomena occur over time scales long compared to the length of

time waste isolation is required. Therefore, projections of repository performance during this period can be made with more confidence than could similar projections for other disposal concepts.

It is important to remember that waste isolation is achieved by the combined effects of several different components of the overall disposal system. The main components are depth of repository below the surface, properties of the host rock, tectonic stability of the area, the hydrologic regime, potential for mining resources from the area, and the multibarrier (engineered and geologic safety features of the system. The multiple barriers include:

The *waste form* itself which is designed to retain its integrity and leach resistance for long times under the anticipated heat, radiation, and chemical conditions that will occur in the repository.

The *waste packaging* that is chosen to reduce the interaction between the emplaced waste and the host rock and any water that may be present.

The low permeability *host rock* that delays the time at which emplaced waste contacts groundwater and the rate at which any isotopes would migrate from the disposal area.

The *shaft sealing* methods that prevent the intrusion of surface water after the disposal site is no longer in use or being actively monitored.

The *monitoring system* that will be in use to verify that site conditions postclosure are the same as those predicted by the computer modeling system and are anticipated to remain that way for the thousands of years of isolation to which the waste is subject.

6.10.1 General Repository Description

Details of the repository design will vary somewhat if spent fuel, high-level waste, and transuranic wastes are disposed together or in separate facilities. This reflects different heat and radiation loads per package of fuel or waste and different container sizes for TRU. The description given here is based on a once-through fuel cycle with spent fuel disposed as waste.

As planned, the geologic repository will be established at a depth of between 2000 and 3000 ft below the land surface. The exact depth will depend upon site characteristics and the characteristics of the disposal medium (rock type in which the waste is placed). In addition to the isolation from the surface, it is important that the rock conditions above, below, and around the waste emplacement be as continuous as possible to retard any movement of radionuclides. Figure 6-18 is an artists's conception of a likely repository design. The depth of waste emplacement (not to scale on this figure) can be put into perspective by realizing that it is similar to the combined heights of the Empire State Building and the Washington Monument.

As was discussed in Chapter 3, the host rock should be strong and

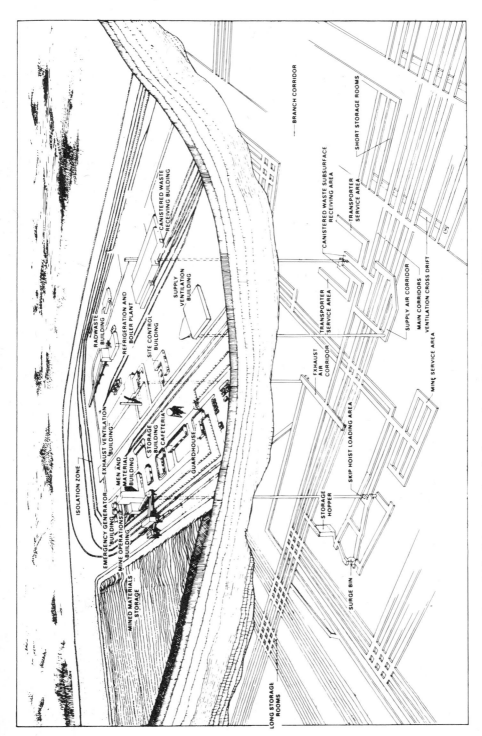

Figure 6-18 Schematic of geologic repository design. Source: (DOE 1980).

ISOLATION ZONE

MINED MATERIALS STORAGE

EMERGENCY GENERATOR BUILDING
EXHAUST VENTILATION BUILDING
MINE OPERATIONS BUILDING
MEN AND MATERIAL BUILDING
RADWASTE BUILDING
REFRIGERATION AND BOILER PLANT
SITE CONTROL BUILDING
STORAGE BUILDING
CAFETERIA
GUARDHOUSE
CANISTERED WASTE RECEIVING BUILDING
SUPPLY VENTILATION BUILDING

STORAGE HOPPER
SURGE BIN
SKIP HOIST LOADING AREA
EXHAUST AIR CORRIDOR
TRANSPORTER SERVICE AREA
MINE SERVICE AREA
VENTILATION CROSS DRIFT
MAIN CORRIDORS
SUPPLY AIR CORRIDOR
CANISTERED WASTE SUBSURFACE RECEIVING AREA
TRANSPORTER SERVICE AREA
SHORT STORAGE ROOMS

BRANCH CORRIDOR

LONG STORAGE ROOMS

impermeable, and have good thermal conductivity and high radiation resistance. The main media investigated by the Department of Energy were salt (bedded and domes), basalt, tuff, granite, and shale.

Each of these materials has particular advantages with respect to desired properties for a high-level waste repository. For example, granite's strength provides better resistance to weathering than other materials, whereas salt's ability to deform and self-seal as a result of the heat transferred from the waste reduces the pathways for fluid flow. The result is that any one of these media could potentially serve as an effective site for a geological repository because other parts of the system would be designed to maximize the rock's advantages and mitigate its less desirable qualities.

The interaction between the host rock, any fluid that might be present, and the waste itself will affect the time at which migration of isotopes might begin and the rate at which it would proceed. The source term of waste available for migration will vary as a function of time as described in Chapter 2 because of the decay of the emplaced isotopes and the buildup and decay of their daughter products.

The needed underground area (excavated using standard room and pillar techniques) is expected to be about 2000 acres. Room and pillar excavation involves removal of rock along corridors or tunnels with structural stability provided by leaving host rock in place at predetermined intervals. Access to the disposal area would be through several shafts. The shafts would initially be used to remove material excavated from the underground disposal area. Separate shafts would be provided for the canistered waste, the ventilation exhaust, and for personnel and materials access. In some rocks, an additional shaft for removing material mined underground may be required. Excavated material would be stored in piles on the surface during facility operation. At closure, about half the excavated material is likely to be used as backfill for the underground mined areas and the access shafts. The remainder would either be stabilized in place or removed to restore the site surface conditions. The actual actions required will depend on the site conditions (both above ground and below ground) and the host rock materials. For example, excavated salt would provide a greater environmental impact due to runoff and wind dispersion than would rocks such as granite or basalt. Surface activities would be restricted during operational and institutional control in an area about four times as large as the excavated underground area to minimize the potential for nearby activities that could have an adverse impact on the performance of the repository.

The amount of waste that can be disposed in a given area is limited by heat transfer between the waste package and the host medium and the heat capacity of the rock itself. That is, the rate of heat transfer must be sufficient to maintain the waste (or spent fuel) temperatures below levels at which waste form and package integrity might be compromised, and the amount of heat absorbed by the host rock must not be so high that fracturing (leading to loss of room and pillar integrity for nonsalt repositories and excessive surface uplift in salt) would result. Because of these limitations a mined repository in salt would require 2.4 times the amount of underground area as a repository in granite or basalt able to accept the same amount of waste (DOE 1980). Since the conceptual design is based on equal underground area, the repository in salt would be able to receive only about 42% of the waste emplaced in granite or basalt. In actual practice, the usable underground area would be determined by the conditions of the host rock formation. It is unlikely that underground areas would be the same for different sites regardless of the host rock.

The rate of heat generation of a given spent fuel or high-level waste canister decreases as a function of time due to radioactive decay. This relationship was illustrated in Figure 2-2. As a result, the total capacity of the repository increases as the age of the emplaced waste increases. The rate and amount of increase vary with type of host rock as shown in Figure 6-19.

As illustrated in Figure 6-18, surface facilities include many structures and areas common to a conventional mining operation. The single largest area is that required to store the rock removed from the underground excavation. The amount of material removed will vary by about a factor of 3 (from about 30 to 90 million metric tons) for the several rock types studied. This is due to the fact that the greater rock strength in granite and basalt relative to salt enables stable excavations to be made with smaller pillars and wider rooms. There is correspondingly more area available for waste emplacement in these rocks than in salt. Other buildings include administration, ventilation supply and exhaust for underground working levels, onsite boiler and refrigeration plant, and maintenance facilities. In addition, structures will be needed for receiving the canistered waste, treating any radioactive waste produced onsite (for example, due to canister decontamination or repackaging), moving personnel and material between the surface and the disposal level, a guardhouse, and a health physics and medical facility. Any surface structure in which radioactive wastes are handled is designed with pressure differentials that direct any leakage from areas of lower to higher contamination potential. That is, the entire structure is maintained at less than atmospheric pressure so that air will leak in rather than radioactive materials leaking out. Exhaust air is filtered prior to release. Planned ancillary facilities include rail connections for shipment of incoming spent fuel and waste containers and other needed supplies (for example, heavy construction machinery and coal to fuel the facility boiler and refrigeration facilities).

Canistered waste is emplaced in holes in the floor of the disposal area. Hole spacing is determined by both thermal and mechanical integrity limits and is expected to be about 6 ft center to center. (There is a possibility that the relatively lower heat-generating boiling water reactor spent fuel canisters would be emplaced horizontally in trenches dug along the sides of the disposal areas in some host rocks.) All waste handling is performed remotely to minimize worker exposure. The conceptual design provides that waste emplaced during the first 5 yr of repository operation would be readily retrievable. This is accomplished by lining the emplacement hole with a steel sleeve and covering it with a removable concrete plug. This period is intended to span the complete mining of the repository working area. Therefore, any conditions that might adversely affect the facility's ability to meet the performance criteria of 10 CFR Part 60 would

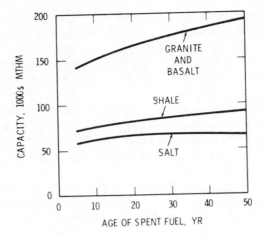

Effect of Spent Fuel Age on Once-Through Cycle Repository Capacities

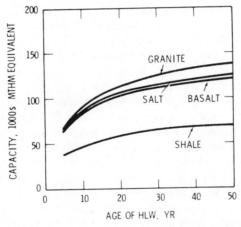

Figure 6-19 Repository capacity for different host rocks as a function of waste age. Source: (DOE 1980).

have been identified while the waste was still in a form that permitted removal using the same equipment as for emplacement. Further, interaction of the emplaced waste with the host rock, particularly waste and local host rock temperatures, can be monitored to document conformance to the predictive modeling. Long-term operation (after the initial confirmatory period) is based on emplacing the waste in unlined holes and backfilling the emplacement rooms with previously excavated material as the rooms are filled. Waste emplaced under such conditions would be recoverable, but at much greater cost in time and money. Underground corridors, transfer areas, and shafts would also be backfilled with excavated material combined with other sealing agents at the end of the repository lifetime.

6.10.2   Waste Isolation Pilot Plant

As described in Chapter 3.4, the Waste Isolation Pilot Plant (WIPP) is a mined geological repository in the bedded salt of the Permian Basin near Carlsbad, New Mexico at which defense transuranic waste will be disposed. The facility is subject to EPA's standards in 40 CFR Part 191 and DOE regulations but an NRC license is not required. Acceptance of waste beginning in late 1988 will enable data to be collected on actual operating conditions at geologic repositories including performance of waste packages, emplacement equipment, and monitoring systems. In particular, the effects of some water in the disposal area (the brine influx condition) will be evaluated before full-scale operation is authorized. Similar conditions have been recognized as occurring and of interest in salt deposits whether for high-level or low-level waste emplacements and have been considered by DOE and its contractors (DOE 1983), but the WIPP facility will provide the first field test in the United States.

6.11   UNDERGROUND INJECTION DISPOSAL WELLS

Use of underground wells for liquid waste disposal is a technology originally developed by the oil industry. Injection of liquid radioactive waste into underground wells for disposal has been investigated as a possible alternative or complement to a geological repository for portions of the waste stream. The two basic variations of the technology that have received the most attention are deep well injection and shale grout injection.

Both high-level waste and remote-handled transuranic waste are candidates for this type of disposal because the waste must be in liquid form, relatively small in volume, and require long-term isolation. It offers the benefit of providing disposal generally at depths greater than the geologic repository, thus reducing the probability of breaching the disposal area by surface activity. It also eliminates the need for solidification and packaging of the waste. However, it is incompatible with the multiple engineered barrier approach to isolation. Isolation results from characteristics of the waste form and geologic formation. Because it is an established technology and would not require extensive equipment and materials development, it is likely to be a relatively inexpensive method of disposing of candidate wastes. It would be particularly advantageous if subsurface characteristics were such that waste generated at a surface facility could effectively be disposed "onsite" without requiring intermediate transportation. This was the case with RH-TRU and other waste disposed by shale grout injection at the Oak Ridge National Laboratory in Tennessee. At Oak Ridge almost 2 million gallons of waste containing over half a million curies of ^{137}Cs, over 35,000 Ci of ^{90}Sr, and other radionuclides was injected over 10 yr. For the case of spent fuel disposal by this method, a surface reprocessing facility would be required.

6.11.1   Design Concepts

Deep Well Injection.   Deep well injection involves pumping acidic liquid waste to great depths (up to 16,000 ft). The receiving area is a porous or fractured rock area such as a depleted hydrocarbon reservoir or zone of natural or induced fractures. Isolation of the injection zone below freshwater aquifers might be accomplished by having intervening impermeable formations such as shale or salt. Thermal capacity of the injection zone would be the limiting criterion both for rate of injection and total amount of material so disposed. Waste concentration would be altered to meet the limiting conditions based on the heat generation criterion mentioned above. In this technology the goal is to maintain the host rock at sufficiently low temperatures (below 100°C) that the waste will not react chemically with the mineral structure of the host rock. Since 60–75% of the heat initially generated in HLW is due to the fission products ^{90}Sr and ^{137}Cs, removal of these isotopes can result in an increase in the amount of long-lived radionuclide bearing waste that the formation could accept. Different disposal methods could be used for the relatively short-lived ^{90}Sr and ^{137}Cs. The fact that the waste remains in liquid form provides the possibility of at least partial recoverability by pumping at a later date, if necessary. Some of the contained waste is expected to adsorb onto the structure of the host rock and would be irrecoverable. Further, depending on local waste chemistry and rock mineral content, there is a possibility of unpredicted reconcentration resulting in thermal "hot spots" or the potential for criticality if plutonium were sufficiently concentrated.

Shale Grout Injection.   Shale grout injection involves pumping neutralized liquid waste or an irradiated fuel slurry mixed with cement, clay, and other additives into an impermeable shale formation. High-pressure water is used to initiate fracturing in the shale (hydrofacture). Continued pumping of the waste grout would extend the fracture and form a thin, horizontal layer that would solidify shortly after injection (on the order of several hours). Additional layers would be formed within the same formation by subsequent injection. Waste would be injected at depths of 1000–1600 ft where the shale formations can be demonstrated to result in horizontal fractures that would retain the waste within the rock formation. The grout mix is designed to achieve rapid solidification and high leach resistance. In combination with the low rates of groundwater flow, high retentive capacity for radionuclides in the minerals in the shale, and distance from the surface, the waste form and shale injection provide a high level of confidence that the waste will remain isolated from the biosphere for long periods of time.

Thermal content would be the controlling factor in the amount of material (number of parallel grout layers) that could be disposed at a given site by shale grout injection just as it was for deep well injection. Because of the relative immobility of the grout, there would not be concern over reconcentration or subsequent diffusion of waste through the rock pores. On the other hand, the waste would not be recoverable without very difficult mining procedures.

6.11.2 General Repository Description

Surface facilities at the site would include receiving facilities to monitor and inspect incoming HLW, spent fuel, and TRU shipments. Spent fuel would be reprocessed and the product acid solution would be the waste form to be disposed. It is at this step that partitioning of the high-heat-producing isotopes of ^{90}Sr and ^{137}Cs could be performed. Waste is stored in stainless-steel tanks prior to emplacement. Waste would be moved from the reprocessing facility to storage tanks and the injection facility through an underground pipeway system consisting of double concentric piping within a concrete shielding tunnel. Leak detectors would be located within the pipe annulus. For the shale grout injection method, additional high-pressure pumps for shale fracturing and grout mixing would also be required. Injection would be performed within a concrete and steel confinement building.

Site size is determined by subsurface conditions. Conceptual designs are based on an initial injection area of 3140 acres. Actual formations utilized might vary from this size. The requirement for additional area between the emplaced waste and site boundary would result in a larger land area requirement. Drilling rigs similar to those used in oil and gas exploitation would be used to drill the injection holes. Sealing is also expected to utilize conventional technology with holes filled with some combination of rock, soil, clay, cement, or other materials that is compatible with site conditions and will provide long-term isolation of the emplaced waste.

6.12 OCEAN DISPOSAL

6.12.1 Ocean Disposal of Low-Level Waste

LLW was disposed at five sites in both the Atlantic and Pacific Oceans from 1946 to 1969. The waste was primarily produced from Department of Energy and defense projects. In general, the material was incorporated into a cement matrix within 55-gallon steel drums. The drums were emplaced on the ocean bottom. Records indicate the estimated number of containers (a total of almost 90,000), undecayed radioactivity content (over 95,000 Ci), the amount at each site, the time period during which the site was used, and the location by latitude and longitutude. In 1970, the United States ceased ocean disposal of radioactive waste. Ocean disposal has been proposed as particularly useful for large volumes of very low specific activity waste such as that produced by decontamination and decommissioning of structures or excavation of slightly contaminated soil from residences. However, it is not expected that this technology will be employed again in this country.

6.12.2 Subseabed Disposal

Subseabed disposal continues to be investigated as an alternative to a land-based geologic repository for long-term isolation of high-level waste, spent fuel, and other waste identified by the Nuclear Regulatory Commission as requiring similar degrees of isolation. The concept involves emplacement of the waste in sedimentary deposits on the ocean floor. It offers the benefits of geologic stability on the order of millions of years, soil conditions with a high sorptive capacity for most isotopes present (iodine and technetium are the exceptions), low biological productivity at the ocean floor, slow interchange between the bottom water and water closer to the surface, and an effective barrier to human intrusion as a result of the distance to the ocean surface.

General System Description. As proposed, the system provides multiple barrier isolation of spent fuel, high-level waste, and fuel cladding hulls. These waste types were chosen for initial study because of the combination of high radiation and heat generation levels in the short term and long-lived isotopes requiring long-term isolation from the biosphere. In addition, the waste concentrations and packaging were amenable to the necessary transportation and emplacement processes. A decision to dispose of other wastes, such as remote and contact-handled transuranic wastes, would primarily reflect comparative economics rather than technical suitability.

The areas under consideration for such disposal are known as mid-plate, mid-gyre sites (MPG). A plate is a portion of the earth's crust that moves as a whole over the lower molten layer. A mid-plate site is remote from both seismic and volcanic processes associated with plate movement and formation. A gyre is a surface circular water mass. The areas under consideration are at the center of these bodies where movement of the overlying water column is relatively slow. Initial investigations are being conducted in both the North Atlantic and North Pacific oceans. The sites have been chosen based on several criteria developed by international scientists in the Subseabed Disposal Program:

1. Geologic and climatic stability—Stability has been observed for periods of over 20,000 yr based on examination of cores taken from these areas.

2. Uniform sediments over wide areas—Cores taken from widely separated locations were analyzed to identify the depth at which magnetic field reversals occurred, a method of dating the individual horizons on the core. Initial results indicate a uniform rate of deposition for periods exceeding one million years. These conditions contribute to the ability to develop and apply models that predict the behavior of emplaced waste with a high level of confidence.

3. Future conditions are predictable—Based on investigation of conditions over several millions of years (determined by examing sediment cores for

parameters such as sedimentation rate and type of material deposited) and knowing the rate and direction of the movement of the plate on which the candidate areas are located, investigators are able to compute the location and sedimentation conditions for various times in the future. The areas under investigation are expected to be subject to continuing buildup of fine sediment that is insensitive to climatic changes such as those due to glacial cycles and resistant to fracturing for times on the order of millions of years.

In addition to the fundamental geologic conditions needed for a disposal site, several other constraints have been applied to reduce the probability of humans coming in contact with waste emplaced in the ocean floor. These include avoidance of areas in which fishing is prevalent, commercial shipping lanes, undersea cables, and areas with potential for resource application (in particular, ferromanganese nodules that might be of interest because of their copper and nickel content). These restrictions are intended to reduce the probability of interaction during transport and emplacement as well as any effect on monitoring instrumentation systems after emplacement.

Several emplacement methods are being investigated and are illustrated in Figure 6-20. These include free fall of the waste from the transport ship at the ocean's surface in specially designed canisters that will increase penetration of the canister into the sediment, control of the canister's descent by a winch to a predetermined height above the ocean floor and free fall from there on, drilling into the sediment and emplacing the waste canisters in the drilled hole, and constructing a trench and emplacing the canisters therein with backfill of disturbed material.

Continuing studies of this disposal method include verification of models developed to predict the physical and chemical performance of the sediments and the waste over long periods of time and with heat and radioactivity contents

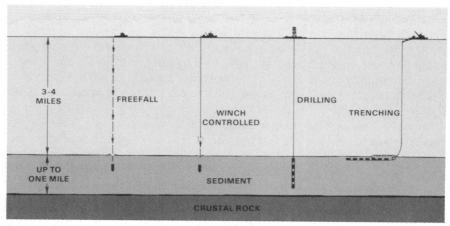

Figure 6-20   Several proposed methods for emplacing waste in a subseabed repository. Source: (DOE 1980).

expected to exist in the emplaced wastes. Once decisions have been reached about the environmental feasibility of the method, engineering studies will be undertaken to determine the best methods for transport, emplacement, and monitoring of the waste and the disposal site. Support facilities would include those needed to ship the waste from the point of generation to a facility at which it would be packaged for the transport to the ocean disposal area. Such demonstrations are not expected to be performed until after the turn of the century.

6.13   SPACE DISPOSAL

Fundamental to the concept of space disposal of radioactive waste is the ability to remove the material from the earth by placing it into orbit around the sun where it would remain for about one million years. The costs of handling and transporting the waste in space are much higher than for land-based alternatives and therefore waste preparation would involve concentrating the waste as much as possible. In effect, this means that reprocessing would be required for disposal of the radioactive materials in spent fuel by this technology. In its environmental review of alternative technologies for high-level waste disposal, DOE characterizes the advantage of space disposal as the ability to place the material where there would be "no long-term risk or surveillance problem as in terrestrial alternatives." The comparison continues, however, that "the risk and consequence of launch pad accident and low earth orbit failure must be compared to the risk of breach of deep geologic repositories" (DOE 1980).

6.13.1   Design Concept

The fundamental system components being considered for space disposal are outlined in Figure 6-21. As indicated in this figure, separated high-level waste is fabricated into a cermet matrix (a homogeneous mixture of ceramic particles in a metallic phase) that would be placed within a shielded canister suitable for loading into a specially designed space shuttle. Once in low earth orbit, the waste would be transferred to a separate vehicle that would transport it and insert it into orbit around the sun. These orbital operations are illustrated in Figure 6-22. It is estimated that transfer from earth orbit to solar orbit would take about 163 days to accomplish.

Several safety questions unique to the space disposal concept are the need to shield the waste material until it is transferred to solar orbit, the need to protect the waste from conditions to which it might be subject during unanticipated reentry into the earth's atmosphere, the ability to develop a waste form and package that would give a high level of confidence that the waste would not disperse under any of the conditions to which it might be subject during launch and transfer, and the ability to track the waste during the launch and transfer operations prior to placing it in solar orbit.

The DOE evaluation indicated no insoluble technical problems with such a

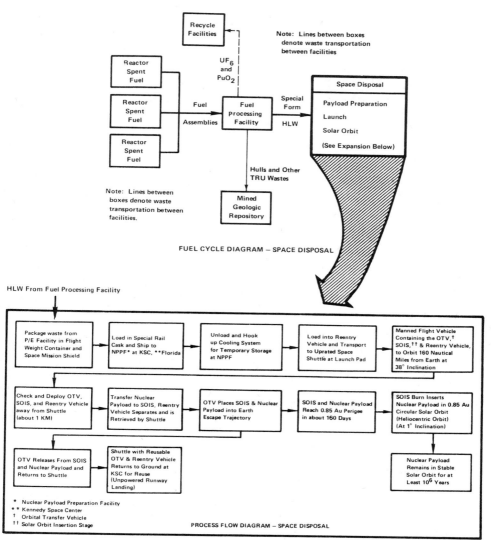

Figure 6-21 Space disposal system components. Source: (DOE 1980).

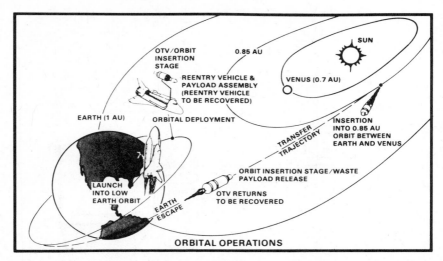

Figure 6-22 Orbital transfer operations for space disposal. Source: (DOE 1980).

system utilizing the cermet waste form, a protective reusable reentry vehicle designed to survive a land or water impact after launch, and reusable equipment such as the orbital transfer vehicle. Launch trajectories would be chosen to avoid having the shuttle fly over land immediately after takeoff to minimize the potential for affecting large numbers of people if an accident occurred immediately after takeoff.

Based on more recent experience with the shuttle program in general and the substantially slowed pace of development and application of launch capabilities, it appears that space disposal will not be a reasonably available alternative for high-level waste disposal until well after 2000. Further, substantial development would have to be performed on the various protective features necessary to avoid having the waste disperse in the atmosphere or on land or water upon either reentry or launch pad failure.

6.14 OTHER DISPOSAL CONCEPTS

Several other disposal technologies have been proposed as alternatives to conventional geologic repositories. These include disposal in the ice sheets, very deep hole disposal, rock melt with liquid or slurry forms of high-level waste, and transmutation of the very long-lived radionuclides into shorter lived or stable isotopes. Each of these technologies was evaluated as a possible alternative to conventional geologic repository disposal for high-level waste (DOE 1980). They were determined to require much more technological development and in some cases to provide lower safety margins than would be achievable at a geologic repository. Thus, they were considered as possible technologies for subsequent application but were not the preferred action for the first or second repository. The major features of a disposal system utilizing one of these technologies and the advantages and disadvantages of each are discussed in the following paragraphs:

6.14.1 Very Deep Hole

Very deep hole disposal involves use of drilling rigs to construct a shaft up to about 30,000 ft deep and emplacement of spent fuel and high-level waste in the shaft. The shaft would be lined throughout its depth and seals would be placed between different waste layers and above the emplaced waste. The incentives for such disposal include

- the waste would be further isolated from the biosphere;
- it would be less likely to come in contact with groundwater;
- should such contact occur, the time over which water transport would occur would result in sufficient radioactive decay that almost no radioactive material would be left when it reaches the biosphere;
- no climatic changes or surface processes are likely to disturb the waste.

Because of the need for very strong rocks with little groundwater contact, it is expected that very deep hole shafts would be sunk in crystalline rock. Most experience with such deep drilling conditions has been accumulated in sedimentary rocks. Substantial technology development would be required to extend currently existing drilling technology to the depths and diameters needed to emplace spent fuel and high-level waste in crystalline rock. No data exist on the long-term performance of anticipated waste form and packaging materials at the temperature, pressure, and chemical conditions that are likely to exist in a very deep shaft.

In addition to the need for technology development, the system does not conform to several of the fundamental objectives of current criteria because there is essentially no method for retrieving the waste after emplacement, and the elaborate multiple barrier and monitored performance features of other technologies are not easily adapted to this technology.

6.14.2 Rock Melt

Rock melt technology involves placing liquid or slurry high-level waste and remote-handled transuranic waste into underground cavities. These two waste streams were chosen because they would not involve accidental criticality risks and they contain substantial heat-producing radioisotopes. The thermal power in the waste acts to evaporate the water present and further actually to melt the surrounding rock. As decay proceeds and thermal power decreases, the rock waste mixture will resolidify (about 1000 yr after emplacement) and form a homogeneous, insoluble solid that will contain the waste for the remaining time needed to protect the biosphere from the contained materials with very long half-lives.

The rock melt concept provides a method for direct disposal of the candidate waste streams without an interim solidification step and the possibility that the eventual soldified form would be more stable than other alternatives. It does not, however, provide multiple barriers between the waste and the biosphere. Site characteristics and performance would be very difficult to monitor. The waste would be essentially nonretrievable. Also, because full solidification would not be achieved until about 1000 yr after emplacement, it is difficult to predict that there will be ongoing responsible authority to monitor facility performance and undertake corrective actions if they are indicated. These factors make development of such a facility very unlikely.

6.14.3   Ice Sheet

Ice sheet disposal would be applicable to high-heat generating wastes that are generally assumed to be packaged and emplaced near the surface of the ice sheets in the Antarctic or Greenland. The method would be particularly applicable to separated high-level waste. Spent fuel could also be disposed by this manner but the rapid decrease in heat generation after removal from the reactor means that relatively fresh (1–2 yr storage after removal) fuel is the best candidate. Emplacement could involve putting the fuel near the surface of the ice such that

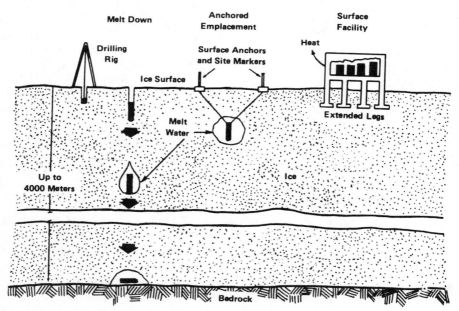

Figure 6-23   Alternative methods for emplacing waste in an ice sheet repository. Source: (DOE 1980).

subsequent heat generation melts the surrounding ice and the waste package sinks further into the ice sheet until it reaches bedrock. Initial investigations indicate that this might occur in 5–10 yr after emplacement. Alternative emplacement methods, illustrated in Figure 6-23, involve anchoring the waste so that it essentially remains suspended in the ice (which would enable the waste to be retrieved at some later time and the condition of the waste canister could be monitored after emplacement) or constructing a surface facility from which decay heat would be transferred to the surrounding atmosphere.

The ice sheet disposal concept has the advantage of providing an extensive heat-transfer area for thermally hot waste materials. It also utilizes land that is unlikely to be particularly attractive for other uses and it can accommodate multiple barriers between the waste and the biosphere. The disadvantages include the fact that the interaction between the waste package and the environment in the long term (in particular, the potential for changes in the rate of movement of the ice mass and possibly related worldwide climatic changes) is not well understood. The waste would be essentially irretrievable and operations would be very difficult under prevailing weather conditions. Further, at the present time, use of the Antarctic ice sheet for radioactive waste disposal is specifically prohibited by the Antarctic Treaty of 1959 and different agreements would have to be negotiated with the other signatories of that treaty (which is subject to renewal in 1989) should such a disposal option be pursued. Similarly, a treaty would have to be negotiated with Denmark (of which Greenland is a territory) if the Greenland ice sheet were to be used.

6.14.4 Transmutation

Transmutation would shorten the required containment period for high-level waste by changing the long-lived transuranic nuclides contained therein into shorter lived or stable isotopes. The degree of confidence with which the behavior of the disposed waste can be predicted increases as the required containment time is reduced. This is accomplished by separating the actinides (radioactive elements with atomic number greater than 88) from the fission products in the high-level waste stream and other sources such as the fuel reprocessing plant and mixed oxide fuel fabrication plant. The separated actinides are then mixed with recycled uranium and plutonium and used as fuel in a nuclear reactor. The interaction between the actinides and the neutrons in the reactor over the course of a year results in about 5–7% of these isotopes being fissioned to stable or short-lived isotopes. It is possible that the process could be repeated until all the actinides had been transmuted and no very long-lived isotopes needed geologic disposal.

Transmutation is attractive because it substantially reduces the demands that need to be put on a geologic repository in terms of duration of isolation required. It is estimated, however, that development of such technology (which is actually a method for treating the waste rather than a separate disposal method) would take at least 20 yr because facilities would be required to perform the actinide waste separation or partitioning and to fabricate fuel elements that contained the separated actinides. This latter requirement would involve a major expansion of the process steps at which highly radioactive material was handled. Development of new equipment to maintain occupational exposures as low as reasonably achievable and to minimize releases of radioactivity to the environment would also be necessary. It should also be noted that no fuel reprocessing capability exists in the United States at the present time nor is any planned in the future so the infrastructure for producing and utilizing the separated actinides would have to be developed from scratch.

REFERENCES

(CN 1988) "Future Low-Level Radioactive Waste Disposal Technology: A Burial Site Operator's Concept." Presented by R. J. Dabolt of Chem-Nuclear Systems, Inc. at Waste Management '88, Tucson, Arizona, March 1988.

(DOE 1980) DOE/EIS-0046F, Final Environmental Impact Statement, "Management of Commercially Generated Radioactive Waste." U.S. Department of Energy, Washington, D.C., October 1980.

(DOE 1983) ONWI-242, "Brine Migration Test for Asse Mine, Federal Republic of Germany: Final Test Plan." Office of Nuclear Waste Isolation Report for the U.S. Department of Energy, July 1983.

(DOE 1987) DOE/LLW-60T, Conceptual Design Report, "Alternative Concepts for Low-Level Radioactive Waste Disposal." Prepared by Rogers & Associates Engineering Corporation for EG&G Idaho, Inc. and the U.S. Department of Energy, June 1987.

(NRC 1979) "Generic Environmental Impact Statement on Uranium Milling." U.S. Nuclear Regulatory Commission, Washington, D.C., 1979.

(NRC 1981) NUREG-0782, Draft Environmental Impact Statement on 10 CFR Part 61, "Licensing Requirements for the Land Disposal of Radioactive Waste." U.S. Nuclear Regulatory Commission, Washington, D.C., September 1981.

(NRC 1982) NUREG-0945, Final Environmental Impact Statement on 10 CFR Part 61, "Licensing Requirements for the Land Disposal of Radioactive Waste." U.S. Nuclear Regulatory Commission, Washington, D.C., November 1982.

(NRC 1983) U.S. Nuclear Regulatory Commission, *Disposal of High-Level Radioactive Wastes in Geologic Repositories.* 10 CFR Part 60, 1983.

(NRC 1984) NUREG/CR-3774, "Alternative Methods for Disposal of Low-Level Radioactive Wastes; Task 1: Description of Methods and Assessment of Criteria." Prepared for the U.S. Nuclear Regulatory Commission by the U.S. Army Waterways Experiment Station. R. D. Bennett, *et al.*, April 1984.

(NRC 1985a) U.S. Nuclear Regulatory Commission, *Criteria Relating to the Operation of Uranium Mills and the Disposition of Tailings or Wastes Produced by the Extraction or Concentration of Source Material from Ores Processed Primarily for Their Source Material Content.* Appendix A from 10 CFR Part 40, Domestic Licensing of Source Material, 1985.

(NRC 1985b) NUREG/CR-3774, "Alternative Methods for Disposal of Low-Level Radioactive Wastes," Volumes 2–5, Belowground Vaults, Aboveground Vaults, Earth Mounded Concrete Bunkers, Shaft Disposal. Prepared for the U.S. Nuclear Regulatory Commission by the U.S. Army Engineer Waterways Experiment Station, October 1985.

(NRC 1987)   U.S. Nuclear Regulatory Commission, *Guidance for Selecting Sites for Near-Surface Disposal of Low-Level Radioactive Waste.* Task WM 408-4, March 1987.

(*W* 1986)   "An Integrated System for the Processing, Storage and Disposal of Low-Level Radioactive Waste." Presented by C. W. Mallory of Westinghouse Hittman Nuclear, Inc. at the International Symposium on Alternative Technology for Low-Level Radioactive Waste Treatment and Disposal, Chicago, Illinois, March 1986.

CHAPTER 7

REMEDIATION AND STABILIZATION OF RADIOLOGICALLY CONTAMINATED SITES

The remediation of radiologically contaminated sites can be broadly divided into two categories: (1) the decommissioning of nuclear plants involving the removal of only contaminated core and structural components and decontamination of building and equipment surfaces to remove radioactivity with the remainder of the plant, buildings, and surroundings being nonradioactive; and (2) the cleanup and stabilization of sites that may contain radioactive contamination in buildings, equipment, waste storage piles, and underlying soil. Decommissioning is a well-structured program that generates a range of radioactive waste forms that are disposed of by routine techniques as described in Chapters 5 and 6. After a brief summary of the waste generation aspects of decommissioning regulations, this chapter will focus primarily on the evolving strategies and technologies for the cleanup and stabilization of sites where contamination is extensive and uncontrolled, resulting in a potential for environmental dispersion.

7.1 HISTORICAL PERSPECTIVE

7.1.1 Decommissioning of Nuclear Facilities

The term "decommissioning" is used to describe the process of taking a plant out of operation and shutting it down permanently. In decommissioning a nuclear power plant after its operating life, it is necessary to remove and/or decontaminate materials, structures, and equipment that are radioactive in addition to the nonradioactive components. More than 95% of the radioactivity, however, is generally located in the components of the plant in and adjacent to the reactor vessel once the fuel and cooling wastes have been removed. In these nuclear plants a comprehensive system of regulations and procedures has been developed to perform the decommissioning, recognizing the potential risk associated with the radioactivity in the materials to be handled.

The radioactivity in the plant is found in two forms: surface contamination consisting of a thin layer of deposited radioactive material ("crud") and induced radioactivity found in the metal and concrete structural material in the core region and surrounding components resulting from neutron activation of these components. The crud is removed by conventional surface cleaning techniques such as brushing and/or scrubbing, sandblasting or high-pressure water jets, chemicals or special cleaning devices, and the resulting liquid and semisolid waste is immobilized and disposed of in the same manner as operational waste. The activated equipment and structures are dismantled or cut away and disposed of in the same manner as routine solid waste from the operational plant. Highly activated core components are dismantled with the aid of remotely controlled equipment and some of this material will be stored for disposal as high-level waste (HLW). Since the major part of the plant is not radioactive, nonrestrictive procedures can be used for decommissioning this portion or it can be used for other industrial purposes.

In general the quantities of waste from a 1000-MWe light water reactor distribute as (ANS 1982) (1) a few hundred cubic meters of highly contaminated material to be ultimately placed in deep underground disposal, (2) a few thousand cubic meters of slightly contaminated material that is disposed of in a manner similar to waste generated in normal reactor operation, and (3) approximately 50,000 m^3 of nonradioactive waste, which can be reused or removed as landfill.

Decommissioning of a nuclear plant can be divided into three stages depending on the amount of radioactive material removed. "Stage 1 decommissioning" involves the removal of all easily accessible radioactive materials with machinery and structures left intact. Stage 1 is used as an interim measure prior to final decommissioning. "Stage 2 decommissioning" involves sealing off those parts of the plant in which the highest radioactivity levels remain for the purpose of allowing for further radioactive decay with time. The less radioactive plant parts are removed after being decontaminated. "Stage 3 decommissioning" involves removing all radioactive materials down to prior background levels to permit immediate use of the remainder of the plant and the site. It is the most extensive and costly decommissioning approach. The choice of stage will depend on comparative economics, site use requirements, and personnel availability.

In the period from 1965 to 1982, approximately 70 nuclear reactors were decommissioned worldwide, including the Elk River, Pathfinder, Piqua, and Peach Bottom plants in the United States. With the large number of plants currently in operation around the world, and with a 30 to 40-yr operating lifetime, it is anticipated that decommissioning of nuclear plants will become a routine activity, as will the disposal of wastes generated during the decommissioning operations.

Although not routine decommissioning operations, the cleanup of the facilities at Three Mile Island (TMI) and Chernobyl after the nuclear accidents represents an extreme form of decommissioning (see Chapter 1).

7.1.2   Remediation of Radiologically Contaminated Sites

Radiologically contaminated sites are sites in which the structures and underlying lands have accumulated excessive levels of radionuclides as a result of past processing or handling operations. In many instances the contamination extends to adjacent properties and may be detectable as elevated radionuclide levels in groundwater, surface water, or airborne pathways. The process of remediation of these sites is distinguished from the previously discussed planned decontamination and decommissioning of nuclear facilities at the end of their useful life in terms of the regulatory and programmatic requirements to clean up the facilities, the nature of the remediation required, and the mechanics of performing the remediation effort. The remainder of this chapter deals with the remediation of radiologically contaminated sites.

The majority of radiologically contaminated sites were either associated with the steps in the paths for uranium and thorium material production during the active years of the Manhattan project, as depicted in Figure 7-1, or in the processing and production of radium and thorium for medical, research, and industrial purposes in the period from 1910 to 1930.

In the former case, the U.S. Army Corps of Engineers Manhattan Engineer District (MED) and its successor, the U.S. Atomic Energy Commission (AEC), conducted programs during the 1940s and 1950s involving research, development, processing, and production of uranium and thorium, and storage of radioactive ores and their processing residues. Those programs were conducted pursuant to the First War Powers Act of 1941 and the Atomic Energy Acts of 1946 and 1954, as amended. Virtually all of this work was performed for the government by private contractors at sites that were federally, privately, or institutionally owned.

Many of these sites became contaminated with low-level radioactivity as a result of this work. When the contracts for MED/AEC activities were terminated, the sites involved were decontaminated according to health and safety criteria and guidelines then in use and applied on a site-specific basis. As we will see, subsequent tightening of these standards has required that further radiological assessments be performed at these sites leading, in many cases, to the need for additional remediation.

The sites that became contaminated as a result of radium and thorium production for commercial and medical purposes have a longer history. Many of these facilities began processing batch quantities of these materials in the early 1900s when the distinctive applications were discovered, and continued production until the applications disappeared. In the absence of specific disposal or health criteria, the residues from these processing operations, contaminated with both radiological and nonradiological constituents, were typically disposed of in on-site pits or surface remains that eventually mixed with the underlying soil.

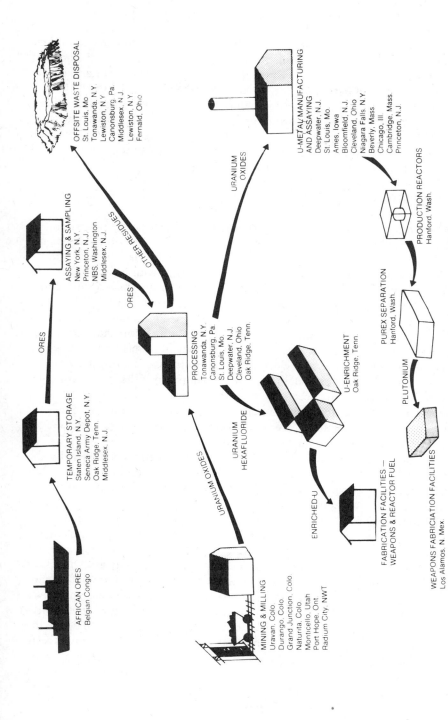

Figure 7-1 Paths for uranium material production during the Manhattan project.
Source: (DOE 1982a).

With time, knowledge of the locations and extent of this contamination was lost and the sites were converted for other purposes. Recent emphasis on the need to rehabilitate sites containing hazardous constituents has resulted in a reassessment of these sites to determine the extent of remediation required.

In addition to these early processing facilities, more recent use of radionuclides in industrial applications (e.g., thorium as an additive to nickel or magnesium for aircraft structural applications) has resulted in slag piles and residues contaminated with low-activity radioactive material at a number of industrial sites.

7.2   CURRENT STATUS OF CONTAMINATED SITES

In the period since the termination of MED/AEC activities, the extent of knowledge on the effects of low-level radiation has progressed and instrumentation for detecting and measuring low-level radiation has become more sophisticated. These factors, in combination with a growing institutional and public concern about the management of hazardous wastes, have resulted in a national effort to identify radiologically contaminated sites and to evaluate the extent of remediation required to restore the sites to (optimally) unrestricted public use. At the same time, the magnitude of this effort has grown as radiological criteria and guidelines governing exposure levels and permissible radionuclide levels in soils after cleanup have become more stringent.

Consequently, significant programs have been initiated at the federal and state levels that will provide the necessary resources to remediate these sites and/or require that the responsible private parties perform the cleanup programs. In many cases private companies are undertaking the remediation of these sites without imposition of governmental directives.

The most far reaching of the remediation efforts is being conducted by the Department of Energy under several program headings. In addition, the program being conducted by the Environmental Protection Agency under authority of the Comprehensive Environmental Response, Compensation and Liability Act (CERCLA) continues to identify radiologically contaminated sites (e.g., Denver Radium Sites) for remediation. These programs are discussed subsequently in this chapter.

Although much progress has been made in identifying and assessing sites potentially requiring some remediation effort, comparatively little actual remediation of these sites has been accomplished to date (1988). This situation derives from a number of factors: the extended litigation in some instances to establish the responsible party(ies), the lack of uniformly accepted standards to define the extent of remediation required, the high costs and lack of government and private funding to achieve an established level of remdiation, the relatively unproven nature of the remediation technology, and a shortage of commercial and government disposal facilities to accept the generally high-volume, low-activity wastes being generated by the remediation programs.

7.3 CHARACTERISTICS OF CONTAMINATED SITES

The similar historical development, processing parameters, and materials handled at a large number of the radiologically contaminated sites have resulted in a set of characteristics broadly common to these sites. The existence of the common characteristics permits the development of remediation approaches that can be readily extrapolated among sites.

Thus, the contaminated sites can be categorized in the following manner:

1. The predominant radionuclides are those in the uranium decay chain consisting of ^{238}U, ^{235}U, ^{226}Ra, ^{230}Th, ^{222}Rn, and ^{210}Pb, or in the thorium decay chain including ^{232}Th, ^{228}Th, ^{228}Ra, and ^{220}Rn.
2. The sites requiring the most extensive cleanup are those containing unstabilized wastes or tailings, which include mining and milling sites, on- and offsite waste disposal sites, and processing facilities.
3. The responsibility for cleanup and stabilization activities lies either with government agencies (i.e., the Department of Energy or Department of Defense), when the facility was either under government control or produced material solely for government use, with commercial organizations, when the production was entirely for private use, or on a shared basis, when the product was used both for private and government purposes.
4. Since the residues from the processing operations also contain metals and processing chemicals, the processing and disposal sites frequently are classified as containing "mixed" wastes composed of both hazardous radiological and nonradiological constituents.

7.4 REMEDIATION PROGRAMS CONDUCTED BY THE ENVIRONMENTAL PROTECTION AGENCY (EPA)

The Comprehensive Environmental Response, Compensation and Liability Act of 1980 (CERCLA), Public Law 96-510, established a program to identify, report, rank, and clean up inactive hazardous waste disposal sites. Since radioactivity is a characteristic of waste that permits it to be classified as hazardous, radiologically contaminated sites are subject to the provisions of CERCLA. The Act created the Hazardous Substance Response Trust Fund ("Superfund") to provide funding for the cleanup of inactive contaminated sites that pose a threat to public health or the environment and where cleanup cannot or will not be accomplished by responsible private parties.

The EPA and individual states have conducted investigations to identify abandoned or uncontrolled hazardous waste facilities. Once these facilities are identified, a Hazards Ranking System, using a model developed specifically for assessing contaminated sites (the Mitre Model), is employed to numerically rank the sites and, thus, permit a quantitative comparison that is used to establish remediation priorities. Appendix C contains the Hazards Ranking System worksheets used in the ranking process for contaminated sites. Initially an Interim Priority List of 160 sites was established, which has subsequently been expanded to more than 700 sites on the National Priority List (NPL). Among these prioritized sites are 34 that are listed as the result of onsite radioactivity (1984). The current listing includes such diverse facilities as the shutdown Maxey Flats, Kentucky commercial low-level waste disposal site, and a grouping of 31 locations in the metropolitan Denver area on which radium was processed in the past, collectively known as the "Denver Radium Site." Since there are approximately 22,000 sites having radioactive materials licenses in the United States, the number of listed radiologically contaminated sites is bound to increase.

The CERCLA remediation program provides that the following sequential steps be conducted on an identified, prioritized site:

1. Assessment of the available data base on the facility and its environs for existing or potential environmental contamination.
2. Confirmation, through a remedial investigation consisting of a program of monitoring and site characterization, to verify the presence or absence of hazardous substances.
3. Engineering assessment (feasibility study) to define and evaluate the potential remediation alternatives through a comparative study of alternative technologies, impact, and costs.
4. Performance of a remedial action program in accordance with an approved remedial action plan for the selected alternative.
5. Verification of the adequacy of the remedial action program through environmental monitoring and documentation of the remediation steps taken.

7.5 REMEDIATION PROGRAMS CONDUCTED BY THE DEPARTMENT OF ENERGY (DOE)

As noted earlier, the progressive application of more stringent radiological criteria for returning contaminated sites to unrestricted use resulted in a reassessment, during the 1970s, of the status of many sites used during the MED project to produce and process uranium and thorium for government use. As a result of this reassessment, a significant number of sites were identified that potentially would require remediation, and programs were defined to accomplish this effort. As the responsible Federal agency, the DOE evolved a Remedial Action Program (RAP), composed of the following components, to clean up these sites:

1. Grand Junction Remedial Action Program (GJRAP)

2. Uranium Mill Tailings Remedial Action Program (UMTRAP)
3. Formerly Utilized MED/AEC Sites Remedial Action Program (FUSRAP)
4. Surplus Facilities Management Program (SFMP)

The parameters of these programs are described in the following:

1. Grand Junction Remedial Action Program (GJRAP)   In the period between 1952 and 1966, several hundred thousand tons of uranium tailings stored at the Climax Uranium Company's tailings pile in Grand Junction, Colorado was removed and used as construction material for a number of residential and institutional projects in the city by incorporation into streets, driveways, foundations of buildings, water pipes and sewer mains, and swimming pools.

In 1966, as a result of increased concern about the effects of radon gas release and groundwater contamination from the dispersed tailings, the use of the tailings was stopped and site investigations were initiated. In 1970, the U.S. Surgeon General issued guidelines for corrective action in Grand Junction based on radon daughter and γ radiation levels (which became the model for subsequent RAPs) and in 1972 Congress appropriated funds for the GJRAP. The GJRAP involves either the removal of tailings or the use of inplace stabilization or ventilation techniques to minimize individual exposure in those structures in which the Surgeon General's guidelines were exceeded. About 1000 structures will be eventually remediated.

2. Uranium Mill Tailings Remedial Action Program (UMTRAP)   Most of the uranium ore produced in the United States from the early 1940s through 1970 was for delivery to the MED/AEC under contracts with private companies on sites located near uranium mines in the west. When processing operations under these government contracts stopped and the mills shut down, millions of tons of uranium mill tailings remained as a residue of the processing operation. At that time, the effects of the low-level radioactivity emitted from the tailings piles were thought to be minimal, and the tailings were not considered to constitute a risk to the public. However, as additional information on the effect of low-level radiation became available, concern about public exposure increased, and increasingly more stringent guidelines were established for those exposure levels. Furthermore, at a number of the shutdown sites, the tailings were not controlled and were spread by wind and water erosion or used in local construction projects (see GJRAP above).

With this background, the AEC successor Energy Research and Development Authority (ERDA) began site investigations in 1975 of inactive uranium mill sites and subsequently identified 25 locations requiring remediation because they were potential environmental and public health hazards. These locations are shown in the map of Figure 7-2 and their operating parameters are provided in Table 7-1 (DOE 1981). Congress then enacted, in 1978, Public Law 95–604, the Uranium Mill Tailings Radiation Control Act, which authorized the Department of Energy (DOE, the ERDA successor) to designate sites needing remedial action,

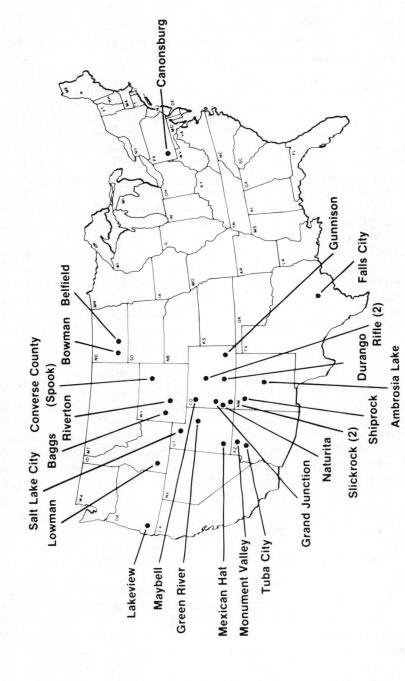

Figure 7-2 Locations of inactive uranium mill sites. Source: (DOE 1982a).

TABLE 7-1 Inactive Uranium Mill Tailings Sites (Engineering Assessment Reports Provided to Congress)

Site	Years Operated	Tons of Tailings (Thousands)	Owner/Operator
Arizona			
Monument Valley	1955–1967	1100	Navajo Nation
Tuba City	1956–1966	800	Navajo Nation
Colorado			
Durango	1943–1963	1555	Ranchers Exploration & Development Corp., Albuquerque, NM
Grand Junction	1951–1970	1900	Shumway, Inc. (purchased portion of site with tailings pile); Bess Investments; State of Colorado
Gunnison	1958–1962	540	Limited Partnership Clarence A. Decker M. Marcus Bishop Roger L. McEachern
Maybell	1957–1964	2600	Union Carbide Corp.
Naturita	1939–1963	704	Ranchers Exploration and Development Corp.[a]; Foote Mineral Company
New Rifle	1958–1972	2700	Union Carbide Corp.
Old Rifle	1924–1958	350	Union Carbide Corp.
Slick Rock (North Continent)	1931–1943	37	Union Carbide Corp.
Slick Rock (Union Carbide)	1957–1961	350	Union Carbide Corp.
Idaho			
Lowman	1955–1960	90	Velsicol Chemical Corp.
New Mexico			
Ambrosia Lake (Phillips)	1958–1963	2600	United Nuclear Corp.
Anaconda-Bluewater	1953–1959	584	Anaconda[b]
Homestake—New Mexico Partners, Milan	1958–1962	1218	United Nuclear Corp.[b]
Shiprock	1954–1968	1650	Navajo Nation
Oregon			
Lakeview	1958–1961	131	Precision Pine (a partnership)
Pennsylvania			
Canonsburg[c]	1942–1957	—	Canon Development Co.

TABLE 7-1 (*Continued*)

Site	Years Operated	Tons of Tailings (Thousands)	Owner/Operator
South Dakota			
Edgemont	1956–1974	2300	Tennessee Valley Authority
Texas			
Falls City	1961–1973	2500	Solution Engineering Co.[d]
Ray Point	1970–1973	490	Exxon Company
Utah			
Green River	1958–1961	123	Union Carbide Corp.
Hite	1950–1958	20	Site covered by Lake Powell
Mexican Hat	1957–1965	2200	Navajo Nation
Monticello	1942–1960	900	DOE
Salt Lake City	1951–1968	1700	Salt Lake County Suburban District— 99 acres[e]
Wyoming			
Converse County	1962–1965	187	D. Hornbuckle—owner of property
Spook Site			Western Nuclear, Inc.— responsible for proper property
Riverton	1958–1963	900	Lome Drilling & Well Service, Inc.; Western Nuclear

Source: (DOE 1981).

[a]Purchased the portion of the site that contains the tailings pile. The remaining portion of the site is owned by Foote Mineral Co.
[b]The inactive mill tailings pile is located on a site under license by the State of New Mexico.
[c]Site was assessed under the Formerly Utilized Sites Remedial Action Program and included by Congress as a designated site in PL 95–604.
[d]Some tailings piles are located on land leased from Lyssy Dairy Farms and Mr. Silvestor Miestroy.
[e]Twenty-nine additional acres are owned by D. Eugene Moenich and David K. Richards.

establish priorities, and enter into cooperative agreements with the host states to establish a RAP at each site. The program is currently (1988) in progress, and involves the performance of the following steps at each prioritized site:

(a) Radiological screening and aerial and ground radiological surveys to characterize site and regional radiological baselines.

(b) Engineering assessments of the existing radiological baselines at the inactive mill tailings sites and on properties in the vicinity of the site leading to definition of existing and potential environmental and public health impact.

(c) Definition and evaluation of remedial action alternatives through a comparative assessment of technologies, health, and environmental impact and costs.

(d) Selection and characterization of a new disposal site for the tailings, if removal is the preferred alternative.

(e) Development of plans and specifications for implementing the remedial actions.

(f) Performance of remedial action programs involving either tailings removal, onsite stabilization, or a combination of both.

(g) Verification of the adequacy of the remedial action program through monitoring and observation leading to a certification that the sites from which tailings have been removed can be released for unrestricted use.

3. Formerly Utilized MED/AEC Sites Remedial Action Program (FUSRAP)   In addition to the inactive tailings piles being remediated under the UMTRAP program, there are other industrial facilities that were involved in the storage of ores, assaying and sampling, processing of U_3O_8, and fuel fabrication for use in the MED/AEC program. Many of these sites also became contaminated with low-level radioactivity as a result of this work. When the work was terminated and the facilities shut down, the sites were decontaminated and remediated based on release and exposure criteria then applicable and, in most cases, released for unrestricted public use. However, as more stringent criteria evolved, the radiation levels at many of those sites were considered to be unacceptable for unrestricted use.

In 1974 the AEC initiated the FUSRAP program to identify these formerly utilized MED/AEC sites and to reassess the amounts of radioactivity present at or near each site. As a result of these historical evaluations and site surveys, 21 sites were initially identified as requiring remedial action (see Figure 7-3) and 14 additional sites were classified as probably requiring remediation. In a number of cases the required remediation extended to adjacent or "vicinity" properties. Table 7-2 (Berlin 1982) lists the definite and probable facilities initially established (1981) and provides estimates of contaminated material volumes. A small number of additional facilities have been added to this list since 1981.

The radioactive material present at FUSRAP sites consists principally of waste by-products from uranium and thorium processing. The by-products contain residual quantities of these elements and their radioactive decay products belonging to the ^{238}U decay series and/or the ^{232}Th decay series. The designated FUSRAP sites can be classified into six major groups (Berlin 1982) based on the types of materials handled and the anticipated radiological constituents remaining at the sites (DOE 1980a; DOE 1980b). The classification system, which is given in Table 7-3, helps establish bounding conditions and permits the majority of FUSRAP sites to be categorized in accordance with their potential radiation sources. The designated FUSRAP sites are categorized in accordance with this classification system in Table 7-4.

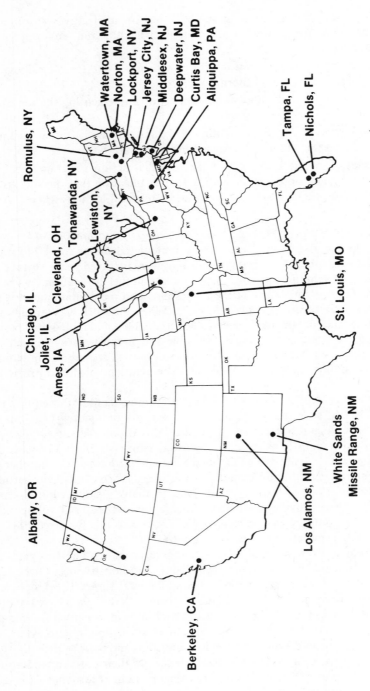

Figure 7-3 Locations of sites requiring remedial actions. Several map locations contain more than one remedial action site. Source: (DOE 1982a).

TABLE 7-2 FUSRAP Sites as of September 1981

	Site	Location	Amount of Contaminated Material (m³)
	Need Established		
1	Kellex Research Facility	Jersey City, NJ	Not applicable[a]
2	Middlesex Sampling Plant and vicinity properties	Middlesex, NJ	60,000[b]
3	Shpack landfill	Norton, MA	3,400
4	Niagara Falls Storage Site vicinity properties	Lewiston, NY	To be determined
5	St. Louis Airport Storage Site vicinity properties	St. Louis, MO	5,000[c]
6	Middlesex landfill	Middlesex, NJ	53,000
7	Linde Air Products	Tonawanda, NY	38,000
8	Bayo Canyon	Los Alamos, NM	2,700
9	E.I. Du Pont de Nemours & Company	Deepwater, NJ	2,100
10	University of California	Berkeley, CA	20
	Need Not Established		
11	Acid/Pueblo Canyon	Los Alamos, NM	1,300
12	Albany Metallurgical Research Center	Albany, OR	2,800
13	Chupadera Mesa	White Sands, NM	To be determined
14	University of Chicago	Chicago, IL	60
	Need Established		
15	Mallinckrodt, Inc.	St. Louis, MO	37,000
16	St. Louis Airport Storage Site	St. Louis, MO	140,000
17	Clecon Metals, Inc.	Cleveland, OH	610
18	Gardinier, Inc.	Tampa, FL	1,500
19	Palos Park Forest Preserve	Chicago, IL	To be determined
20	Conserv, Inc.	Nichols, FL	100
21	Seneca Army Depot	Romulus, NY	310
22	Guterl-Simonds Steel Division	Lockport, NY	270
23	Seaway Industrial Park	Tonawanda, NY	30,000
24	Ashland Oil Company	Tonawanda, NY	37,000
25	Pasadena Chemical Company	Pasadena, TX	To be determined

TABLE 7-2   (*Continued*)

Site	Location	Amount of Contaminated Material (m^3)
Need Not Established		
26 W. R. Grace & Company	Curtis Bay, MD	20,000
27 Harshaw Chemical Company	Cleveland, OH	7,000
28 Iowa State University	Ames, IA	50
29 National Guard Armory	Chicago, IL	20
30 Olin Chemical Corporation	Joliet, IL	230
31 Universal Cyclops	Aliquippa, PA	40
32 Ventron Corporation	Beverly, MA	80
33 Watertown Arsenal	Cambridge, MA	200
34 Ashland Oil Company No. 2	Tonawanda, NY	To be determined
35 Staten Island	Port Richmond, NY	To be determined

Source: (Berlin (1982)).

[a]Decontamination was completed in 1980.
[b]Based on material already excavated.
[c]Contamination associated with the offsite ditches.

After the FUSRAP sites were identified and surveyed, the sites were prioritized for cleanup, and the remediation program initiated. Remedial action was started in 1979 at sites in Jersey City and Middlesex, New Jersey. The program, which is scheduled for completion in the 1990s, generally follows the same sequence of steps at each site from detailed characterization of radiological condition, assessment of existing and potential health impact, definition and evaluation of the viable remedial approaches, implementing the selected remedial actions, and ultimately certifying the satisfactory radiological condition of the site. Remediation may involve removal of wastes to permit use without restriction, or stabilization with restriction on future use of the site.

4. Surplus Facilities Management Program (SFMP)   The SFMP involves the decontamination and decommissioning (D&D) of some 500 radioactively contaminated government-owned facilities or locations that have no current use or have been retired. Many of these facilities are situated on DOE laboratory sites or at DOD installations. They include such diverse systems as reactors, fuel reprocessing equipment, laboratory facilities, tanks, pipelines, waste treatment systems, and areas containing uranium and thorium residues. Figure 7-4 shows the locations of the DOE surplus facilities included in this program. Many of these facilities contain more than one contaminated site, with approximately 75% of the contamined sites located on the Hanford Reservation in Richland, Washington. Major decommissioning activities under this program have been initiated at the Weldon Springs site near St. Louis, Missouri, the Niagara Falls

TABLE 7-3 Classification System for FUSRAP Sites

Group	Title	Principal Radionuclides	General Description
I	Ore/concentrate sites	Uranium-238 decay series	Large quantities of uranium ore or concentrate were handled at these sites. In some instances, the ore/concentrate residual was buried at the site or has contaminated large quantities of soil at the site and in the nearby area. These sites are located in New York, New Jersey, and Missouri and account for approximately 90% of the total volume of FUSRAP contaminated material.
II	Salt/metal processing sites	Uranium-238 decay series	Activities at these sites involved handling uranium salts and fabricating uranium metal. These sites characteristically contain contaminated equipment, structures, and some soil and debris.
III	Phosphoric acid uranium recovery sites	Uranium-238 decay series	These sites were used for extraction of uranium from wet-process phosphoric acid production. The contaminated material consists principally of some soil, processing and drying equipment, and, in some cases, structures contaminated with phosphogypsum. These sites also contain phosphogypsum waste piles.
IV	Thorium processing sites	Thorium-232 decay series	These sites were used principally to process thorium ores and to fabricate thorium metal. The contaminated material consists principally of buildings, equipment, and some soil.
V	Nuclear weapons and/or disposal sites	Fission products and possibly transuranics	These sites were used for weapons testing and/or disposal of wastes from weapons programs. The principal contaminants are fission products and certain transuranics.
VI	Miscellaneous	Tritium and uranium-238 decay series	This group consists of all sites not categorized into one of the five preceding groups.

Source: (Berlin 1982).

TABLE 7-4   Classification of FUSRAP Sites

	Site	Location	Total Volume of Contaminated Material (m^3)
	Group I: Ore/Concentrate Sites		
2	Middlesex Sampling Plant and vicinity properties	Middlesex, NJ	400,000
4	Niagara Falls Storage Site vicinity properties	Lewiston, NY	
5	St. Louis Airport Storage Site vicinity properties	St. Louis, MO	
6	Middlesex landfill	Middlesex, NJ	
7	Linde Air Products	Tonawanda, NY	
15	Mallinckrodt, Inc.	St. Louis, MO	
16	St. Louis Airport Storage Site	St. Louis, MO	
21	Seneca Army Depot	Romulus, NY	
23	Seaway Industrial Park	Tonawanda, NY	
24	Ashland Oil Company	Tonawanda, NY	
34	Ashland Oil Company No. 2	Tonawanda, NY	
35	Staten Island	Port Richmond, NY	
	Group II: Salts/Metal Processing Sites		
1	Kellex Research Facility	Jersey City, NJ	9,600
9	E. I. Du Pont de Nemours & Company	Deepwater, NJ	
14	University of Chicago	Chicago, IL	
22	Guterl-Simonds Steel Division	Lockport, NY	
27	Harshaw Chemical Company	Cleveland, OH	
28	Iowa State University	Ames, IA	
29	National Guard Armory	Chicago, IL	
31	Universal Cyclops	Aliquippa, PA	
32	Ventron Corporation	Beverly, MA	
	Group III: Phosphoric Acid Uranium Recovery Sites		
18	Gardinier, Inc.	Tampa, FL	1,800
20	Conserv, Inc.	Nichols, FL	
25	Pasadena Chemical Company	Pasadena, TX	
30	Olin Chemical Corporation	Joliet, IL	
	Group IV: Thorium Processing Sites		
12	Albany Metallurgical Research Center	Albany, OR	23,000
17	Clecon Metals, Inc.	Cleveland, OH	

TABLE 7-4   (*Continued*)

Site		Location	Total Volume of Contaminated Material (m³)
26	W. R. Grace & Company	Curtis Bay, MO	
28	Iowa State University	Ames, IA	

Group V: Nuclear Weapons and/or Disposal Sites

8	Bayo Canyon	Los Alamos, NM	4,000
10	University of California	Berkeley, CA	
11	Acid/Pueblo Canyon	Los Alamos, NM	
13	Chupadera Mesa	White Sands, NM	

Group VI: Miscellaneous

3	Shpack landfill	Norton, MA	3,600
19	Palos Park Forest Preserve	Chicago, IL	
33	Watertown Arsenal	Cambridge, MA	

Source: (Berlin 1982).

site at Lewiston, New York, and the Mound Laboratories Plutonium Facility in Miamisburg, Ohio.

7.6   INDUSTRIAL REMEDIATION PROGRAMS

A number of industrial firms, having contaminated locations at their facilities, have undertaken their own RAPs. The majority of these sites contain residues from processing or manufacturing operations, which were originally emplaced when the regulations either did not cover the particular radiological constituent or where the regulations have been tightened in recent years to impose new disposal constraints. The impetus to undertake remediation and/or upgrading the disposal sites has generally been the threat of future action by the EPA under CERCLA, or the need to meet current NRC standard to relicense a facility.

A significant example of the need for remediation based on the imposition of new standards are the cleanup of onsite residues, in the form of landfills or waste piles, containing varying concentrations of thorium. Thorium was used as an additive to metals (e.g., thoriated nickel) in small quantities to enhance certain desirable characteristics such as strength or durability. At the time (1940s–1970s) no regulations existed governing the management of wastes containing these thoriated materials, and they were generally handled as solid wastes and disposed of accordingly. The NRC subsequently, in 1981, issued a Branch Technical Position, SECY 81–576, establishing graded disposal criteria as a function of

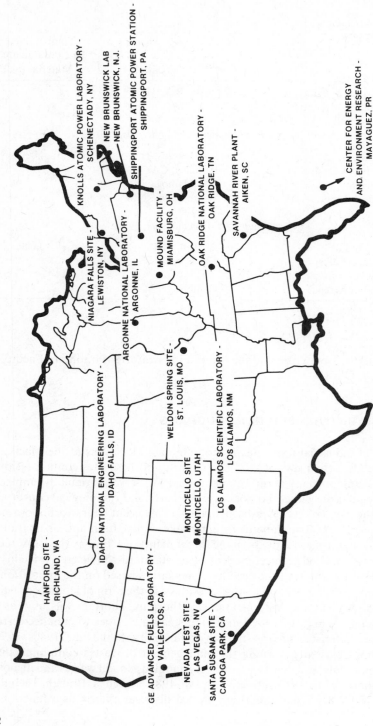

Figure 7-4 Locations of Department of Energy radioactively contaminated surplus facilities. Source: (DOE 1982a).

thorium concentration in the waste. The enforcement of these criteria creates the need for the remediation of these sites.

Although the remediation process does not have the same formal data collection, assessment and reporting requirements as the EPA and DOE programs, the remediation approach is quite comparable since the end result is to require cleanup and stabilization to ensure comparable short- and long-term protection of the environment and the health and safety of the population. This standard dictates a remediation program consisting of a thorough site and vicinity characterization of radiological sources, evaluation of existing and potential impact through dispersion, definition and comparative assessment of remediation alternatives, selection and performance of a remediation approach, and subsequent validation of success of the remediation process.

7.7 REMEDIATION STANDARDS AND CRITERIA

Remediation criteria and standards must be defined prior to assessing the viability of the selected remediation alternatives so that the effort required to achieve these standards can be quantified. However, considerable controversy exists as to the appropriate basis for the standards, and the resultant numerical limits. The controversy in part involves whether environmental cleanup standards, requiring remediation to remove contaminated material above specified levels in soils, should be used or whether health-related standards, requiring remediation to the extent that the contaminated material must not potentially cause exposure limits to be exceeded, should be applied. Since the choice of appropriate standards directly and significantly affects cleanup costs, this issue has frequently become quite heated between government advocates of use of cleanup standards versus industrial advocates of health-related standards.

The historical evolution of the regulatory standards has essentially focused on the remediation of sites contaminated with the constituents of the uranium and thorium decay chains because of the predominance of these radionuclides at these facilities. Thus, the numerical limits encompass the control of these radionuclides specifically in addition to dealing with the general case of all radiological constituents.

The health-related limits that may be applied to a remediation program are

1. Exposure levels (doses) for building interiors primarily used as a measure of the extent of decontamination required, and/or limits of occupational exposure to workers.
2. Environmental exposure levels (doses) beyond the facility boundary for the individual and population. These doses are used to protect the general public against acute or chronic effects of exposure.
3. Maximum permissible concentrations (MPCs) of radiological contaminants in air and water for members of the general public to provide protection similar to (2).

The cleanup standards that may be applied to a remediation program are

1. Concentration limits for radionuclides (e.g., ^{226}Ra) in soil after cleanup to be used as a measure of soil removal required.
2. Emissions of gaseous radionuclides (i.e., ^{222}Rn) from soil surface to be used as a measure of the extent of surface stabilization required.

A body of criteria and regulations has evolved that provides the basis for the limits being used in the site remediation programs for radium-contaminated sites. The applicable provisions are as follows:

7.7.1 Criteria for Radium-Contaminated Sites

Grand Junction Criteria. The Grand Junction Criteria were developed by the U.S. Surgeon General to guide remedial actions with respect to structures built with, on, or adjacent to uranium mill tailings in Grand Junction, Colorado. These criteria, which were the first remedial standards promulgated for radiologically contaminated sites, provide action levels (Table 7-5) for remediation in terms of the health-related standards of external γ radiation levels and indoor radon daughter concentration levels (above background) found within dwellings constructed on or with uranium mill tailings.

U.S. EPA 40 CFR Part 192. Standards for Remedial Actions at Inactive Uranium Processing Sites. The Uranium Mill Tailings Radiation Control Act of 1978 (UMTRCA) charges the EPA with the responsibility for promulgating radiological standards for inactive mill sites. These standards were issued by the EPA in 1983 (EPA 1983a) and subsequently adopted by the DOE for the UMTRAP and FUSRAP programs. They are summarized in Table 7-6 for ^{222}Rn and daughters in the air, and ^{226}Ra in soil and water. They provide both health and cleanup criteria.

TABLE 7-5 Grand Junction Criteria

External γ Radiation (mR/hr)	Indoor Radon Daughters Concentration (WL)	Recommendation
(a) > 0.1	> 0.05	Remedial action indicated
(b) 0.05–0.1	0.01–0.05	Remedial action may be suggested[a]
(c) < 0.05	< 0.01	No remedial action indicated

Source: (AEC 1972).

[a]When DOE-approved data on indoor radon concentrations are available, use radon daughter concentration of > 0.01 WL (Working Level) for dwellings and schoolrooms, $\geqslant 0.03$ WL for the structure. When DOE-approved data are not available, use external γ level of $\geqslant 0.05$ mR/hr and radon daughter concentrations of $\geqslant 0.01$ WL for dwellings and schoolrooms, $\geqslant 0.15$ mR/hr for other structures.

TABLE 7-6 U.S. EPA Standards for Remedial Actions at Inactive Uranium Processing Sites[a]

Radionuclides	Guideline	Action	Condition
^{222}Rn decay products	> 0.03 WL[b]	Required	Within structures
	0.02–0.03 WL[b]	Based on case-by-case evaluation	Within structures
^{222}Rn	> 20 pCi/m^2-s[c]	Required	Over surface of controlled property
	> 0.5 pCi/L[c]	Required	Boundary of controlled property
^{226}Ra	> 5 pCi/g[c]	Required	Average in 15 cm of soil
	> 15 pCi/g[c]	Required	Averaged over 15-cm-thick layers more than 15 cm below surface
^{226}Ra	> 5 pCi/liter[d]	Required	Drinking water
γ Radiation	> 0.02 mR/hr[c]	Required	Within structures

[a]Includes background.
[b]Includes background.
[c]Above background.
[d]EPA limit for combined ^{226}Ra and ^{228}Ra (40 CFR Part 141, National Interim Primary Drinking Water Regulations) regardless of source.

U.S. NRC 10 CFR Part 20. Standards for Protection Against Radiation. These Standards, which are periodically updated, provide both occupational and general public exposure limits for exposure to radionuclides. The limits pertinent to remediation of sites contaminated by constituents of the uranium decay chain are (NRC 1982)

1. ^{222}Rn concentrations in air are limited to 3 pCi/L [0.03 working level (WL) of radon progeny] for an unlimited exposure period (24 hr/day) to a member of the general public (nonradiation worker in an unrestricted area)
2. Gamma exposures are limited to 500 mrem/yr to the whole body for a nonradiation worker in an unrestricted area.

U.S. EPA 40 CFR Part 190. Radiation Protection Standards for Nuclear Power Operations. These Standards apply to radiation doses received by members of the public in the general environment and to radioactive materials introduced into the general environment as the result of operations that are part of the nuclear fuel cycle. The regulations relevant to uranium processing facilities are.

"Annual dose equivalent does not exceed 25 millirems to the whole body, 75 millirems to the thyroid, and 25 millirems to any other organ of any member of the public as the result of exposures to planned discharges of radioactive materials, *radon and its daughters excepted*, to the general environment from uranium fuel cycle operations and to radiation from these operations" (EPA 1982). These standards would not be relevant where the release of radon gas is an environmental concern since radon and its daughters were excluded by the standard in estimating the annual dose limits.

In addition to the above-discussed standards, each of the states has promulgated radiation control regulations that generally are the same as the comparable federal standard.

In general, the quoted standards would be applied in the following manner to sites contaminated by residue produced in uranium fuel cycle operations.

Exposure Levels for Building Interiors. An exposure level of 0.02 WL (above background) from radon decay products would be considered as a limit above which the need for remedial actions exist. The 0.02 WL value is based on the EPA's standard for reduction of radon decay products under the Standards for Remedial Actions at Inactive Uranium Processing Sites. It is also consistent with the Grand Junction Criteria, where a remedial limit of 0.03 WL (including background) is required for structures other than dwellings or schoolhouses, which translates to 0.02 WL above background.

A whole body γ dose of 57 μR/hr (0.5 rem/yr), based on the limits in the NRC regulations 10 CFR 20, is considered appropriate for exposure of a controlled indoor population of nonradiation workers. The more restrictive value of 20 μR/hr for indoor γ radiation in the UMTRAP regulations is relevant to an uncontrolled indoor population of nonradiation workers.

Exposure Levels for Outdoor Air Immersion. A ^{222}Rn concentration of 0.5 pCi/liter in the air outside the boundary of the controlled property is considered to be an appropriate limit. This value also derives from the EPA Standards for Remedial Actions at Inactive Uranium Processing Sites and the DOE RAP standards.

Radon Emission (Flux) from Soil. A radon flux limit of 20 pCi/m²-sec (above background) from the surface of the soil or stabilization cover generally is used as the relevant standard for site cleanup activities. This limit is based on the levels stipulated in the EPA Standards for Remedial Actions at Inactive Uranium Processing Sites and adopted for the DOE RAPs.

Concentrations of Radium-226 in Soil. A limit on soil radium content after remediation of 5 pCi/g (above background) averaged over the top 15 cm of soil and 15 pCi/g averaged over any 15 cm of soil below this has been generally accepted as the criteria. These limits are also derived from the EPA Remedial Action and DOE RAP standards.

These site remediation and health-related limits cannot be applied independently; they are interrelated. For example, the radon flux limit from the soil surface is a function of the ^{226}Ra concentration in the underlying soil and other variables (e.g., porosity, moisture content). Even if the ^{226}Ra soil concentration limits are achieved, a diffusion analysis must be performed to validate that the flux from the surface does not exceed the 20 pCi/m^2-sec standards. Further, the outdoor and indoor radon concentrations in the air are related to the radon flux from the soil surface. A dispersion analysis in air considering the local climatological, topographic, and structural parameters is required to validate that the radon concentration remains below the appropriate standard even if the flux limit from the soil surface is achieved.

For those sites requiring remediation that are contaminated with radiological constituents other than radium and its daughters, different remediation criteria will apply. These sites can be generally categorized as those containing (1) residue of material from uranium or thorium processing operations, and (2) waste material from nuclear power plants and/or other fuel cycle facilities having low levels of fission products and/or actinides.

7.7.2 Criteria for Uranium- and Thorium-Contaminated Sites

Many of the sites formerly used for processing thorium and uranium that are currently contaminated with processing residues contain large volumes of relatively low-activity material. The NRC has recognized that alternatives other than removal of this contaminated soil for offsite disposal would be justifiable based on public health considerations, a weighing of costs and benefits, and the unavailability of commercial burial sites for these wastes. They have identified (NRC 1981) acceptable options for disposal or storage of the contaminated soil, and have defined maximum permissible concentrations in the residues for ^{232}Th and various forms of uranium for each option. These concentrations and options are summarized in Table 7-7.

7.7.3 Criteria for Sites Contaminated with Other Radiological Constituents

This category of contaminated sites is quite limited, applying primarily to the low-level (LLW) burial sites now designated for remediation that accepted nuclear power plant or other industrial or institutional wastes containing fission products, or actinides, and the land underlying or around nuclear research facilities. In general, if these sites are to be remediated under NRC license, or DOE jurisdiction, the relevant standards will be the exposure limits and maximum permissible concentrations in air and water established in 10 CFR 20, Standards for Protection Against Radiation. In addition, where the facility is part of the nuclear fuel cycle, the standards in 40 CFR Part 190, the EPA's Radiation Protection Standards for Nuclear Power Operations, will apply to doses received by the public in the general environment around the facility.

TABLE 7-7 Summary of Maximum Concentrations of Thorium-232 and Uranium Istotopes

Kind of Material	1[a]	2[b]	3[c]	4[d,e]
Natural thorium (^{232}Th + ^{228}Th) with daughters present and in equilibrium	10	50	—	500
Natural uranium (^{238}U + ^{234}U) with daughters present and in equilibrium	10	—	40	200
Depleted uranium				
Soluble	35	100	—	1000
Insoluble	35	300	—	3000
Enriched uranium				
Soluble	30	100	—	1000
Insoluble	30	250	—	2500

[a] Option 1 permits the disposal of acceptably low concentrations of the listed material (based on EPA cleanup standards) with no restriction on the burial method. NRC expects that licensees will minimize the possibility of soil contamination during operations and, if such occurs, reduce contamination to ALARA levels.

[b] Option 2 permits the disposal of certain low concentrations of the listed material with no land use restrictions and no continuing NRC licensing of the materials providing (1) the buried materials will be stabilized in place and not transported away from the site, (2) topographical, geological, hydrological, and meteorological characteristics of the site combine to make the site acceptable, (3) burial depth will be a minimum of 4 ft below the surface, and (4) concentrations will be such to limit individual doses to 170 mrem/yr.

[c] Option 3 permits the disposal of low concentrations of natural uranium ores in areas zoned for industrial use and where the title documents for the land are amended to state that the land contains buried radioactive material and with a covenant that it may not be used for residential building and providing (1) the burial criteria outlined for option 2 are met and (2) concentrations will be such to limit equivalent exposure to 0.02 WL or less. There will be no continuing NRC licensing of the material.

[d] Option 4 permits the disposal of land-use concentrations of the listed materials in areas zoned for industrial use and where the title documents for the land are amended to state that the land contains buried radioactive material and with a covenant that it may not be (1) excavated below stated depths in specified areas unless cleared by appropriate health authorities, (2) used for residential or industrial building, or (3) used for agricultural purposes and providing that (a) the burial criteria outlined in option 2 are met and (b) concentrations will be such to limit individual doses to 500 mrem/yr and, for natural uranium, limit exposure to 0.02 WL or less. There will be no continuing NRC licensing of the burial site.

[e] When concentrations exceed those specified in option 4, the thorium and/or uranium may be stored onsite under an NRC license until a suitable offsite disposal site is available. Radiation doses to individuals during the storage period may not exceed 10 CFR Part 20 limits and will be maintained at ALARA levels.

7.7.4 Impact of Provisions of Superfund Amendments and Reauthorization Act of 1986 (SARA)

Where the remediation of a contaminated site is conducted under CERCLA designation on the NPL, the provisions of SARA, which represents a major overhaul to the original CERCLA law, will affect the standards applied to the remediation process. Section 121 of SARA requires that the remediation of the site "must attain standards and criteria of other federal laws and of state laws if these standards are legally applicable or relevant and appropriate under the circumstances of the release." Thus, the requirements of the federal laws, such as the Safe Drinking Water Act (SDWA), the Clean Air Act (CAA), the Toxic Substance Control Act (TSCA), and the Resource Conservation and Recovery Act (RCRA), as relevant, will apply to contaminant levels onsite and any effluent releases after cleanup. Furthermore, Section 121 expressly calls for remedial actions to attain recommended maximum contaminant levels (RMCLs) established under the Safe Drinking Water Act and water quality criteria established under the Clean Water Act where these requirements are relevant and appropriate.

Section 121 also restricts the situations in which an alternate concentration limit (ACL) can be used as the cleanup standard. An ACL cannot be used instead of a legally applicable standard if, in applying the ACL, a point of exposure is assumed beyond the facility boundary, except for the cases where (1) groundwater enters a surface water at a "reasonable" distance from the facility, (2) there is no statistically significant increase in hazardous constituents in the surface water, and (3) enforceable measures exist to preclude human exposure to groundwater between the facility and points of entry to surface water.

Section 121 also allows the EPA to waive the application of federal and state standards where (1) the remedial action under consideration represents an interim measure leading to a complete remedial action that will attain the standards, or (2) compliance with the standards will result in increased risks to human health and the environment, or (3) compliance with the standards is technically impractical, or (4) the selected remedial action will attain a cleanup standard that is equivalent to the applicable standard, or (5) the use of an applicable standard will not provide a balance between protection of the site in question (for a fund-financed cleanup) and other NPL sites.

The short-term result of the application of these SARA provisions is likely to be further confusion in selection of cleanup standards at Superfund sites and, because of the acceptance by other federal agencies of the EPA guidance, at other DOE sites undergoing cleanup. The potential application of RMCLs, water quality standards, and other state standards could result in very expensive cleanups in many cases and encourage potentially responsible parties (PRPs) to attempt an ACL demonstration. However, the SARA law limits the use of ACLs and, until significant precedents are established, it is difficult to predict whether EPA approval could be obtained.

NPL sites at which Records of Decision (RODs) or consent orders have been signed prior to enactment of SARA, establishing the extent of site cleanup to be undertaken, are not subject to the new cleanup standards, but will be remediated under the standards established in the original CERCLA law.

7.8 GENERIC REMEDIAL APPROACH

Although each of the responsible federal agencies has established formalized remediation programs for radiologically contaminated sites that differ somewhat procedurally and in structure, the general sequence of steps associated with site identification, prioritization, and characterization, the identification and evaluation of remediation alternatives, selection of a preferred alternative, planning and performance of the remediation program, and validation of its success is the same. For example, the EPA uses a structured Remedial Investigation (RI) and Feasibility Study (FS) to encompass the program leading up to selection of the preferred remediation alternative, whereas the NRC and DOE generally accomplish the same tasks under the traditional NEPA process involving the preparation of an Environmental Impact Statement (EIS).

The generic remedial approach for radiologically contaminated sites, as described in the following sections, will sequentially encompass the following;

- Identification of contaminated site and prioritization for cleanup.
- Planning of remediation program components.
- Characterization of the site.
- Definition and assessment of remediation and disposal alternatives.
- Selection and support of preferred alternative.
- Performance of the remediation program.
- Postremediation monitoring and maintenance.
- Final disposition of site.

7.8.1 Identification of Contaminated Sites and Prioritization for Cleanup

The initial assessment that a site is radiologically contaminated at some level is generally a matter of public knowledge, derived from information on past site activities, prior environmental or effluent monitoring programs, or, in extreme cases, measurements of radiation levels in local structures and residences. To provide sufficient basis to permit an assessment of the extent and severity of contamination and thus establish a priority for remediation, it is necessary to collect the available data, organize it, and evaluate the implications of the contamination in terms of potential impact on public health and the environment. Generally, at this stage this will be accomplished without instituting major field sampling programs though selected essential measurements may be made. In addition, a history of facility operations and/or disposal practices is compiled to assist in pinpointing contaminant sources and past or potential dispersion mechanisms.

The available data base is then evaluated to establish, on a comparable basis with other sites in the program, the necessary priorities within the program for remediation of the site. This comparative evaluation is accomplished in different ways under the various RAPs. Under the UMTRAP, FUSRAP, and SFMP, direct measurements of radiation levels on and in the vicinity of the site, in conjunction with an assessment of the existing and potential exposure levels to the population based on site demographics and accessibility to dispersion pathways, provided the basis for priority ranking of sites for cleanup. Thus the Vitro tailings pile located in downtown Salt Lake City received the highest priority for remediation under the UMTRAP program.

The radiologically contaminated sites identified for potential remediation under CERCLA are evaluated for ranking on the NPL in the same manner as the sites with nonradiological contamination. The model developed by the MITRE corporation (the "MITRE" model) is used to quantify the level of hazard associated with the contaminated site. This Hazards Ranking System permits the establishment of a uniform approach to the evaluation of all sites throughout the United States that have been recommended for consideration by individual states, and reduces the assessment of a range of parameters to a quantifiable basis, thus permitting a numerical comparison and ranking.

The Hazards Ranking System worksheets are provided in Appendix C. The net numerical ranking (Sm) used for consideration for inclusion on the NPL is developed from an evaluation of the existing and potential health impact along the groundwater, surface water, and air pathways. Currently a cutoff score of 28 is being used to list a site on the NPL for remediation. Although the Hazards Ranking System has proved to be a useful tool for establishing priorities for remediation under CERCLA, it has inherent flaws that reduce its accuracy. Since it fails to consider all the factors that contribute to the extent of health impact along a pathway (e.g., ease of source mobilization and dispersion), it may tend to produce a nonrepresentative result, either better or worse than the existing situation. Furthermore, the system may not give sufficient negative weight to a situation of severe impact along one pathway and negligible impact along the other pathways. The EPA continues to look into potential improved approaches in ranking contaminated sites for remediation.

7.8.2 Planning Remediation Program Components

Before initiating any aspect of the remediation program, once the program has been authorized, it is essential that the program components be fully scoped, and a work plan developed for conducting the program. To accomplish this, it is generally necessary for existing data as to on- and offsite contamination levels, site and local characteristics, and history of operations to be compiled and assessed. It is then possible to judge additional data collection needs and to obtain a preliminary indication of the extent of required remediation and of the feasible remediation alternatives for this site. The products of this planning process will include

- A work plan identifying the remedial program objective, and breaking down the remediation program by phases and tasks within each phase, and providing schedules and estimated costs for each task. Incorporated into the work plan should be the site and vicinity sampling and measurements plan to obtain the data to fully characterize the source, and an initial determination of the remediation alternatives for further evaluation.
- A summary of the assessed existing data base.
- Sampling protocols, health and safety plans, data management plans, quality assurance plans, and a community information plan.

Although each of the federal agencies responsible for the RAPs, and the industrial firms who remediate their own sites under NRC jurisdication, will produce their planning documents and data summaries in different forms, the requirements for generating this material will exist in each case. It should be noted that the preparation and subsequent review process by the agencies and interested public groups (when appropriate) could be lengthy, involving 6–12 months before approval is obtained to proceed with field investigations.

Although the DOE, and industrial firms operating under NRC license, would now proceed to the next phase of the program using their own remediation contractors, the recently enacted SARA legislation has changed both the emphasis and the mechanics of the EPA's remediation programs under CERCLA. It is now required that the identified Potentially Responsible Parties (PRPs) be given the opportunity to decide whether they want to perform the subsequent investigations and site remediation, rather than have the EPA perform the process using their contractors and assess the PRPs for the costs. If the PRPs choose to perform the work themselves, they must first obtain EPA approval of their proposal to perform the remediation in accordance with the work plan. The provision was enacted to attempt to speed up remediation of hazardous sites under CERCLA by removing day-to-day administrative and management responsibility for remediation programs from an overburdened EPA staff and by eliminating time-consuming and costly litigation in favor of cooperative agreements that would be less costly for PRPs to accomplish.

7.8.3   Characterization of the Site

Site characterization involves the collection and analysis of data on the site (and vicinity properties) needed to provide input for the assessments that comprise the investigation of remedial alternatives. The data collection program is intended to fill in the gaps in the data base originally developed during the identification and prioritization phase, and is conducted in accordance with the sampling and measurement plan. The site investigations should result primarily in a determination of the type and extent of the radiological contamination to permit an evaluation of the source concentrations and volumes of contaminated material. It should also confirm the viable exposure pathways to the environment and local human and animal receptors.

The extent of field investigations required to fully characterize a site is highly variable and will depend on the extent and accuracy of the existing data base, how explicit the definition of source strength and extent is, the number and complexity of the dispersion pathways from the source to the local receptors, and the potential severity of the environmental and public health impacts. If the need for immediate and extensive remedial activities is obvious, only sufficient data will be collected to establish the site features contributing to the problems, and permit a thorough assessment of remedial alternatives.

In general this site characterization program will require a minimum of 1 yr of sample collection and analysis and compilation and assessment of the data base. The characterization program is not an independent phase, but is generally conducted concurrently with the evaluation of remedial alternatives to permit feedback as to evolving input data requirements to assess the remedial alternatives. Under the DOE RAPs, the site characterization is performed as part of the preparation of the EIS, and the results are incorporated in the EIS. Under a CERCLA designated cleanup, the EPA performs the site characterization under the Remedial Investigation (RI) and the accumulated data base will be incorporated into an RI report or, alternatively, an Information Document. NRC licensees performing site characterization will generally do the work as part of their development of the site remediation plan. In each instance, the objectives of the data collection program and the mechanics of its performance are essentially the same.

The field sampling and measurements program covers the following individual investigations:

1. Characterization of radiologically contaminated wastes stored or disposed of onsite above or below ground. These may be found in tanks, drums, lagoons, impoundments, piles, pits, or a variety of wood or cardboard containers. The wastes may also be dispersed on the surface or mixed with other media (i.e., soil, water).

2. Hydrogeologic investigations to assess the horizontal and vertical distribution of radiological constituents in the underlying groundwater. This information will, in turn, permit the evaluation of the short- and long-term potential for contaminant dispersion in the groundwater both on- and offsite, and also provide a basis for evaluating the suitability of a site for long-term waste containment and isolation from the environment.

3. Surface and subsurface soils investigations to assess the location and extent of contamination from each significant constituent. In many cases, the prior mixing of the wastes with the soil makes the soil characterization a critical aspect of defining the source strength and extent.

4. Surface water investigation to evaluate the extent of contribution from the

source to contamination of local surface water bodies. The sampling program should attempt to distinguish between contributions from runoff, deposition, or cross-contamination from groundwater sources. In addition, both in the case of the surface and groundwater investigations, samples should be taken from up-gradient and down-gradient of the contaminant source to isolate the source contribution. Surface water investigations will also include collection and analysis of sediment samples, since the sediment frequently acts as a preferential concentrator of contaminants.

5. Air investigations to determine the tendency of airborne particulates and gases to be released into the atmosphere, the on- and offsite migration and deposition patterns, and particularly the concentrations at significant locations such as site boundaries and local residences. In conjunction with the contaminant measurements program, it is also essential that local wind patterns (directions and stability classes) be determined to provide input to the airborne dispersion modeling program.

Where wastes contain constituents of the uranium or thorium decay chains it is necessary that the airborne releases of radon or thoron gas, respectively, be measured in terms of fluxes from the contaminated surface and concentrations at significant receptor points. The values establish the starting point for cleanup activities that will reduce the fluxes to essentially background levels.

6. Local flora and fauna analysis to permit determination as to whether the contaminants have entered the food chain, and to assess the tendency of various species to concentrate or eliminate individual contaminants. In some cases, it is necessary to supplement the field investigations with controlled bench or pilot-scale studies. These studies may be performed to simulate a mobilization or dispersion mechanism, or the complex chemical interactions between the waste form, surrounding matrix, or soil pathways, and/or the effectiveness of certain technologies in preventing migration or providing the required level of isolation. These pilot studies are often defined as feedback is obtained from the assessment of remedial alternatives.

7.8.4 Definition and Assessment of Remediation and Disposal Alternatives

This phase of the remediation program encompasses the identification and development of the remedial/disposal alternatives and their evaluation leading to the selection of a preferred alternative. In selecting the range of alternatives for evaluation, the primary objective is that the alternative must either prevent or minimize the release of contaminants so that they do not disperse and thus cause a substantial threat to public health or the environment. Furthermore, the alternative should be technically feasible, must meet regulatory standards, and must be cost effective, all factors considered.

The initial scoping of remediation/disposal alternatives will be broad, encompassing the full range of response actions considered feasible for the site conditions. This scoping process will generally occur concurrently with the performance of the site investigations and will benefit from the developing data base. The more definitive the data base becomes, the easier it is to examine the feasibility of an alternative, and decide whether to include it in more detailed evaluations. Conversely, the identification of an alternative may help direct the accumulation of data during the site investigation phase.

The viable remediation and disposal alternatives, as currently assessed for radiologically contaminated sites, are discussed in Sections 7.9 and 7.10.

Before the remedial/disposal alternatives are subjected to detailed comparative assessments, they are subjected to a broad screening that considers whether (1) the technologies incorporated in the general response action are feasible and applicable to the specific site, (2) the response actions would cause obvious unacceptable public health and environmental impact, and (3) the costs of implementation of the alternative are significantly higher (i.e., order of magnitude) than the other alternatives without providing clear benefits in terms of minimizing public health or environmental impact and/or increasing the long-term reliability of the response action. Using these "go–no go" criteria, the list of alternatives is narrowed to those requiring detailed assessment to judge their acceptability.

In general, this remaining list of final alternatives will encompass the range from a "no-action" scenario, which is used as a base case to compare the other alternatives against, to a full removal of waste and soil to agency-specified cleanup criteria. Between these two extremes lie alternatives that are dependent on varying extents and combinations of waste and soil removal, site stabilization through use of surface seals (covers or caps) and/or other engineered barriers, and minimization of exposure through removal of the receptor and/or prevention of access to the facility.

Although neither the DOE nor NRC formalizes the type or number of final remedial/disposal alternatives to be evaluated, the EPA under CERCLA does require that the evaluation of alternative in the Feasibility Study include, as a minimum, at least one alternative from each of the following categories:

- Alternatives for treatment or disposal at an offsite facility approved by EPA.
- Alternatives that attain applicable and relevant federal public health or environmental standards.
- As appropriate, alternatives that exceed applicable and relevant public health or environmental standards, or
- Alternatives that do not attain applicable or relevant public health or environmental standards but will reduce the likelihood of present or future threat from the hazardous substance.
- A no-action alternative.

Once the final alternatives are defined and the sequence of steps in each is specified in detail, a series of analyses is performed to characterize each alternative as to technical aspects, public health and environmental impact, ability to meet institutional constraints, and capital and operating costs. The objective is to quantify these parameters wherever feasible to permit direct comparison of alternatives. Where the data base is insufficient to permit quantification, comparison is achieved through "discussion" or "argument."

The analyses performed of the remediation/disposal alternatives would therefore encompass the following components:

Technical Analyses. Each remedial/disposal alternative is first evaluated to determine whether it can be readily implemented at the selected site. Once the constructability of the alternative is established, its performance in terms of effectiveness in reducing the source concentration, immobilizing the source, or preventing movement along dispersion pathways is assessed, and the useful life is evaluated. Its reliability, in terms of meeting operation and maintenance requirements, and the ability to withstand a range of postulated failure modes are determined. Analysis of technical aspects must also cover the effectiveness of methods for mitigating known adverse environmental and public health impacts. Since the "as low as reasonably achievable (ALARA)" concept is applicable to the cleanup of radiologically contaminated sites, the remedial concepts should be capable of achieving reductions below regulatory limits consistent with maintaining a cost-effective approach.

Public Health Analysis. This major component of the evaluation requires that each alternative be assessed as to its ability to protect the public health during each stage of implementation of the remedial/disposal concept and subsequent to remediation. Much of the data collected during the site investigations will be used as input to the public health analysis. In general, the sequence of tasks performed to determine the extent and duration of human exposure to site contaminants is as follows:

- The onsite radiological source, or mixed radiological and chemical source, in the waste and/or soil and water is characterized as to type, concentration, and extent of constituents. These data are developed through sampling programs of the onsite and vicinity properties.
- The potential migration and exposure pathways to both onsite and offsite populations are identified for each alternative. These would include air, surface and groundwater, and direct radiation pathways.
- The site and regional characteristics that influence the mobilization and migration of the contaminated material are determined. The significant characteristics are site climatology in terms of wind direction and stability (χ/Q), surface topography, local and regional land use in terms of location of constituents in the food chain, surface water location and flow rates and directions, and groundwater levels, flow rates, and directions.
- The population(s) potentially at risk are identified, and the demographic characteristics are established for these population(s).

- The potential concentrations and doses at the receptor locations are calculated for the no-action and other alternatives. Ideally these calculations are performed using one of the sophisticated computer models developed specifically for assessing radiological contaminant migration, generally under NRC auspices. Appendix B contains a brief description of the models currently available to perform the analyses. The input data to the calculations include the previously determined source concentrations and release rate, site characteristics, pathway characteristics, and receptor parameters. The key receptor locations at which concentrations and doses are evaluated typically include.

1. The site boundary or fence in the predominant wind direction.
2. The nearest residence to the site, and the nearest residence in the predominant wind direction.
3. Occupied onsite buildings used by either nuclear or nonnuclear workers.
4. The nearest population center to the site.
5. Wells used as drinking water or irrigation sources.
6. Water intakes from surface water bodies subject to deposition, runoff, or mixture with groundwater.

- The doses are converted into health effects in terms of increased potential cancer incidence and/or increased morbidity. The calculated concentrations, doses, and health effects can then be compared to appropriate regulatory standards. Appropriate standards for radiological contaminants include concentration limits in air and water and permissible exposure levels specified in 10 CFR Part 20, exposure limits at remediated sites in 40 CFR 192, and health risk levels based on current federal policy as to acceptable risk levels that underlie the environmental regulations.

If the public health assessment demonstrates that modification of the remedial/disposal alternative will succeed in significantly reducing or eliminating exposures, and the modification is both technically feasible and cost effective, the public health assessment should be repeated with the modification incorporated into the remediation process.

Environmental Analysis. The environmental analysis will also consider the effect of dispersion of the contaminants from the onsite sources along environmental pathways on and off the facility. This assessment, however, focuses on impact on environmental receptors rather than human receptors. The no-action alternative provides the base case with which other alternatives are compared. In performing an environmental assessment, the factors generally considered include (1) effects on environmentally sensitive areas, (2) whether environmental standards are now or potentially violated, (3) whether the potential long- and short-term effects vary, and (4) what resources are irreversibly committed as a result of application of the alternative. Assessment of the no-action case permits identification of the existing and likely environmental impacts and their significance. It should also permit a quantification of the value of the areas that are contaminated or threatened with contamination. Evaluation of the other alternatives should demonstrate how this situation can be improved.

The alternatives may produce both beneficial and adverse environmental impact. The beneficial impact would include a reduction in ambient residual concentration with a resultant improvement in the biological environment and an improvement in the environmental resources available to the population and in the quality of life. Adverse impact on biota, environmental quality, and resource availability needs to be categorized as to reversible impact subject to mitigation, and irreversible ones that cannot be alleviated. Where impacts can be mitigated, the approach to achieve this mitigation, its effects, and costs should be stipulated.

Institutional and Community Analysis. This is a broad category covering factors such as community, general public, and bureaucratic attitudes and involvement in the alternative evaluation and selection process, involvement of other agencies in the decision-making and implementation processes, the impact of other federal and state standards on the selection and comparison of alternatives, effects of local conditions such as surrounding land use, accessibility to roads and rail lines, and availability of support services on the viability of alternatives, and, conversely, the impact of the remediation process on access to, and availability of, local resources. The institutional analysis would also consider these factors in the region along waste transport routes.

Although it is unlikely that any institutional factors will become the controlling influence in the selection of a preferred alternative, after the initial screening of alternatives, it is essential that they be fully considered to ensure that the concerns of all interested parties are represented in the decision-making process, and to enable planning to commence to alleviate any concerns and issues associated with the viable alternatives.

Cost Analysis. As would be expected, the relative cost of the various alternatives is a significant factor in assessing their appropriateness. This is particularly true if there are only small variations in factors such as public health and environmental impact among a number of generally acceptable alternatives. The cost analysis should include the following steps:

1. *Estimation of capital, operating, and maintenance costs.* Capital costs are generally one-time costs associated with the initial remediation of the site and disposal of the waste. Capital costs cover

   (a) Engineering of site remediation including preparation of plans, designs, specifications, and other documentation.
   (b) Onsite remediation activities including decontamination, building

demolition, earth moving and packaging, emplacement of engineered barriers, and vehicle loading.

(c) Equipment purchase or rental for performing remediation and transport of waste, including instrumentation and auxiliary equipment to support the program.

(d) Vicinity land purchase and remediation.

(e) Quality assurance, health physics, administration, liaison, and other outside support services during remediation and disposal.

(f) Transport and disposal of waste.

(g) Permits and licenses, and legal support.

Operating and Maintenance costs cover

(a) Postclosure monitoring and maintenance program (labor and consumable materials).

(b) Energy and other utility services.

(c) Site administration and reporting.

(d) Security and surveillance.

(e) Insurance, taxes, legal and accounting fees, consulting, technical support, and other purchased services.

It is appropriate to include contingency funds for both the capital and operating items to cover the uncertainty in the estimates and unknown items.

2. *Determination of the present worth of the costs.* The cost items should be analyzed, as appropriate, and then related to present worth to permit a direct comparison of the cost elements and total costs for each alternative.

3. *Performance of sensitivity analysis.* This analysis is performed to determine the sensitivity of the cost estimate to changes in key parameters. Although not always performed, the analysis provides a valuable tool to assess the potential uncertainty in the cost estimates as both a weighing factor in comparing alternatives and a planning tool to allow for sufficient contingency.

Upon completion of the analyses to assess each of the remediation/disposal alternatives the results are summarized and tables prepared providing a direct comparison between parameters for each alternative. The objective at this point is to present the material in a manner that aids in the decision-making process leading to the selection of the preferred alternative.

7.8.5 Selection and Support of Preferred Alternative

The final selection of the preferred alternative is made by the responsible regulatory agency in consultation with other involved parties. Although the agency may have employed outside consultants to assist in the data gathering and analysis, evaluation of alternatives, and comparison of alternatives, or the licensee or named principal responsible party may have been responsible for these activities, the responsible regulatory agency must make the final decision as to preferred alternative and be prepared to support that decision as its own.

To assist the agency in the decision-making process, the results of the comparative analyses must be presented in a manner that permits a ready determination of the differences between alternatives leading to the choice of the preferred alternative. This can be accomplished through judicious use of comparative tabulations of the results of the analyses, accompanied by precise text enumerating the similarities and differences among alternatives. It can also be accomplished by the use of a benefit–cost (risk) analysis in which the elements of benefit and cost are weighed based on perceived significance in the final decision process and a summarization of the weighting is prepared to permit ready comparison between alternatives. Since it is not possible to quantify all elements of benefit and cost, comparative weighting of the unquantifiable elements is not always possible, and it is necessary to supplement the weighting with discussion. Experience has shown that the agencies tend to rely upon the comparison of factors derived from the analyses of technical, public health, and environmental impacts as well as cost elements to provide sufficient basis to select a preferred alternative rather than use of a system such as cost–benefit analysis to quantify the decision-making process.

The primary basis for the selection of the preferred remedial/disposal alternative, as embraced by each of the responsible federal agencies, is the short- and long-term attainment of the governing public health and environmental exposure standards. This presumes a situation in which one alternative, which achieves technical, cost, and institutional factors comparable to the other alternatives, shows a clearly lower level of potential exposure to the public or onsite workers. This situation rarely occurs in the remediation of radiologically contaminated sites. Generally, because of the relatively low concentrations of radionuclides on the site, the calculated exposure levels for a number of the remedial alternatives will be well below governing exposure limits, thus permitting other factors to be more heavily considered in the selection of the preferred alternative. These factors include a weighing of the other evaluation parameters, particularly cost; however, political reality also requires that nonevaluation parameters be taken into consideration.

Thus the responsible agencies have evolved a current position on the remediation of radiologically contaminated sites that encourages the consideration of permanent onsite disposal or containment as a preferred alternative. This position has evolved from the following realities:

1. Existing commercial LLW burial sites are reluctant to accept the high-volume, low-activity waste (soil) generated from the remediation of radiologically contaminated sites, or are prevented from doing so by license constraints. Furthermore, the commercial burial sites will impose surcharges on the burial charge should they agree to accept this type of waste.

2. Attempts to find secure local landfills or storage facilities for the material have proven to be frustratingly unsuccessful. Local communities and their political representatives have emotionally rejected accepting waste that was not generated in their community. A prime example of this situation is the inability of the State of New Jersey to find a permanent disposal site for radium-contaminated soil removed from residential areas of Glen Ridge, West Orange, and Montclair. The drums of soil were first rejected by the commercial burial sites, and have been turned away by local communities in New Jersey.

3. Although the waste could be disposed of at a future LLW disposal facility developed under a state compact agreement, the unavailibility of the compact sites for at least 6–8 yr would require temporary storage or no action for that period. This remedy is generally unacceptable to the local residents and political entity.

4. It has been recognized that removal of the waste from the site and disposal in an offsite location do not necessarily guarantee long-term containment of the material in a manner that is significantly better than onsite remediation and containment. Since the elevated radioactivity of the waste often persists for hundreds of thousands of years, and the current state of the art cannot demonstrate long-term isolation for that period, a new engineered offsite disposal facility may not provide major improvements in isolation potential over an engineered onsite solution. Futhermore, the process of excavating the waste and soil, moving it onsite, transporting it offsite, and reburying it provides potential migration pathways.

The net result of these considerations is to place increased emphasis, where public health exposure standards can be met, on permanent onsite disposal of waste in an engineered facility. This approach would obviously not be possible where the contaminated site is in a densely populated area, where insufficient land is available onsite to permit development of a disposal facility and achieve required buffer-zone standards, or where the contaminant pathways cannot be eliminated through remediation and/or incorporation of engineered barriers.

Both the process of development and assessment of remedial disposal alternatives and the selection of a preferred alternative are subject to public participation and review through formally structured community relations programs conducted by the federal agencies. These programs are intended to provide the public, as represented by the local community and public interest groups, with descriptions of the alternatives and the evaluatory process in layman's language and elicit constructive participation to help ensure that the final decision reflects the valid concerns of the public.

The process of defining and evaluating the alternatives is documented by the agencies or their contractors. The documentation will take the form of an Environmental Assessment or Environmental Impact Statement, depending on the potential impact of the remediation process, when the project is performed under DOE auspices (i.e., UMTRAP or FUSRAP site). The EPA requires that a Feasibility Study be prepared documenting the process. NRC licensees may prepare an assessment of alternatives as input to an Environmental Impact Statement prepared by the agency.

The final step in this evaluation process is the preparation, by the agency, of a Record of Decision (ROD), documenting the entire data collection and alternative assessment. The ROD becomes the formal record for both public and bureaucratic review and provides the basis for any subsequent litigation undertaken by the agency.

7.8.6   Performance of the Remediation Program

Once the preferred remediation alternative is selected, and concurred in by the involved parties, a detailed plan is developed by the remediation contractor for the performance of the remediation process. The plan may be documented as a Closure Plan if prepared for the NRC or DOE, or a Remedial Action Plan if prepared for the EPA under CERCLA. This plan will typically include the following components:

1. Decontamination, excavation and demolition, packaging, handling, transport, and disposal procedures, and drawings and specifications prepared in sufficient detail to submit fixed price subcontractor bids and perform the remediation steps.
2. A work plan incorporating descriptions and schedules of remediation tasks and manpower loading.
3. Equipment, instrumentation, and material specifications.
4. Engineering cost estimates for each phase of remediation.
5. A health physics plan for the protection of the remediation and transport workers.
6. A quality assurance plan providing for audit, inspection, record keeping, and control of day-to-day operations.
7. Standards for assessing completion of remediation in conjunction with procedures to perform final radiation surveys.
8. A postclosure program covering maintenance, monitoring, and surveillance procedures.
9. A public information program to be conducted during the remediation process.

The above described plan is prepared by the engineering firm hired by the agency, the PRP, or the involved industrial firm to perform the remediation program. Once the plan is approved, the engineering firm implements the remediation phases, with the agency personnel or another contractor providing oversight to assess compliance with the remediation plan. A rigorous quality assurance program is conducted to validate adherence to procedures and specifications. Upon completion of the remediation program, radiation surveys are conducted to demonstrate that concentrations and exposure levels have been reduced to within the applicable standards.

7.8.7 Postremediation Monitoring and Maintenance

The postclosure period is designed to validate the adequacy of the remediation program in either eliminating the contamination source or successfully isolating it from the environment over the long term. It has three primary components: monitoring of all environmental dispersion pathways to demonstrate that the remediation program has been successful in either removing the contaminant source or providing the necessary isolation to reduce pathway concentrations to essentially background levels and hold them at these levels over the long term, a "passive" maintenance program consisting of routine site maintenance activities, and a site surveillance program designed to control access and prevent unauthorized entry.

The postclosure monitoring, maintenance, and physical surveillance programs are defined at time of closure and are conducted by either the site owner, if available as the responsible party, or a contractor to the regulatory agency performing the remediation program. The regulators will continue compliance activities through the postclosure period to determine that the closure objectives are being met. The postclosure phase does not have a stipulated time limit but will extend until the regulators are assured of site stability from assessment of monitoring data. Site stability must be achieved without an active maintenance program being carried out to repair or upgrade engineering barriers; routine, passive maintenance is acceptable.

7.8.8 Final Disposition of Site

When the remediated site is deemed acceptable by the regulatory agency, there are several options available for future use depending on the types and volumes of contaminants remaining on the site after remediation, and the potential for public exposure through migration of these contaminants. If all radiological contaminants have been removed from the site, or the residual contaminants reduced to within background levels, the site may be returned to full public use. If limited amounts of contamination remain, but are contained and not accessible, the site may be used for limited commercial activities such as industrial warehouse facilities.

In the case of certain remediated facilities such as those under the UMTRAP or FUSRAP or the closed-down commercial LLW burial grounds, the site will be taken over by the DOE or host state for perpetual ownership without permitting public use. In these instances the site still contains significant amounts of radioactive material that is isolated by use of engineered barriers.

When the facility being remediated by the owner is licensed by the NRC, the license is not terminated until satisfactory performance is demonstrated over an extended period of monitoring and passive maintenance. The extended demonstration period is referred to as the institutional control phase. There is generally no major distinction between the postclosure phase and the institutional control phase except for the degree of maintenance performed on the remediated site.

7.9 REMEDIATION ALTERNATIVES

Remediation alternatives for the cleanup of a radiologically contaminated site generally encompass the same range of viable approaches, whether the remediation is being accomplished in accordance with the DOE, NRC, or EPA defined remediation process. Although the terminology and the requirements for inclusion of a specific alternative in the assessment process may differ, in general the following alternatives are defined and evaluated at each site:

1. No action—the site remains essentially as it was when reported and characterized.
2. Complete removal of contaminated soil and structures to relevant cleanup standards.
3. Stabilization of the site with a clean cover material after removal or decontamination of contaminated structures.
4. A combination of partial soil removal and application of a stabilization cover.
5. Onsite treatment of the contaminated wastes to reduce toxicity and mobility to acceptable levels.
6. Relocation of onsite or nearby receptors (residents, workers) and restriction of future use of the site.

Remedial alternatives 2–5 generally incorporate either waste removal, containment, or treatment techniques, individually or in combination, to accomplish the required cleanup and achieve regulatory criteria. In describing the remedial alternative, reference is made to certain of the more accepted techniques. The various onsite containment and treatment techniques are described in Section 7.11.

The characteristics of each remedial alternative, their advantages and disadvantages are as follows.

7.9.1 No Action

The no-action alternative is characterized by a continuation of the status quo, with no action being taken to decontaminate or remediate the site. It is based on a demonstration that short- and long-term offsite individual and population exposures do not exceed health-related regulatory standards and that validation of this situation, perhaps coupled with control of onsite activities, would be sufficient to satisfy regulatory constraints. The nature of the continued onsite operation would depend on the type of work conducted and extent of worker exposure (i.e., nuclear or nonnuclear worker) ranging from full operations at one extreme to warehouse storage and controlled access at the other. Periodic environmental monitoring would be continued at the site boundary and offsite and a routine maintenance program conducted to maintain site conditions and prevent erosion and material dispersion.

The advantages of this alternative are

1. It represents the lowest cost alternative.
2. It causes the least disruption to the environmental setting and the local population.
3. Since there would be no contaminated soil movement, or removal from the site, no short-term impact in terms of increases doses would be incurred by onsite workers or the offsite population.

The disadvantages of this alternative are

1. The no-action alternative fails to remove the contaminated source that was the basis for the original designation of the site for remediation. Even if an assessment of the existing and potential public health and environmental impact demonstrates that this alternative meets short- and long-term standards, the continued presence of the contamination will likely violate the EPA onsite cleanup standards, and require that the responsible agency insist on more substantive remediation to ensure that the contamination is removed or contained. Further, once a site is included on the NPL or designated by another agency for cleanup under an RAP, there is no bureaucratic mechanism for delisting it without a remediation process occurring.
2. Where the potential exists for spread of the contamination from the soil to underlying groundwater, the no-action alternative fails to remove the contaminant source.
3. It generally will require an extended future program of environmental monitoring to validate that the contamination is contained onsite.
4. Once a site is designated by the responsible agency as requiring remediation, the local populace and officials will require that some remedial action be taken before the site is considered acceptable.

The viability of the no-action alternative has become a controversial issue, particularly in the case of Superfund-designated sites having some radiological contamination. In some instances the PRP is able to demonstrate, through dispersion modeling and monitoring, that the contaminated soil, if left in place, will pose no significant public health or environmental risk. However, the agency (EPA and State) requires that stipulated soil cleanup standards (e.g., 5 pCi/g ^{226}Ra in top 15 cm of soil) be met through either soil removal or use of covers.

Thus, the question of which standard, health-related or cleanup, should apply is joined. Since substantial differences in cost are involved, a resolution of this contentious issue will be required before the place of the no-action alternative can be assessed. Further, the SARA amendments clearly demonstrate (Section 121) preference for site cleanups that permanently and significantly reduce the volume, toxicity, and mobility of wastes through use of practical treatment technologies when they are available. Offsite transport of wastes and contaminated soil is to be considered the least favorable option when these treatment technologies can be alternatively used.

7.9.2 Complete Removal of Contaminated Soil and Structure to Relevant Cleanup Standards

Under this alternative, the contaminated soil is removed until levels that achieve regulatory cleanup standards are met and the soil is disposed of in an offsite disposal location. A variation would require, where radon gas emanation from the soil occurs, that the soil be removed until a surface concentration rate of $20 \, pCi/m^2$-sec is achieved. Generally, structures on the site over the contaminated soil region would be disassembled and disposed of in the same manner as the soil. In some instances, an interim step would permit onsite storage until a permanent offsite disposal location is qualified.

In general the following steps are undertaken for this remediation alternative from the point where the radionuclide concentrations in the soil, air, and water and the surface radiation levels on equipment and building surfaces are characterized:

1. Remove all noncontaminated equipment from the building.

2. Raze the contaminated buildings and structures. To avoid the spread of contaminated material and increased occupational and environmental exposures during this activity (and subsequent operations) provide onsite health physics monitoring and use personnel protective equipment (i.e., masks, clothing, respirators) on an as-needed basis to limit exposure. Disassemble buildings by components, rather than using techniques that generate large volumes of fugitive dust. Institute a dust-control program.

Alternatively, where buildings and equipment are to be preserved, decontaminate the buildings and equipment by removing all surface contamination until surface activity and measured exposure rates in the building are within regulatory limits. Collect and package any liquid and solid decontamination materials.

3. Excavate and remove soil to sufficient depth and breadth to achieve regulatory concentration limits in the remaining soil (e.g., ^{226}Ra concentration would be reduced to $5 \, pCi/g$ above background averaged over the top 15 cm of soil, and $15 \, pCi/g$ averaged over any 15 cm of soil below this).

4. Apply a stabilizing chemical or surface sealant over excavated regions to control dusting. The soil is either left in this condition or a layer of topsoil is emplaced and a vegetative cover of indigenous plants is grown.

5. The excavated soil and building rubble are then removed to a designated area onsite for interim storage. The storage area should be underlain with an impervious pad of natural or artificial material to prevent seepage and the surface of the storage pile covered to prevent infiltration of water.

Alternatively, the soil and building rubble may be moved directly to a staging area for loading on to transport vehicles for offsite shipment. The transport mode may consist of

(a) Bulk shipments using an ore transporter with the material wetted and covered with a tarpulin to prevent dusting.

(b) Packaged in 55-gallon drums and/or crates and shipped in a van.

(c) Bulk shipment by covered rail car from a local rail siding.

Waste disposal alternatives are discussed in Section 7.10.

6. Subsequent to stabilization of the remediated site surface, a poststabilization environmental monitoring and active maintenance program is conducted to validate and ensure the adequacy of the remediation measures. This is followed by a period of additional monitoring and passive maintenance before the site is either used on a nonrestricted basis or dedicated to state or federal agencies.

The advantages of this alternative are

(a) The source of the contamination is fully removed from the site, and thus this source of public health and environmental impacts is eliminated after excavation and transport is completed.

(b) Unrestricted future use of the site is generally possible.

(c) No long-term monitoring, maintenance, or surveillance programs are required.

The disadvantages of this alternative are

(a) It is the most expensive remedial alternative as a result of the combination of extensive soil and rubble removal, high volume transport over extended distances, and disposal at either a commercial burial site or new site developed for the waste.

(b) If a commercial burial site agrees to accept the waste, the waste occupies a volume that could be used for higher activity, lower volume waste material, thus conserving scarce burial space in the commercial burial sites.

(c) Offsite disposal is the least preferred option under the SARA law of 1986, and has been a difficult concept to sell to the public near the disposal site and along the transport routes for all RAPs.

(d) It is difficult to accurately determine the extent of contaminated soil removal required to reach regulatory limits because of the high degree of variability in contaminant concentrations in the soil.

(e) Even when dust suppression methods are employed, increased impact from airborne dispersion during the excavation and soil movement periods generally occurs.

(f) Some relocation of residents of adjacent buildings may be required during the excavation and soil movement periods because of the potential for increased doses.

This option, involving the excavation and removal of the contaminated soil and building rubble to offsite disposal locations, has been the preferred agency approach for remediating contaminated sites. However, strong objections by the public in the vicinity of the proposed disposal locations expressed through their local representatives, constraints against disposal of high-volume low-activity waste in the limited space in existing commercial sites, and the fact that the contaminant source is just being moved (at some increased risk) rather than eliminated have led to reassessment of the viability of this remediation approach. The constraints against its use in the SARA Act and a reemphasis on onsite treatment and containment in other RAPs are direct results of these factors.

7.9.3   Stabilization of the Site with a Clean Cover Material

The premise behind this alternative is that a cover can be used to reduce airborne dispersion of particulates and gases, prevent surface erosion of the soil, and redirect precipitation to prevent percolation of water through the soil and thus reduce radiation levels both onsite and offsite to achieve regulatory standards over the long term. The cover employed could be a natural material of low permeability such as bentonite or montmorillonite clay, an artificial material such as asphalt, concrete, or reinforced plastic, or a combination of materials. Existing structures over the contaminated soil region would be either disassembled and disposed of in an offsite disposal facility or decontaminated and left standing.

In general, the following steps are undertaken for this remediation alternative after characterizing the radioactivity in the soil, air, and water and on building and equipment surfaces:

1. Remove all noncontaminated equipment from the buildings.

2. Raze the contaminated buildings and equipment applying the same precautions as described under the "complete removal" option above.

3. Stabilize and cover all areas of the site where the radioactivity in the soil exceeds the regulatory standard. Use sufficient thickness of material to reduce the surface emanation rate to within regulatory limits where radon gas release is to be controlled. Cover thickness is evaluated in advance

based on a diffusion model analysis and verified during and after emplacement.

4. Subsequent to stabilization, an environmental monitoring and active maintenance program is conducted to validate the adequacy of the remediation measures. This is followed by continued monitoring and passive maintenance until stability of the site in terms of maintaining releases constant and below regulatory limits over the long term is demonstrated. The site would then be released for use.

The advantages of this alternative are

1. Use of a stabilizing cover is generally considerably less expensive than excavation and removal offsite.
2. There is no increased impact to the public health or the environment from dispersion of contaminated material since the soil excavation, movement, and transport phases are eliminated.
3. A barrier is provided between the contamination and the environment, thus eliminating or minimizing the potential for the action of natural forces and resultant dispersion of the soil.
4. No additional burden is placed on offsite disposal facilities to accept waste material.

The disadvantages of this alternative are

1. The source of potential environmental contamination is not removed but is in part isolated. Thus, the possibility of future dispersion of contaminants exists if the cover integrity is not maintained. Long-term surveillance of the cover integrity is therefore required.
2. Although the cover reduces the percolation of water through the soil and thus reduces the potential for formation of leachate and its contamination of underlying aquifers, it does not eliminate the potential for groundwater contamination since the migration pathway to the aquifer is not interrupted. Thus, at a minimum, long-term monitoring of the underlying soil and groundwater is required.
3. If the groundwater shows evidence of contamination prior to undertaking remediation, or if concern about this pathway becoming contaminated exists, the use of a cover in itself would not solve the problem since a pathway for contaminant migration to the groundwater may already be established. The use of more invasive measures such as pumping the contaminated aquifer down-gradient of the site and treating the water, construction of an enveloping grout curtain to further contain the contaminated soil, more extensive groundwater monitoring in the form of closely spaced wells, or a combination of these measures may be required.

Even with these limitations on the use of the technique, application of stabilizing covers is likely to find increased use in the future. There are many sites to be remediated where potential groundwater contamination is not a significant factor, and where the cover can serve to stop airborne and surface dispersion of the soil. With improvements in the techniques for manufacturing artificial covers and in the emplacement of both natural and artificial covers, and with increasing legislative and public constraints against offsite shipment, the use of stabilizing covers as a total or partial remedial alternative will be encouraged.

7.9.4 Combination of Partial Soil Removal and Application of a Stabilization Cover

This alternative would ideally be suited for sites at which localized areas of high radionuclide concentrations exist relative to a larger region of soil with low levels of contamination. The soil would selectively be removed from the areas of high concentration (hot-spots) and either stored onsite for an interim period in a secure location or shipped for offsite disposal. Depending on the contamination levels remaining after hot-spot removal, the site would then be treated with a stabilizing agent and revegetated, or covered with an artificial or natural cover of sufficient thickness to achieve regulatory limits for upper layer soil concentration and emanation rate from the surface.

In general, the following steps are undertaken for this remediation alternative after characterizing the radioactivity in the soil and on building and equipment surfaces:

1. Remove all noncontaminated equipment from the buildings.
2. Raze the contaminated buildings and structures or, alternatively, decontaminate the buildings and equipment applying the same precautions as described under the "complete removal" option above.
3. Remove the soil to sufficient depth and extent to eliminate surface "hot-spots" where significantly higher than average local radionuclide concentrations exist.
4. Stabilize and cover all areas of the site where the radioactivity in the soil exceeds the regulatory standard. Use sufficient thickness to reduce surface emanation rates to within regulatory limits where radon gas release is to be controlled. The cover thickness is evaluated in advance based on a diffusion model analysis and verified during and after emplacement.
5. The excavated soil and rubble are then removed to a designated area onsite for interim storage or, alternatively, moved to a staging area and loaded onto transport vehicles for offsite shipment, as described under the "complete removal" alternative.
6. Subsequent to stabilization, an environmental monitoring and active maintenance program is conducted to validate the adequacy of the remediation measures, followed by continued monitoring and passive

maintenance until stability of the site is demonstrated. The site would then be released for a designated use.

The advantages of this alternative are

1. If the removal of "hot-spots" in the soil reduces the existing or projected health-related impacts (doses) to occupational workers and the public to within regulatory limits, it would then only be necessary to satisfy relevant cleanup standards. Since, as previously noted, the imposition of cleanup standards is a controversial issue that may be subject to negotiation as to extent, the satisfaction of health-related standards would go a long way toward alleviating the problematic nature of the site.
2. The alternative will be less costly than full removal of contaminated soil.
3. Increased impact during remediation and transport is minimized since soil disturbance and movement are minimized.
4. A barrier is placed between the remaining contaminated soil and the environment, thus eliminating the potential for airborne or surface dispersion of the soil through the action of natural forces.

The disadvantages of this alternative are

1. The source of potential environmental contamination is not entirely removed, with the remainder of the contaminated soil contained by the cover. Thus cover integrity must be maintained, and long-term surveillance is required to validate the continued cover integrity.
2. Long-term monitoring of the underlying soil and groundwater will also be required to ensure that groundwater contamination is not occurring from migration of leachate from the contaminated soil. If the monitoring indicates that potential contamination of the groundwater is likely, other measures will be required (see discussion of "stabilization" alternative above).
3. The alternative requires the disposal or storage of some contaminated soil that is relatively high in activity compared to the typical material from this type of site. Thus, the problems attendant with offsite disposal, such as unavailability of a commercial disposal site and public antipathy, will be experienced to some degree.

The use of this alternative is likely to find acceptance in those instances in which localized regions of high-activity contamination can be significantly reduced with a relatively small removal of soil, and the site can then be stabilized without further offsite activities. It may thus be possible to convert a highly contaminated site into a readily manageable one with this approach consistent with the growing trend toward avoidance of major offsite shipments of waste.

7.9.5   Onsite Treatment of Contaminated Wastes to Reduce Toxicity and/or Mobility to Acceptable Levels

In addition to the aforementioned alternatives based primarily on waste removal or containment techniques, another approach to managing contaminated wastes (soil and water) at sites undergoing remediation is the treatment of the waste material to reduce its toxicity and/or mobility. This approach, used either singly or in combination with containment techniques, permits onsite stabilization of the wastes to be accomplished. Treatment technology, which is discussed in Section 7.11.2, can be used to eliminate or mitigate the hazard-producing characteristics of the contaminated soil, groundwater or leachate, or surface water. The majority of these biological, chemical, and physical waste treatment techniques are not new, but their application to radiologically contaminated sites is still in the developmental stage.

In general, the following steps are undertaken for this remediation alternative after characterizing the radioactivity in the soil, air, and water and on building and equipment surfaces:

1. Remove all noncontaminated equipment from the buildings.
2. Raze the contaminated buildings and structures or, alternatively, decontaminate the buildings and equipment applying the same precautions as described under the "complete removal" option above.
3. Based on prior laboratory analyses and characterization of the contaminated soil, groundwater, or surface water, select appropriate treatment technologies and validate their applicability through bench-scale testing. The correct selection of the technologies and the order of application (i.e., process train) are the key factors in achieving cost-effective treatment.
4. Mobilize the necessary equipment and personnel onsite to perform the treatment. Contaminated soils and water will either be treated *in situ* or removed for treatment and then replaced. The treatment process will be conducted using necessary protective health-physics and dust-suppression measures. Residues from the treatment process, in the form of sludges or liquids containing concentrated radioactivity and/or treatment process by-products, will be collected and packaged for offsite disposal or interim onsite storage.
5. Based on the success of the treatment process in reducing radioactivity in the soil and water, further remedial steps will be undertaken. These could consist of (a) application of chemical stabilizers and vegetative cover, (b) emplacement of artificial or natural covers to contain the remaining waste, or (c) removal of selective contamination not amenable to treatment. The use of treatment techniques is intended to minimize or eliminate the need for these steps.
6. Subsequent to stabilization, an environmental monitoring and active maintenance program is conducted to validate the adequacy of the

remediation maneuver, followed by continued monitoring and passive maintenance until stability of the site is demonstrated. The site would then be released for a designated use.

The advantages of this alternative are

1. The hazardous characteristics of the site may be reduced to an acceptable level based on existing regulatory standards without the need to excavate and remove soil or to use containment techniques, both approaches that generate their own ancillary problems.
2. Treatment should result in permanent elimination of the hazardous source onsite, eliminating the need for long-term monitoring.
3. Very high-activity wastes can be reduced in toxicity onsite before undertaking other remedial approaches, thus making the handling and disposal of the treated wastes more manageable.
4. Treatment techniques are generally cost effective by comparison to removal and containment.

The disadvantages of this alternative are

1. Most treatment techniques produce a residue of concentrated radionuclides of elevated activity levels that requires offsite disposal in either a commercial burial site or government-controlled facility. The residue may be a liquid, semisolid (sludge), or solid and may require further treatment (i.e., solidification, immobilization) prior to disposal.
2. Some treatment techniques will generate effluents that potentially may increase the exposure level of both onsite workers and the local population during the treatment process, and require the incorporation of costly controls to minimize their dispersion.
3. A number of promising treatment techniques have been demonstrated only on a laboratory or pilot scale, and are yet to be applied to actual site cleanup processes. Thus, the parameters of their performance have not yet been demonstrated, and some uncertainty exists as to the ability to achieve projected reductions in toxicity and mobility consistent with competitive costs and schedules.

There is considerable promise associated with the refinement and application of process trains of treatment techniques to contaminated sites. The use of these techniques will undoubtedly grow because the avoidance of waste removal and transference to another location and the uncertainty associated with the viability of long-term containment will be overriding considerations. Thus, proven cost-effective treatment techniques will increasingly serve as the first option considered for remediation of radiologically contaminated sites.

7.9.6 Relocation of Onsite or Nearby Receptors and Restrictions on Future Use of the Site

Under this alternative, the receptor of the impact from the contaminated site, being either the onsite workers or local population at the end of the existing or potential dispersion pathways, is removed rather than the contaminant source being removed, contained, or treated. Thus, the dispersed contaminant will not cause a public health impact. Under these conditions it will be necessary to restrict the future use of the site and affected nearby locations and maintain those restrictions in perpetuity or until another remediation measure is adopted.

In general, the relocation alternative would be considered for only highly contaminated sites having severe current or potential impact, not readily amenable to mitigation. It would be accomplished through

1. Determination by measurement or analysis of the high levels of radioactivity onsite, along the dispersion pathways, and at the receptor locations, accompanied by confirmation that these concentrations were not subject to mitigation.
2. Evacuation of the site and surrounding affected areas possibly accompanied by razing of previously occupied structures.
3. Establishment of restricted access to the site and surroundings and maintenance of surveillance to ensure that the access restrictions are not violated.

The advantage of this alternative is

1. Public health impact to onsite workers and the affected population is eliminated.

The disadvantages of this alternative are

1. The source of the contamination is not eliminated or mitigated. No action is taken to remediate the site.
2. Surveillance, monitoring, and maintenance are required in perpetuity to prevent access and ensure that contaminant dispersion does not affect other receptors.
3. The lives of the local populace are severely disrupted.
4. It fails to meet the objectives of the DOE RAPs, CERCLA, and SARA.

Evacuation of receptors would be considered as a "remedy" only in extreme situations, where all other approaches are not currently feasible, and where exposure levels are unacceptably high. It is not a viable solution for the large majority of radiologically contaminated sites, and would be an interim measure even in extreme situations until viable remediation techniques are available, or radiation levels have been reduced to manageable levels. The contaminated site at Chernobyl after the accident is an example of a situation in which evacuation of receptors is an appropriate interim measure until more permanent cleanup techniques can be employed.

Contaminated offsite or "vicinity" properties, if contiguous with the site undergoing remediation, will generally be subjected to the same type of cleanup process as the primary site if the public exposure conditions are comparable. Vicinity properties that are not contiguous need to be evaluated on a case-by-case basis, based on individual site characteristics and pathways to humans and the environment.

Where sites are contaminated with mixed radioactive and hazardous chemical wastes, comparable remedial alternatives are defined and evaluated. The remediation objectives would then require that public health and cleanup standards be achieved for both classes of constituents. The types of containment and/or treatment options might vary depending on the predominant constituents of the waste material, but the methodology would be essentially the same.

7.10 WASTE DISPOSAL ALTERNATIVES

For those remediation alternatives involving the removal of contaminated soil and building rubble, disposal alternatives must be defined and assessed in conjunction with the remediation alternatives. Although there are hypothetically a range of potential disposal options, in practice a limited number of options are practical to consider because of a combination of factors such as the status of the disposal technology, public and governmental attitudes, availability of secure offsite disposal locations, and the need to achieve regulatory standards. Thus, the disposal alternatives that represent viable options for current and near-term disposal of radiologically contaminated waste are

1. Onsite disposal in a secure, engineered landfill.
2. Onsite temporary storage.
3. Onsite permanent storage/disposal.
4. Offsite disposal in an existing commercial burial site.
5. Offsite disposal in a new burial site.
6. Offsite disposal in conjunction with other remediation wastes.

The characteristics of each of these disposal alternatives and their advantages and disadvantages are as follows.

7.10.1 Onsite Disposal in a Secure Engineered Landfill

Under this disposal option, contaminated soil and building debris are buried, essentially in bulk form, in an onsite landfill engineered to meet regulatory standards. Thus, the landfill would incorporate containment features similar to an offsite facility, and may have to undergo comparable characterization and qualification associated with licensing these types of LLW disposal facilities under 10 CFR 61. This alternative can be considered for only a very limited number of sites where (1) sufficient land is available to locate the landfill and maintain the required buffer zone to the site boundary, (b) the performance objectives and prescriptive requirements of 10 CFR 61 or comparable standards can be achieved, and/or (c) the volume of contaminated material is relatively small. The sites likely to achieve these conditions are those requiring only minimal soil excavation or located in relatively isolated regions.

The advantages of this disposal alternative are

1. It eliminates the need for offsite transportation of the waste with the attendant impact and risks associated with transport and the increased public concern that it generates.
2. Onsite disposal eliminates the need to use scarce burial volume at existing commercial disposal facilities or to create new offsite disposal facilities for this material.
3. Onsite bulk disposal is a relatively low-cost alternative compared to offsite bulk disposal or to storage/disposal of packaged wastes in vaults or other engineered structures.

The disadvantages of the alternative are

1. It is highly unlikely that industrial sites, or other contaminated facilities located in urban or suburban areas, will possess sufficient land, or have the necessary hydrogeologic characteristics to develop an engineered landfill that meets regulatory standards. It was usually the lack of suitable site characteristics that initially contributed to the creation of the problem. Thus, this disposal option is best suited for selected mine or mill sites possessing adequate land and suitable characteristics, but where the existing waste is not adequately contained.
2. Onsite disposal would tend to disrupt ongoing operations in those cases in which an onsite process facility continues to function.
3. Continued long-term monitoring of the site should be required to validate the containment characteristics of the engineered landfill.

7.10.2   Onsite Temporary Storage

When a permanent offsite disposal option is not immediately available, and onsite permanent disposal is not a suitable alternative, it is necessary to store the waste until an offsite disposal facility can be established. Under current conditions, where existing commercial disposal facilities are not available for the bulk of the contaminated material generated in RAPs, onsite storage may be the only available option. Typically, soil and rubble in bulk form are placed on an impervious pad made of clay, asphalt concrete, reinforced plastic, or a combination of these materials, and the waste is stabilized with a cover of similar materials and possibly revegetated or covered with a tarpaulin if storage is of short-term duration. Alternatively, the waste is packaged to provide better control over its movement and storage, and to create an additional barrier against dispersion into the environment. Monitoring is required to ensure that isolation of the wastes from the environment is maintained, and site surveillance will be necessary to prevent access to the storage area.

The advantage of this alternative is that it provides a temporary way to place the waste in a contained location until a permanent disposal option is available. The availability of temporary storage will permit the remediation program to proceed.

The disadvantages of this alternative are

1. Temporary storage is an interim measure; a permanent solution is still required.
2. Temporary storage measures may be required for an extended period since it may not be possible to accurately predict the availability of permanent disposal facilities at the time the material is placed in storage.
3. Although temporary storage concepts provide better isolation of the waste from the environment, they do not achieve the same degree of long-term isolation that licensed disposal facilities do. Thus, the site operator will incur additional labor costs for the monitoring, maintenance, and surveillance activities required over the life of the storage facility to achieve comparable levels of containment.

7.10.3 Onsite Engineered Permanent Storage/Disposal

Under this alternative, contaminated soil and rubble are emplaced in engineered facilities (e.g., precast concrete structures, surface or subsurface vaults), usually after being packaged in drums or crates, which are designed for long-term storage or permanent disposal. These "enhanced shallow land burial" or "alternative burial" concepts, which are described in Chapter 6 (Section 6.8), serve to improve the long-term isolation characteristics of the site and thus convert a marginal or unsuitable location into a viable disposal site while offering the option of retrievability for certain of the design concepts. Although these alternative disposal concepts are not limited to onsite waste management and are in fact the basis for the LLW facility design requirements established by a number of the LLW compact sites, they are an attractive option for the onsite disposal of waste from RAPs.

The advantages of this alternative are

1. It permits the use of sites for permanent storage/disposal that would be otherwise unsuitable by creating conditions that meet regulatory standards

(e.g., 10 CFR 61 performance objectives and prescriptive requirements) and thus opens up many sites for consideration for onsite disposal. This, in turn, meshes with the SARA objectives and those of other governmental agencies.

2. It eliminates the need to use offsite commercial or new disposal facilities, thus eliminating transportation impact, and conserving existing disposal space for higher activity waste, or waste from generators having a more critical need for the available volume (e.g., hospitals, research institutions).

3. It provides a secure means for permanent disposal, yet permits ready access to the waste in the future if a more attractive long-term storage/disposal concept is developed.

4. The concept is relatively inexpensive compared to the transport and disposal of waste at an offsite commercial disposal facility.

The disadvantages of this concept are

1. Engineered permanent storage/disposal involves the use of design concepts that, while attractive on paper, have not been proven over the long term to provide necessary containment characteristics to achieve ALARA releases. Thus, it will still be necessary to provide long-term monitoring to demonstrate that the approach does achieve improved isolation over properly designed shallow land burial sites or surface impoundments.

2. The original source of the radiological contamination, while in secure locations, remains onsite, which will be objectionable to a segment of the population.

3. The concept requires substantial land area to accommodate the engineered enclosure containing the emplaced packages of waste. This will limit the number of sites that are suitable for the disposal approach.

4. It is considerably more expensive to provide for engineered storage/disposal than for shallow land burial or surface impoundments.

7.10.4 Offsite Disposal in Existing Commercial Burial Site

Under this concept, contaminated soil and rubble are brought to an onsite staging area, either packaged or left in bulk form, loaded onto transport vehicles (i.e., truck or train), and transported to a commercial disposal site licensed for burial of LLW. There are three disposal sites in the United States currently receiving commercially generated LLW. These are the facilities at

- Richland, Washington, operated by U.S. Ecology.
- Beatty, Nevada, operated by U.S. Ecology, and
- Barnwell, South Carolina, operated by Chem-Nuclear Systems Inc.

There are, however, existing restrictions that limit the availability of these sites for the type of high-volume, "low"-activity waste typically generated by site remediation programs. The Barnwell site is not licensed to accept radium-contaminated waste. In addition, because of current constraints imposed by the host states or by the site operator, there is reluctance to accept the high-volume wastes that use disposal volume that could be reserved for higher priority wastes. The State of Nevada has recently refused to permit shipments of contaminated soil to the Beatty site, and the Richland site operators will consider each shipment on a case-by-case basis. A high premium (surcharge) is likely to be charged for the limited volume of RAP waste that is accepted.

A permanent prohibition against use of the three existing commercial disposal sites will undoubtedly be in effect after 1992. The provisions of the Federal Low-Level Radioactive Waste Management Amendments Act permit the host states to restrict the acceptance of waste to only that material generated within the compact region after that date. Because of diminishing available burial space, and a desire to avoid shipments of wastes from outside the region, it is likely that even small volumes of RAP waste will not be accepted after that date. Since the large majority of RAPs will not have excavated waste available for shipment until after the cutoff date, this disposal option is effectively precluded from consideration in those cases.

As previously noted, the SARA provisions also would place low priority on the use of offsite disposal concepts and encourage the use of onsite treatment or disposal as the preferred approach.

The advantages of this disposal alternative are

1. The onsite source of radiological contamination is removed, and the attendant problems of existing or potential offsite impacts are eliminated.
2. The need for long-term monitoring, active maintenance, or surveillance is either eliminated or substantially reduced. There will, in general, be no long-term restrictions on use of the site.
3. The waste is removed to a facility specifically designed and operated for disposal and containment of radioactive waste.

The disadvantages of this alternative are

1. Offsite disposal in a commercial facility is generally the most expensive disposal option as a result of the high transportation and burial costs.
2. The population along the transport route is subjected to somewhat elevated exposures during routine transport and to the effects of potential accidents. In addition the total occupational exposure, in terms of person-rem, is generally highest for the combination soil removal, handling, transport, and disposal process.
3. Most significantly, the constraints and obstacles described above for this disposal option will, at best, limit the applicability to special situations and,

at worst, prevent its consideration for any high-volume, low-activity waste generated in the cleanup of radiologically contaminated sites. ·

7.10.5   Offsite Disposal in New Burial Site

This alternative is similar in concept and mechanics to disposal in an existing commercial facility, with the significant difference that a new disposal facility would be selected, qualified, constructed, and licensed under the current body of regulations, particularly the provisions of 10 CFR 61 governing LLW disposal. There are potentially two different mechanisms for the establishment of a new LLW burial site to accept waste from remediated sites;

1. Incorporation of the waste into an LLW site developed under a regional compact program.
2. Development, by the remediator, of a single-use site for the disposal of the waste.

Use of a Regional Compact Site. A number of the regional compacts of states, as part of their assessment of anticipated disposal volumes, are projecting the receipt of wastes generated from remediation programs within the region. Thus, the facilities will be designed to accommodate remediation wastes, with the disposal units engineered for the projected volumes and the type, activity, and mobility of the radiological contaminants. The advantage of such an approach, in addition to those described for an existing commercial site, is that the new facility would be built, operated, and closed in accordance with current standards and would thus provide an enhanced long-term isolation potential. However, additional disadvantages would include the inevitably higher costs for use of the new commercial site and the need for temporary storage of the wastes for sites undergoing near-term remediation until a new facility is licensed for operation.

Development of a Single-Use Site for the Disposal of the Waste. This disposal alternative has been considered a viable option for disposal of waste generated by cleanup of UMTRAP or FUSRAP sites and certain CERCLA-designated sites. It is particularly attractive where onsite disposal cannot be considered, and where readily qualifiable offsite disposal locations are within a short transport distance. Examples of such situations include the removal, under the UMTRAP program, of the Vitro tailings, from Salt Lake City to a newly qualified site in a desert area in Utah, and the attempt by the State of New Jersey to use a quarry in Vernon, New Jersey to dispose of radium-contaminated soil from the remediation of residential areas in Montclair, West Orange, and Glen Ridge, New Jersey. The latter approach was dropped because of strong opposition on the part of local Vernon residents.

Development of a single-use disposal site by the party performing the remediation program involves the selection, qualification, operation, and closure of the site in accordance with NRC and state licensing requirements governing new sites for disposal of LLW or by-product materials. The process is rigorous, and sequential, and would likely extend over a period of 5–6 yr from initiation. The major advantage of this disposal option, in addition to those described elsewhere for a new site, is that the remediating party also controls the offsite disposal of the material, and is, therefore, not subject to uncertainties over control of schedule and cost inherent in other disposal concepts. In addition, the disposal site would be located in general proximity to the remediation site to minimize transportation impact.

The concept does present a number of disadvantages. It puts the remediating party into the radioactive waste disposal business, albeit temporarily, and requires increased personnel involvement in the construction and operation of the facility. The remediating party may continue to retain liability for the waste emplaced in the disposal facility, and would have to provide monitoring, maintenance, and surveillance during the postclosure observation phase.

7.10.6   Disposal in Conjunction with Other Remediation Wastes

Under this concept, remediation waste would be disposed of in a site designated for disposal of waste from other remediation projects or from major processing operations. The types of waste disposal facilities that could be considered for this disposal approach are

1. Overburden piles generated from mining operations.
2. Uranium tailings impoundments at commercial mills.
3. An UMTRAP or FUSRAP site.
4. A CERCLA site.

Overburden Piles Generated from Mining Operation. The incorporation of bulk remediation waste into an existing pile of overburden material stripped off a uranium mine site appears to be an attractive option because the negotiated costs could, in principle, be considerably less than those for other offsite disposal options. However, this concept in most cases would not represent a viable option. Overburden piles are typically composed of material ranging in composition between natural soils and the lower cutoff concentration for processing uranium ores. They are not licensed as either by-product or source materials and do not fall under NRC or comparable state radiological regulations associated with licensed facilities. Therefore, it is highly unlikely that a mine operator would consider, or be permitted to consider, the incorporation of outside radiologically contaminated wastes into the overburden pile with the resultant changes in regulatory status, and thus closure and stabilization costs, associated with such a change.

Uranium Tailings Impoundments at Commercial Mills. The use of an active or inactive uranium tailings impoundment at a commercial mill presents a more realistic possibility than the use of an overburden pile. This option would be of interest for those remediation projects in which the contaminants are parts of the uranium or thorium decay chains, and the mill sites are in reasonable proximity to the remediation site. The remediation wastes would usually represent a small fraction of the total volume of material in the impoundment, and would not affect the total impact of the facility on the environment. In addition, bulk waste material could be emplaced in the tailings impoundment, either intermixed with the tailings or encapsulated in pods in the tailings.

Furthermore, the tailings impoundments already possess by-product material licenses for possession of the tailings material. Although it would be necessary to amend the existing license to permit acceptance of the outside wastes, there would be no need to modify either operating practices or closure and stabilization concepts. If this situation prevailed, and no other significant barriers were imposed by the regulatory agencies, a number of mill operators would undoubtedly consider the additional revenues, obtained in the form of a negotiated disposal fee, to be an attractive inducement for accepting the waste. From the standpoint of the party disposing of the wastes, the negotiated costs should be considerably less than for other offsite disposal options such as disposal at an existing or new commercial disposal facility. In addition, this option provides a currently available alternative, eliminating the need to delay remediation projects or store remediated wastes.

A variation on the use of a tailings pile at a commercial site involves the incorporation of the waste in a pile where the tailings are commingled from those produced under government contract for the MED/AEC programs and those generated when the product was used for commercial purposes. These commingled tailings piles are located at 13 sites in six western states. Table 7-8 lists the sites, facility owner, location, operating status, and tailings quantities.

Under the requirements of Public Law (P.L.) 96–540, Section 213, the Secretary of Energy was required to develop a plan for a cooperative program to provide assistance in the stabilization and management of the commingled tailings. This legislation grew out of a recognized inequity in the Uranium Mill Tailings Radiation Control Act of 1978 (P.L. 95–604) that authorized federal assistance to stabilize tailings at the 24 inactive uranium processing sites that produced concentrate solely for government use, but did not authorize any assistance for those sites in which commingled tailings existed.

The DOE does not now have any statutory or contractual legal responsibility for the decommissioning or restoration of the commingled uranium tailings pile sites. These facilities are commercial licensees of the NRC or an Agreement State and, as such, the facility operators are required to decommission the site and dispose of the tailings pursuant to the current NRC regulatory requirements. However, a plan was developed by the DOE (DOE 1982b) for stabilization and management of the commingled uranium mill tailings and submitted to Congress in June 1982. The plan provided estimates of the total cost for reclamation of the commingled tailings sites, and estimated the federal government share of the costs for several cost-sharing approaches. Although providing the specific information

TABLE 7-8 Commingled Tailings Sites Parameters

Mill/Location	Estimated Capacity (TPD)	Ownership during Deliveries under AEC Contact (Period)	Parent Ownership	Mill Status
Cotter Corporation Canon City, Co	1500	Cotter Corp. (1958–1965)	Commonwealth Edison Co.	Active
Union Carbide Uravan, CO	1300	Union Carbide Corp. (1949–1970)	Union Carbide Corp.	Active
Anaconda Minerals Co. Bluewater, NM	7000	Anaconda Copper Co. (1953–1970)	Atlantic Richfield Co.	Shutdown, 3/82
Homestake Mining Co. Grants, NM	3500	Homestake-NM Partners to 1961 Homestake-Sapin Partners to 1968 United Nuclear-Homestake Partners after 1968 (1957–1970)	Homestake Mining Co.	Active
Kerr-McGee Nuclear Corp. Ambrosia Lake, NM	7000	Kermac Nuclear Fuels Corp. (1958–1969)	Kerr-McGee Corporation	Active
TVA Edgemont, SD	750	Mines Development, Inc. (1956–1968)	Tennessee Valley Authority	Shutdown, 1974
Atlas Minerals Division Moab, UT	1500	Uranium Reduction Co. to 1962 Atlas Minerals (1956–1970)	Atlas Corporation	Active
Dawn Mining Company Ford, WA	600	Dawn Mining Co. (1957–1965)	Newmont Mining Corporation	Active
Federal-American Partners Gas Hills, WY	950	Federal-Radorock Gas Hills Partners (1959–1969)	Federal-American Partners	Shutdown, 11/81
Pathfinder Mines, Inc. Gas Hills, WY	2800	Utah Construction and Mining Company, later known as Utah International (1958–1970)	Pathfinder Mines, Inc.	Active
Petrotomics Company Shirley Basin, WY	1500	Kerr-McGee-Getty Partnership (1962–1966)	Getty Oil Company	Active
Union Carbide Corp. Gas Hills, WY	1400	Union Carbide Corp. (1960–1970)	Union Carbide Corp.	Active
Western Nuclear, Inc. Jeffrey City, WY	1700	Lost Creek Uranium Western Nuclear Corp. to 1959 Western Nuclear, Inc. (1957–1969)	Phelps Dodge Corp.	Shutdown, 6/81

Source: (DOE 1982b).

requested by Congress, the DOE made no recommendation as to the advisability of implementing legislation authorizing an assistance program. Congress, to date, has taken no further action, and thus no program yet exists to provide assistance to the facilities containing commingled tailings. Since none of the facilities listed in Table 7-8 has undertaken to decommission their sites, the issue of availability of government assistance has not yet been joined. However, a number of shutdown facilities will have to face up to initiating closure activities in the next few years.

If the commingled tailings piles were to be available for incorporation of remedial wastes, and federal government assistance was available, the disposal option could be considered. DOE participation in the negotiations with the facility owner would undoubtedly be required. The cost structures for incorporation of the remediation wastes should reflect the DOE's sharing of closure and reclamation costs, and thus be somewhat reduced from the option of using a tailings pile at a site not possessing commingled tailings. For example, the DOE analysis (DOE 1982a) of the various cost-sharing plans show the government assuming about 40% of the total site reclamation cost. Reductions of comparable magnitudes should be feasible for negotiated arrangements to dispose of outside waste.

The major disadvantage associated with incorporation of the remediated waste in a tailings pile at a commercial mill involves liability questions, namely, uncertainty as to whether the generator of the waste would be relieved of responsibility for the waste after disposal, and the question as to what liabilities the mill owner assumes. In addition, as with any offsite disposal, the shipment of the wastes would incrementally increase the doses received by the population along the transport route and those incurred by onsite waste handlers.

An UMTRAP or FUSRAP Site.   This disposal option would involve the shipment of the wastes from the remediated site to an UMTRAP or FUSRAP site undergoing remediation for incorporation with the onsite wastes.

As described earlier, the UMTRAP and FUSRAP programs are in various stages of planning, engineering, and remediation. The DOE has assigned a priority to each of the designated sites, and is proceeding with the step-by-step process of preparing EISs and preliminary engineering design reports, final designs of remedial action plans, and construction of onsite or offsite disposal facilities. To data, only selected "high"-priority sites have reached the cleanup stage, and the programs have been drawn out by reductions in funding and difficulties in confirming disposal options. The DOE has informally indicated a willingness to consider the incorporation of remediation wastes at an UMTRAP or FUSRAP disposal facility if the projected schedule and costs for the planned RAP were not affected by the incorporation of the remedial site wastes. If arrangements were to be made for the use of an UMTRAP or FUSRAP disposal facility, they would have to be accomplished before final design of the disposal facility was initiated to permit inclusion of the offsite remedial wastes in the remediation plan, and to allow for early shipment of the waste to the disposal site.

To implement this option, a commitment of availability of a specific site would have to be obtained from the DOE, and an arrangement negotiated allocating costs, future liabilities, schedule, and mechanics of the disposal operation. The negotiated costs should be less than other offsite disposal options since only incremental expenses of handling the remedial wastes would be involved. In addition, once agreement of the parties is achieved, no further licensing or regulatory constraints would affect the disposal of the waste.

However, before the use of an UMTRAP or FUSRAP disposal facility can be assumed, written arrangements would have to be made with the DOE and NRC (or state), and the availability of a specific site confirmed. Schedule uncertainties would have to be resolved. Because of the slippage in the RAP schedules, it is difficult to confirm dates of availability of these sites.

A CERCLA-Designated Site. The use of a site on the NPL that is undergoing a cleanup process that will result in the wastes being stored or disposed of onsite presents the possibility for incorporating other remedial wastes under certain limited circumstances. Although there will be a reluctance on the part of local authorities and the public to increase the volume of onsite contaminated material, the option may be feasible for the following situations:

- The site is used as a staging area where temporary storage of the wastes from other sites is performed until a permanent disposal facility is available. This approach would be feasible when the sites are in relatively close proximity to each other, and the wastes contain similar radiological constituents. In 1986, such an approach was proposed by the EPA (EPA 1986) for the interim storage of waste from other locations at the Mentor Co. Site, one of the 31 locations comprising the "Denver Radium Site." It was subsequently dropped because of local opposition.

- There are a number of vicinity sites in the area of the NPL-listed sites that were contaminated from material dispersed from the primary site. When remediation of the primary site involves onsite disposal of the wastes, waste material from the vicinity sites will generally be incorporated into the disposal facility.

- An engineered onsite disposal facility has been developed at a remediated NPL-listed site that has available disposal volume, and where the inclusion of selected offsite wastes does not increase the potential for contaminant mobilization and dispersion.

7.11 TECHNOLOGY OF REMEDIATION AND CLOSURE

The remedial response alternatives described in Section 7.9 are based on the use of certain state-of-the-art technologies for the removal of the contaminated material and disposal in a secure facility, onsite conditioning and treatment, and/or imposition of engineered barriers to contain the contaminated material or

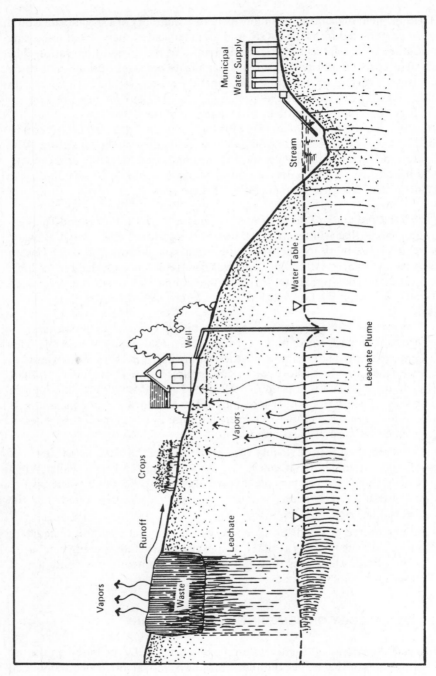

Figure 7-5 Environmental pathways from a generic waste site.

Municipal
Water Supply

Stream

Water Table

Well

Leachate Plume

Vapors

Crops

Runoff

Leachate

Vapors

Waste

eliminate contact with potential transfer media. These technologies are described in this section. Since radioactive waste management treatment and disposal technologies have been covered in detail in Chapters 5 and 6; this section will serve primarily to reference those technologies to the appropriate remedial alternative. Remedial technologies not previously described are dealt with in greater detail.

The intent of the application of the remedial technologies is to remedy the hazards due to the ingestion or potential ingestion of the radiologically contaminated materials from the site through the available environmental pathways (see Figure 7-5). To accomplish this, the remedial program may employ a number of technologies in combination to provide an integrated approach that minimizes the shortcomings of individual approaches. Physical removal of contaminated waste and soil to another site provides a long-term solution at the uncontrolled site, but transfers the contaminant source to the new site where long-term monitoring will be required. In addition, the transfer process creates potential exposure concerns during removal and transport. Containment of the contaminated source will reduce exposure to an acceptable level, but may require long-term maintenance and monitoring to assess the integrity of the containment systems. Treatment of the contaminated material at the site may reduce the volume or toxicity of the waste to within acceptable standards without requiring removal of the material. But treatment technologies are somewhat developmental as applied to the radiologically contaminated sites, generally are costly, and are applicable only in a limited number of situations. Thus, a combination of these approaches may be most effective in achieving long-term control of the hazard. As noted previoualy, the increasing concern about the inability to find satisfactory offsite disposal locations for radioactive waste from remediated sites is also acting as a stimulus for the application of onsite remedial technologies. Combinations involving the use of containment systems built around engineered barriers and treatment systems applied over an extended period may prove to be the most effective way to eliminate the problem; the containment to limit short-term environmental dispersion and potential elevated exposure and treatment to ultimately eliminate the radiological source.

7.11.1 Removal and Disposal of Contaminated Waste and Soil

Waste material and contaminated soil are the primary sources of problems at contaminated sites. Once this material is mobilized it may provide the contamination that is dispersed through air, surface water, and groundwater pathways. Removal of the waste and contaminated soil offers a long-term, permanent solution at the uncontrolled location. It will likely also permit the site to be adapted for alternative uses.

The waste material generally is found in the form of landfilled drummed and/or packaged material, and unconsolidated solid bulk material (e.g., process residue) and components, impounded semisolids (e.g., tailings, sludge) and liquids, and surface-stored piles (e.g., slag heaps, overburden, other process residue) or drummed wastes.

The sites have become problems because the packaging or land barriers have failed and the contaminants have been mobilized. The underlying and adjacent soils have become contaminated because the contaminants have been leached from the waste and seeped into the soil or because the waste has been mixed with the soil. At times, the dispersion of waste constituents over an extended period has been extensive enough to contaminate adjacent vicinity properties.

Excavation. Removal techniques may be used to transfer the waste and soil to offsite treatment and/or disposal facilities or to newly established onsite disposal facilities sited in accordance with 10 CFR 61 requirements. The major removal technology is excavation using either a backhoe or dragline crane to dig up the waste and contaminated soil. The backhoe, a hydraulically powered digging unit mounted either on tracked or wheeled vehicles, is usually preferred since it is more maneauverable and offers more accurate digging bucket placement than the dragline (EPA 1983a). A dragline operates by dragging a bucket into the surface of the ground. A backhoe can be used to remove compacted as well as unconsolidated material, to remove drums when equipped with a sling, and backfill and grade an excavated site or a drained surface impoundment. A dragline generally is used for the removal of consolidated materials. Although the use of this equipment for excavation at construction sites is well demonstrated and routine, in applying excavation techniques to radiologically contaminated sites certain additional precautions must be observed;

1. Equipment operators and other onsite workers must be protected against inhalation of contaminated dust particulates and radon gas (if radium or thorium is present), direct radiation exposure, or ingestion of contaminated soil. To accomplish this rigorous dust suppression, measures such as use of water or chemical sprays and reduced speed of vehicles and earth-moving operations are employed. In addition, exposed personnel may be required to wear protective clothing and respirators if exposure levels are elevated. A health physics plan will be prepared in advance, and implemented through personnel, area, and environmental monitoring during excavation and backfilling operations, and restrictions on onsite working, eating, and other practices that could increase exposure levels. In addition, decontamination of equipment will be required before it is removed from the site.

2. If the site contained buried waste, particularly in drums or packages that may have failed, subsidence may have occurred, or the load-bearing capacity of the overlying soil reduced. These factors must be considered before deploying heavy equipment on the site. In addition, drums must be handled carefully to avoid further leakage with the resultant spread of contamination, or the possibility of explosions or fires from sparks igniting the waste or gaseous vapors.

A significant result of these precautions is to extend the excavation process considerably beyond the time required for routine construction operations. This must be considered in the development of remedial alternative cost estimates and work plans.

Semisolids such as sludge bottoms in surface impoundments or saturated soils can be removed by use of dredging techniques such as centrifugal pumping and hydraulic pipeline dredges, both of which are proven techniques. Since the wastes cannot be packaged for disposal in a licensed burial site with a water content exceeding 1%, dewatering and/or immobilizing treatment would be required prior to drumming. Bulk waste would also have to be dewatered prior to shipment.

Packaging and Handling. After the waste and soil are excavated they are conveyed to a staging area and either packaged or loaded in bulk in ore carriers (or equivalent) for offsite shipment, or for movement onsite to a storage or disposal area. The waste and soil will be shipped in bulk form only when it is high volume, low activity, and generally homogeneous, and does not pose a significant hazard during handling and shipping. The material will be loaded onto open back trucks, sprayed to increase the surface moisture content, and covered with a tarpulin to minimize dispersion of fugitive dust during transport.

The preferred approach for the majority of excavated radioactive waste and soil is to package the material in 55-gallon drums for shipment or storage. While this method is more expensive and time consuming than shipment in bulk, it affords greater protection for remedial workers and the public, and more flexibility in that the drums can be handled, moved, stored, and loaded as units typically without concern about dispersion into the environment.

When corroded or breached drums, or drums whose contents are suspected to be dangerous, are uncovered during remediation they are generally transferred to the staging area, samples taken and analyzed, and incompatible drums segregated to prevent accidents. The contents of damaged drums may be transferred to new drums or placed in an overpack prior to transport.

Packaging of radioactive waste is discussed more fully in Section 5.8.

Transport. Waste and soil from remedial programs are almost always transported to the storage or disposal facility by truck because of the lesser costs and greater accessibility to these facilities than can be provided by rail shipment. Most radioactive waste storage and disposal facilities developed for receipt of remediated waste do not have rail sidings, as also is the case with many of the sites being remediated. In addition, the availability of a range of truck types and sizes to accommodate a variety of shipment types and distances, in conjunction with the general inflexibility of the railroads to adjust for the needs of radioactive waste shippers, further accentuates the advantages of truck shipment.

Transport of radioactive waste and soil from remediation projects generally requires the same types of radioactive waste shipments. The relevant aspects of radioactive waste transport are discussed more fully in Section 5.8.

Disposal. The subject of radioactive waste disposal technology has been comprehensively discussed in Chapter 6. Radioactive waste and soil from remedial projects are disposed of as LLW using land burial techniques (Section 6.8) or surface impoundments (Section 6.9). Since the remediation process has progressed to the disposal stage in only a few of the government-sponsored and industrial remediation programs there is no currently significant backlog of experience to demonstrate the adequacy of the technology to achieve long-term isolation. In many instances public and local government opposition to siting of these disposal facilities has resulted in the waste and soil being temporarily stored.

Where disposal of the waste and soil has occurred, as in the case of a number of the UMTRAP tailings piles (e.g., the Vitro pile in Salt Lake City) and the residue from the Grand Junction Remedial Action Program, the material has been transported in bulk to a new land burial site in a low-populated area. The site was selected, characterized, and qualified in accordance with DOE criteria that to a large extent follow the NRC standards established for licensing of new disposal facilities for by-product material (tailings). In particular, the burial units are designed to contain the material through use of engineered barriers such as liners and caps, water management and diversion systems are incorporated into the design to prevent infiltration and leachate formation, and monitoring capabilities are put in place to validate the long-term success in achieving containment of the material.

New disposal facilities for radioactive remedial wastes, whether the material is derived from NRC licensed facilities, Superfund sites, or other government-sponsored programs, will be designed to meet NRC criteria for LLW burial sites under 10 CFR 61 or tailings disposal facilities under 10 CFR 40 Appendix A depending on the classification of the waste. These facilities may be developed as state compact facilities, secure disposal facilities on industrial sites and re-mediated sites, or DOE-owned facilities for wastes from DOE-sponsored programs.

7.11.2 Treatment and Conditioning Technologies

Treatment and conditioning technologies are employed at radiologically contaminated sites in a limited manner. Since these technologies are also used predominantly for as-generated radioactive wastes, they are discussed in detail in Chapter 5, and will only be summarized here. These technologies are employed at contaminated sites to (1) decontaminate surfaces of buildings and equipment, (2) eliminate or mitigate the hazard-producing aspects of the waste, soil, leachate, or contaminated groundwater, (3) remove contaminants from vented gases prior to release to the atmosphere, and (4) solidify or immobilize the waste or soil prior to packaging and shipping.

Decontamination of Surfaces of Buildings and Equipment. Decontamination technologies involve the removal of deposited radioactivity from contaminated equipment and other solid waste forms by physical and/or chemical methods. They are used at contaminated sites to clean these surfaces so as to restore structures and equipment to a condition to be reused or, prior to demolition, to reduce the potential radioactivity in the waste. Decontamination activities usually involve the use of four processes—hands-on manual decontamination, chemical decontamination, electropolishing, and ultrasonic cleaning.

Elimination or Mitigation of Hazards-Producing Aspects of Media.
Radiologically contaminated soil and solid waste can be treated on- or offsite with the objective of detoxifying, separating, and/or concentrating the material. Since economics of the treatment process is a governing factor, treatment technologies are generally restricted to onsite processes where the cost of packaging, transport, and licensing of a new facility can be avoided. In each instance, the process will leave a contaminated residual that must be disposed of, but since the residual is generally a small fraction of the volume of the original contamination (although typically of higher radionuclide concentration), offsite disposal opportunities are available for the residual that were not available for the larger volume of material. The physical/chemical processes available for use at radiologically contaminated sites for wastes and soil that can be removed and treated include

- Sectioning or cutting of dry solid waste components that have large void volumes within their dimensional envelopes (e.g., tanks, boxes) to eliminate the voids and thus minimize the volume required for final disposal.
- Combustion of ignitable organic materials (e.g., paper, plastics, rubbers, ion-exchange resins, and solvent) to reduce the volume and weight of the material and to convert the material to inert or less reactive forms. Combustion technologies that are being developed for treatment of radioactive wastes and soil include incineration, pyrolysis, and acid digestion. The various combustion processes can reduce the volume of combustible wastes by factors of 20 to 100, depending on the process, the composition of the waste, and the as-generated waste density. The resulting residue may be packaged as is, or immobilized and then packaged.

Although incineration has not been used to treat contaminated wastes at uncontrolled sites, the growing emphasis on onsite stabilization and disposal of these wastes, coupled with improvements in the technology, has increased interest in use of this technology. Incineration is a proven organic waste treatment technology. In general, a rotary-kiln incinerator will be the best type for the waste mixtures found at uncontrolled sites. Mobile incinerators, currently under development, may prove to be most attractive for onsite treatment of radioactive waste having a significant organic material content.

- Separation and recovery of the radioactive constituent in the waste with disposal of the depleted residual onsite. This technique is applicable to wastes and soil containing a separable radionuclide that is predominant in the material, and whose removal would make the residual disposable without significant constraints. One example of a separation technique suitable for treatment of radioactive wastes includes the use of barium chloride as an additive to semisolid (slurry) or liquified tailings or waste containing radium. The resultant barium–radium–chloride compound formed precipitates out and can be removed and recycled leaving a residue that has in excess of 99% of the radium removed. This technique, which is commonly employed in precipitation ponds to treat slurried uranium mill tailings, can also be used at radium-contaminated uncontrolled sites or at offsite processing facilities. Another example is the separation and recovery of thorium from waste (Anderson 1959) that involves the grinding and hot (90°C) water leaching of the thorium-contaminated waste, leaching of residual solids with hot (100°C) H_2SO_4, and precipitation of the thorium from the pregnant sulfate leach liquor using sodium fluoride. Although volume reduction is minimal, since about only 2% of the volume is removed as thorium, the recovery efficiency is 85% which will usually permit the residual to be recycled or disposed of onsite under SECY81-576.

A variation of the separation and recovery technique applied as *in situ* extraction (or solution mining) involves the introduction of a solvent liquid into the waste or soil mass. The contaminants in the waste are removed by the solvent and collected for disposal or treatment on the surface from wells placed to intercept the solvent plume. Although this technique is yet to be applied to an uncontrolled site, solution mining has been used to recover uranium from selected formations in Wyoming and Texas and could readily be applied to sites contaminated with constituents from the uranium and thorium decay chains.

- *In situ* vitrification involves the fusing of the waste into a glassy, stable matrix by heating it in place. One developmental technique of this nature passes an electrical current through the wastes to produce high temperatures that fuse the waste. *In situ* techniques such as vitrification must be carefully applied to avoid the risk of volatilizing other toxic constituents or damaging underground utility lines.

Liquid and semisolid waste (slurries and sludges) can be treated to remove contaminants and thus permit the treated liquid to be disposed of, or to concentrate the liquid (and contaminants) and reduce the volume to be disposed of. Liquids, in the form of contaminated groundwater or leachate, can be pumped to the surface for treatment, or can be treated *in situ*. Generally, the treatment of liquids or semisolids will produce an effluent that can be discharged and a residual sludge or liquid in which the radioactive constituents are concentrated. The processes that are readily available for use at radiologically contaminated sites for treatment of liquids and semisolids include

- Chemical precipitation, or separation, can be used to remove soluble metals from contaminated groundwater or impounded liquids. This is done by adding chemicals to create insoluble forms of the metal, which are then separated by gravitational settling. As noted previously, this technique is used to remove radium from liquid streams.

- Ion-exchange techniques can be used to remove inorganic salts from a liquid waste stream but, because of their relatively high cost, would usually be used as a polishing step for low-solid-content liquids. Ion exchange consists of selective removal of contaminants from liquids onto a resin that also generates secondary wastes in the form of expended ion-exchange media. There are portable systems available.

- Evaporation can be used to reduce the volume of radioactive liquids and to remove water from a solution. Evaporation has been employed to reduce the volume of contaminated water pumped from trenches at the Maxey Flats LLW burial site. Natural evaporation has been employed, where climatological conditions are suitable, to reduce liquid volume in surface impoundments.

- Filtration technology, which removes suspended solids from a solution by forcing the liquid phase through a porous medium and collecting the solid phase on the filter medium, can be used to treat liquid waste streams prior to impoundment, or collected leachate or groundwater prior to further treatment. A variety of filter media is available including screens, cloth, sand, and diatomaceous earth.

- Holdup-for-decay is a conditioning process that has been used for selected low-level liquid wastes at DOE facilities. The concept of time delay in the discharge of liquids to permit decay is useful when the radioactive constituents have short half-lives. Under these conditions the contaminated leachate, groundwater, or surface water can be collected and held in storage tanks or surface impoundments, and then released to the environment when activity levels have dropped below regulatory release limits.

- Activated carbon adsorption is a physical treatment that can be used to remove complex mixtures of organic contaminants from leachate or groundwater. The contaminants are bound to the carbon by physical and/or chemical forces (EPA 1983b).

It is generally necessary to apply several treatment techniques in sequence (process train) to remove complex mixtures of contaminants from collected leachate or groundwater. Although this approach is commonly used to treat liquid wastes at nuclear fuel cycle facilities, to date it has not been applied extensively to radiologically contaminated sites.

In addition to the techniques available for treating liquid wastes, semisolids (slurries and sludges) are also subject to treatment to remove the water prior to immobilization or disposal. These treatment techniques can generally be categorized as dewatering concepts, in which pumped aqueous solutions are subject to physical processes (e.g., filtering, and screening) to separate the liquid phase, and/or a combination of flocculation/precipitation and sedimentation. Flocculation is the process whereby small unsettleable particles suspended in a liquid are made to agglomerate into large particles. Precipitation would be used to remove the larger particles from solution. Sedimentation is a purely physical process by which these particles can also be made to settle out by means of gravitational and inertial forces. Although flocculation, precipitation, and sedimentation are individual process steps, they are often combined into a single overall treatment process. Again, although these techniques are available for use at uncontrolled sites, their primary application has been for treatment of semisolid process streams at nuclear fuel cycle facilities.

Removal of Contaminants from Vented Gases. Engineered approaches for the control of airborne emissions from contaminated sites are discussed in Section 7.11.3. When the gases are collected and vented to the environment they are usually treated prior to release to remove the hazardous constituents. The treatment techniques include

- Adsorption to fix the hazardous component on an adsorbing material, usually activated carbon.
- Absorption to dissolve constituents in the gas by passing the gas through a liquid or spraying a liquid through the gas in spray towers.
- Incineration of organic gases such as methane using auxiliary fuel to achieve the high temperature necessary to destroy the hazardous components.

Treatment of vented gases at a radiologically contaminated site is generally not practical. Control of radon release is generally accomplished through soil or waste removal or site capping. Generation of other gases has not represented a major problem except when a significant quantity of organics has been present in buried waste. Where burial trenches have been vented to remove these organic gases the radionuclide concentrations have been generally low enough to permit release without treatment. A more significant problem associated with decomposition of organics such as cartons and boxes is the formation of voids and the resultant subsidence in the trench covers.

Solidification (Immobilization) of the Waste. Solidification systems are used to immobilize inorganic waste within an inert matrix. Solidification is required for liquid wastes to achieve regulatory limits on water in packages. It is also used for the stabilization of semisolid and solid wastes. There are a number of binder materials commercially available, the selection of which will be dictated by the characteristics of the waste and the relative economics of the competing techniques. Although most waste and soil from a remediated site are shipped offsite in unpackaged bulk form, packaged essentially water free and not stabilized, there are situations in which liquid and/or semisolid wastes will require stabilization and packaging prior to shipment.

Among the available binder materials, cement has the longest record of experience and is applicable to a wide range of waste compositions. The cement is either used as a binder by itself or mixed with a material such as fly ash or cement kiln dust.

The waste stream is slurried into this mixture and allowed to set, creating a volume typically twice that of the original waste volume. Thermoplastic binding using asphalt is a competing technique that, although generally more expensive than cementation, provides certain improvements in leaching characteristics and greater volume reduction factors. Solidification technologies are discussed in greater detail in Chapter 5.

A variation on straight solidification involves the use of integrated volume reduction/solidification systems that achieve a volume reduction factor higher than that achievable by separate evaporation and solidification operations.

7.11.3 Control through Use of Onsite Engineered Technologies

Control technologies are used to both contain the existing region of contamination and prevent further dispersion of the contamination from this source. The control technologies function to (1) prevent airborne dispersion of particulates and gases through use of surface barriers, (2) contain the leachate or groundwater plume through use of in-ground barriers, or (3) prevent contamination of surface or groundwater by diverting flow away from the contaminated region.

Air Control Technologies. Airborne releases of radioactive particulates from uncontrolled sites occur primarily as a result of wind erosion of the surface of the waste pile or contaminated soil. Human activity that disturbs the surface may also release particulates into the air. The release of ^{222}Rn gas from soil or waste containing the ^{226}Ra precursor is a major problem at sites in which the waste material contains constituents of the ^{238}U or ^{232}Th decay chains. Since a large number of sites requiring remediation are contaminated with uranium and/or radium residues, the control of radon release into the air receives a good deal of attention.

The gaseous radon that is generated from decay of ^{226}Ra diffuses through the soil and waste and is released from the surface. The diffusion rate is a function of source strength, the characteristics of the soil such as moisture content and permeability, and local climatological conditions such as temperature, pressure, and wind speed. Radon is an inert gas, with a half-life of 3.8 days, which decays to produce daughter products of ^{218}Po, ^{214}Pb and ^{214}Bi often referred to as radium A, B, and C. The daughter products are short-lived α emitters that attach themselves to dust particles and may be inhaled. These daughters may be a cause of bronchogenic cancer to exposed individuals such as miners, mill workers, or potentially people in poorly ventilated structures in close proximity to certain radiologically contaminated sites.

The state-of-the-art control technology to prevent airborne release of particulates and/or radon gas from the *in situ* waste is to place an engineered cap over the waste or soil composed of material or materials of low diffusion characteristics. Natural deposits of bentonite or montmorillonite clay, if found in the region of the site, provide an ideal natural cover because of their high density, low permeability and diffusivity, and excellent resistance to erosion and surface traffic. In the absence of clay-based material, other natural local soils of greater thickness are generally used either singly or in combination to achieve comparable reductions in radon emanation rate.

The standards applied by the federal agencies, as defined in 40CFR192 "Standards for Remedial Actions at Inactive Uranium Processing Sites," limit the average radon emission from the surface of tailings piles to no more than 20 pCi/m²-sec (above background) or 0.5 pCi/liter in air outside the disposal site. This standard has been applied by the DOE to stabilized tailings under the UMTRAP program and residue and wastes under the FUSRAP program, and by the NRC to remediation of licensed facilities.

The analytical methodology used to establish the relationship between ^{226}Ra source concentration, ^{222}Rn flux from the soil and cover surfaces, cover thickness, and cover material characteristics relates to the configuration shown in Figure 7-6.

The steady-state diffusion equation governing radon diffusion from radium-contaminated soil or waste is

$$Dd^2C/dx^2 - \lambda C + R\rho\lambda E/P = 0 \qquad (7\text{-}1)$$

where C = radon concentration in pore space (pCi/cm³)
D = diffusion coefficient for radon (cm²/sec)
λ = decay constant of radon (2.1×10^{-6} sec^{-1})
R = specific activity of radium in soil (pCi/g)
ρ = dry bulk density of soil (g/cm³)
E = radon emanation coefficient
P = total porosity

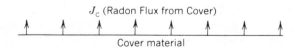

J_c (Radon Flux from Cover)

Cover material

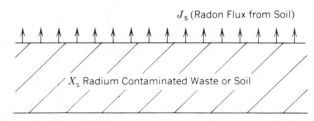

J_s (Radon Flux from Soil)

X_s Radium Contaminated Waste or Soil

Figure 7-6   Configuration of cover placed over material emitting radon gas.

The radon emission (flux) from the surface of a bulk material is related to the radon concentration in the pore space of the material through Fick's law.

$$J = 10^4 \, DP \, dC/dx \tag{7-2}$$

where J = surface radon flux (pCi/m^2-sec)
X = thickness of bulk material (cm)

Equation (7-1) and (7-2) can be solved for the cases where (1) there is no cover over the radium-contaminated waste or soil and (2) there is a cover applied to the waste or soil to reduce the radon flux at the surface.

1. Bulk Waste or Soil: Boundary conditions:

$$dC/dx(\text{at } X = X_s) = 0, \qquad C(X = 0) = 0$$

The solution to (7-1) and (7-2) is

$$J_s = 10^4 R\rho E \sqrt{\lambda D_s} \tanh\sqrt{\lambda/D_s}\, X_s \tag{7-3}$$

where J_s = radon flux from waste or soil surface (pCi/m^2-sec)
s = waste or soil region

2. Covered Waste or Soil: Boundary conditions: Flux is continuous from soil to cover, or

$$\frac{C_s}{1 - (1 - k)m_s} = \frac{C_c}{1 - (1 - k)m_c}$$

and

$$P_s D_s \frac{dC_s}{dx} = P_c D_c \frac{dC_c}{dx}$$

It is also assumed that the radium source term in the cover is negligible. The solution to (7-1) and (7-2) is

$$J_c = \frac{2J_s e^{-b_c X_c}}{[1 + \sqrt{a_s/a_c}\tanh(b_s X_s)] + [1 - \sqrt{a_s/a_c}\tanh(b_s X_s)]e^{-2b_c X_c}} \tag{7-4}$$

where $b_i = \sqrt{\lambda/D_i}(i = c \text{ or } s)(\text{cm}^{-1})$
$a_i = P_i^2 D_i[1 - (1 - K)m_i]^2(\text{cm}^2/\text{sec})$
$m = 10^{-2}\rho M/p$ = fractional moisture saturation
M = moisture content (dry weight percent)

$$K = 0.26 \frac{\text{pCi/cm}^3 \text{ in water}}{\text{pCi/cm}^3 \text{ in air}}$$

c = cover material

When using this relationship to solve for the required cover thickness, X_c, for a given flux criterion, J_c, Eq. (7-4) can be rearranged to obtain

$$x_c = \sqrt{D_c/\lambda} \ln\left[\frac{2J_s/J_c}{(1 + \sqrt{a_s/a_c} \tanh b_s X_s) + (1 - \sqrt{a_s/a_c} \tanh b_s X_s)(J_c/J_s)^2}\right]$$
(7-5)

The following example illustrates the application of this methodology. Representative tailings pile parameters are

$$R = 400 \text{ pCi/g}, \qquad \rho = 1.5 \text{ g/cm}^3, \qquad E = 0.2$$

$$D_s = 1.3 \times 10^{-2} \text{ cm}^2/\text{sec}, \qquad P_s = 0.44, \qquad M_s = 11.7\%, \qquad X_s = 300 \text{ cm}$$

Typical cover material parameters are

	$D(\text{cm}^2/\text{sec})$	P	$M(\%)$	m
Clay	0.0078	0.30	6.3	0.4
Natural overburden	0.022	0.37	5.4	0.25

The radon flux from the surface of the bare tailings is, from Eq. (7-3),

$$J_s = 198 \text{ pCi/m}^2\text{-sec}$$

The cover thickness over these tailings to reduce the surface flux to the federal standard of 20 pCi/m²-sec is, from Eq. (7-5), for a clay cover,

$$X_c = 118 \text{ cm} \qquad \text{or} \qquad 3.9 \text{ ft}$$

Similarly, for a natural overburden cover, $X_c = 179$ cm or 5.9 ft.

The establishment of a stabilizing cover over the radium-contaminated soil also serves to reduce γ exposure above the remediated site. Each foot of cover material, regardless of the material composition, will reduce the γ exposure at the surface by approximately one order of magnitude (factor of 10). Although the calculation of cover thickness for a clay-based material will usually show that a relatively thin cover (2–3 ft) is suitable to reduce radon emanation to within the 20 pCi/m²-sec standard, the regulatory agencies will typically require that a thicker cover be used, generally a composite of the clay with an overlayer of a local soil, to provide a soil base to establish a vegetative cover, and ensure long-term stability of the cover under natural erosion and man-made factors. The final cover thickness will generally be sufficient to reduce γ radiation exposure at the surface to well within the accepted standard of 57 μR/hr, which derives from the 10 CFR 20 limit of 0.5 R/yr for exposure of nonradiation workers.

The 3.9 ft of clay cover calculated for the representative tailings pile will also reduce the γ radiation level at the surface of the cover by a factor of $10^{3.9}$. Thus, for a representative γ dose of 8000 mrem/yr from the bare tailings, the dose at the surface of the clay cover will be approximately 0.8 μrem. For the 5.9 ft of natural overburden, the surface dose will be reduced to approximately 0.008 μrem.

Other gases may also be released from the contaminated site. If organic materials such as cellulose, cardboard, and animal carcasses are present, the organic compounds may continuously volatilize and produce gases such as methane and hydrogen sulfide that diffuse from the surface. Furthermore, chemical process residues contained in the waste and/or soil (mixed waste) could volatilize or react to produce gaseous emissions. These gases, in addition to being potentially hazardous, also may carry radioactive particles into the air. Continued generation of gases under an emplaced cover could lead to cracking or perforation of the cover if pressure buildup exceeds the covering material's working limits, or to lateral flow through the soil if the cover retains its integrity.

Remedial approaches used to treat these gaseous emissions in addition to site covers include (1) removal of the source of the gaseous release by excavation of the waste, (2) increasing the moisture in surface and underlying soil layers by wetting or irrigation, which both reduces the rate of gaseous emission at the surface and increases the rate of biological decay, and (3) stripping the entrained and dissolved gases from liquid wastes in surface impoundments or increasing the gases' solubility by adding chemicals.

When a cover is emplaced on the site, and no means to control organic or chemical gas production is provided, it is necessary to collect the gases and vent them to the atmosphere, first treating them to remove radioactive particles and other hazardous constituents. Collection and venting can be accomplished by using ditches, pipe vents, and barriers. The ditches are filled with rock or gravel and located in or around the site to permit the gases entering the trench to flow to a central collection point or to a surface outlet and then the atmosphere. When pipe or trench vents are used, perforated pipe is laid through or around the site to interrupt the gases and provide a pathway to a controlled release point. The incorporation of gas barriers in the site with a collection system helps to channel the gas flow toward the collection or release point. These barriers are similar to the groundwater barriers discussed in the following subsection. If the natural pressure in the collected gases is not sufficient to drive the gases through the pipes or ditches, pumps may be required.

Fugitive airborne emissions of particulates can also occur during the remediation process from the excavation of waste and soil materials and from vehicular traffic on exposed surfaces and haul roads. Fugitive dust emissions from roadways are generally controlled through use of oil or calcium chloride, or by periodic wetting of the surface. When dust release occurs from excavation activities, surface spraying with water to control the release is usually performed. Chemical stabilizers can also be used to bind the surface and create a thin crust.

When sites are remediated using a surface cover, a stabilizing cover is placed on top of this barrier. Vegetative covers using indigenous species provide a natural permanent control of dust release. When a vegetative cover cannot be maintained, physical stabilizers such as rock, bark, wood chips, or gravel can be applied to the exposed surface. In selecting plant species for vegetative covers on radiologically contaminated sites, it is essential to avoid species with deep-growing root systems that penetrate into the contaminated soil and pick up the radioactive constituents from the soil. It is also necessary to avoid species that preferentially concentrate radionuclides in their leaves and other edible material that could enter the food chain through consumption by grazing animals. Plants

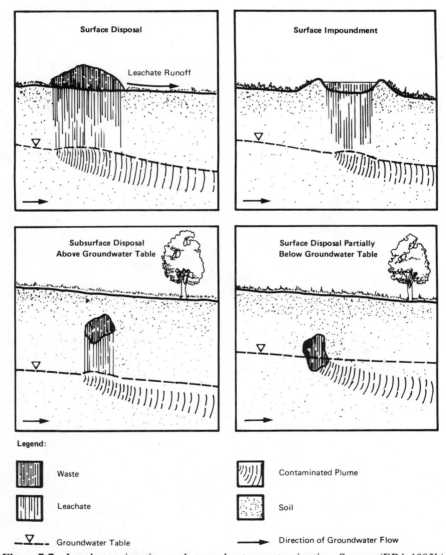

Figure 7-7 Leachate migration and groundwater contamination. Source: (EPA 1983b).

should also be selected with the site end use in mind and consistent with the need to avoid active maintenance during the postclosure phase.

Groundwater Control Technologies. Groundwater contamination results from the migration of leachate from the radioactive waste or contaminated soil. Water percolating through surface waste piles, landfills and trenches, surface impoundments, and underlying contaminated soil can pick up suspended particles to form a leachate. The leachate migrates from these sources to the groundwater through a number of paths, as shown in Figure 7-7 (EPA 1983b). Human exposure to the leachate can occur in a number of ways:

1. Wells used as a drinking water supply, crop irrigation, food processing, etc. can become contaminated from the leachate plume in the groundwater.
2. The leachate plume moving through the groundwater can flow into an intersecting surface water body that may serve as a potable water supply or support biota life.
3. The leachate generated from surface piles can run off on the surface and contaminate nearby surface waters, or enter the food chain by contaminating crops.

The permissible concentrations of radionuclides in groundwater (and surface water) that may be directly consumed or enter the food chain are generally taken as the federal limits in the Drinking Water Standards. These limits are 15 pCi/L for α concentrations, 50 pCi/L for β concentrations, and 5 pCi/L combined ^{226}Ra/^{228}Ra concentrations.

Groundwater control technologies involve the prevention or minimization of leachate formation through reduction of surface infiltration, which is accomplished through use of impermeable surface barriers and/or by management and diversion of surface water flow away from the waste disposal units; engineered barriers located up-gradient of the waste that prevent the groundwater from contacting the waste, or down-gradient of the waste to contain the leachate-contaminated water; and the use of leachate and groundwater collection and removal systems to intercept the leachate plume and move it to a discharge outlet where the contaminated water can be treated and disposed or stored. These technologies are often used in combination with each other.

The barrier and containment control technologies that are now state-of-the-art include the use of slurry and other walls, grout curtains, surface seals (covers or caps), groundwater pumping, and subsurface drains.

Slurry walls (see Figure 7-8) are "fixed underground physical barriers formed by pumping slurry, usually a soil or cement, bentonite (clay), and water mixture into a trench as excavation proceeds" (EPA 1983b). The slurry is then allowed to set if a cement–bentonite material is used or backfilled with a suitable engineered material if a soil–bentonite slurry is used. Slurry walls are used to contain or divert contaminated groundwater, divert uncontaminated groundwater around the waste mass or contaminated soil, or provide a hydrologic barrier for

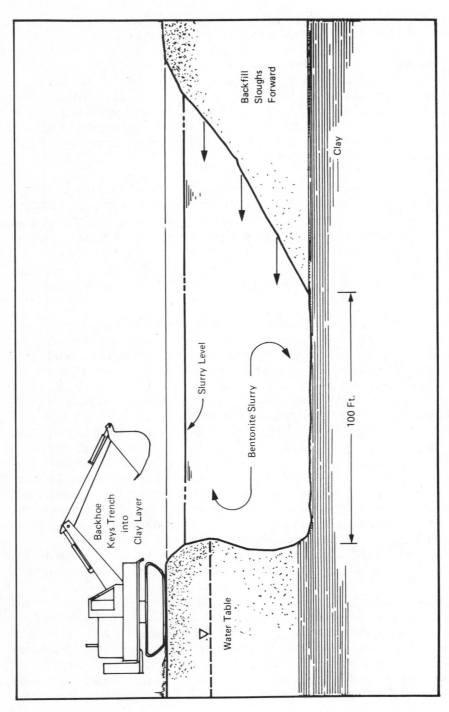

Figure 7-8 Construction of a bentonite slurry wall. Source: (EPA 1983b).

a groundwater treatment system. Slurry walls are a proven technology that continues to be improved as applications increase. It is important to verify that the wall is contiguous and, if possible, extends over to bedrock of comparable low permeability.

Grout curtains are also used to contain or divert contaminated groundwater, or divert uncontaminated groundwater around the waste mass or contaminated soil. They are also fixed underground physical barriers, but are formed by injecting grout into the ground through pipes at a series of well points in a pattern typically of two or three adjacent rows. They are useful under certain site-specific conditions, but it is difficult to determine whether a contiguous wall has been formed, and the maximum depth is limited by depth of the injection well. A range of grout materials is available whose selection is dependent on compatibility with site conditions such as soil permeability and grain size, groundwater flow rate, chemical constituents of the groundwater, and grout strength required. Obviously cost will be a major consideration. Typically portland cement or other particulate grouts are used whenever possible (90% of all grouting in the United States), with chemical grouts such as sodium silicate used for special situations.

Surface seals, in the form of caps or covers, are impermeable barriers placed over a waste disposal site or contaminated soil that serve to reduce surface water infiltration, water and wind erosion, and dust and gaseous emission. Although, in theory, a surface water control technology, they are included here because infiltration of surface water is a primary cause of leachate formation. Seals can also be used as a base to support vegetative growth, and for postclosure use as roadway subfoundations. The technique has been frequently used for closure of properly designed disposal sites, such as stabilizing uranium tailings piles or LLW disposal trenches. Its long-term stability and effectiveness in reducing leachate formation and migration at uncontrolled sites have yet to be demonstrated. Many combinations of impermeable materials, such as soils and clay, asphalt concrete and soil cement, or reinforced membranes of plastic or rubber, can be used to create the impermeable cover, with each layer providing a different function. Generally the surface seal will consist of a barrier layer of low permeability sandwiched between buffer soil layers to prevent cracking, wear, drying, or other damage to the barrier and to support vegetative growth. Filter layers to prevent fine particles of the barrier from penetrating the coarse buffer layer and a gas channeling layer of sand or gravel directly above the waste may also be included. Two representative cover seals are shown in Figure 7-9. Table 7-9 ranks the performance of a range of soils as to their effectiveness as surface seals, and Table 7-10 describes the advantages and disadvantages of available man-made or refined products to be used as impermeable barrier layers.

Similar types of materials are used as liners for newly constructed impoundments and landfills to contain the leachate in the disposal unit and prevent migration into the groundwater. This technology is not strictly a remediation approach since it can be used only on a newly developed disposal site or a new location for the contaminated waste. It is discussed in Chapter 6.

Groundwater pumping is a technique that can be used to lower the water table

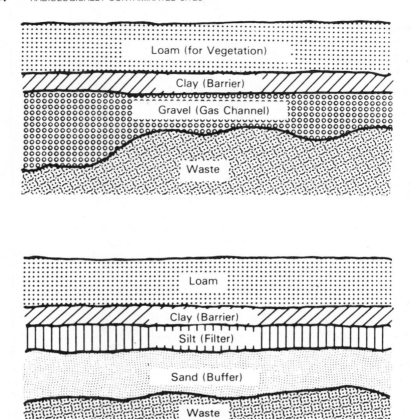

Figure 7-9   Typical surface seals. Source: (EPA 1983b).

and contain a leachate plume. It is a relatively expensive conventional technology that can be used in cases of extensive groundwater contamination or rapid plume movement in conjunction with impermeable barriers to prevent contamination from spreading into valuable water resources. Conversely, pumping may be the most practical approach to remove the leachate when groundwater conditions are stagnant. The concept involves emplacing a well system at the proper location to intercept the plume front, and pumping the groundwater to the surface where it may be treated and discharged. There are two variations to the concept, depending on the depth to groundwater. Deep well systems, which are similar to monitoring wells in construction, are used in aquifers located in depths up to several hundred meters. In shallow unconfined aquifers (up to 10 m deep), a well point system consisting of a series of riser pipes secured at the bottom and connected to a common header pipe and a centrifugal pump is employed. A typical well point dewatering system is shown in Figure 7-10 (EPA 1983b). Well spacing and location in both cases depend on individual site conditions and plume

TABLE 7-9 Ranking of USCS Soil Types According to Performance of Cover Function[a]

USCS Symbol	Typical soils	Trafficability	Water Infiltration		Gas Migration		Erosion Control		Crack Resistance	Support Vegetation
			Impede	Assist	Impede	Assist	Water	Wind		
GW	Well graded gravels, gravel–sand mixtures, little or no fines	E	F	G	F	E	E	E	E	F
GP	Poorly graded gravels, gravel–sand mixtures, little or no fines	E	P	E	F	E	E	E	E	F
GM	Silty gravels, gravel–sand–silt mixtures	G	F	F	F	G	G	G	G	F
GC	Clayey gravels, gravel–sand–clay mixtures	G	G	F	G	F	G	G	G	G
SW	Well-graded sands, gravelly sands, little or no fines	E	F	G	F	G	E	E	E	F
SP	Poorly graded sands, gravelly sands, little or no fines	E	P	E	F	G	E	E	E	F
SM	Silty sands, sand–silt mixtures	E–G	F	G	F	G	F	G	E	E
SC	Clayey sands, sand–silt mixtures	G–F	F	F	G	F	F	F	G	E
ML	Inorganic silts and very fine sands, rock flour, silty or clayey fine sands, or clayey silts with slight plasticity	G–F	G	F	G	F	P	F	G	G
CL	Inorganic clays of low to medium plasticity, gravelly clays, sandy clays, silty clays, lean clays	F	E	P	E	F	P	F	F	F
OL	Organic silts and organic silty clay of low plasticity	F	—	—	—	—	P	F	F	G
MH	Inorganic silts, micaceous or diatomaceous fine sandy or silty soils, elastic silts	F	G	F	—	—	F	F	F	G
CH	Inorganic clays of high plasticity, fat clays	F	E	P	E	F	F	F	F	F
OH	Organic clays of medium to high plasticity, organic silts	P	—	—	—	—	F	—	F	F
Pt	Peat and other highly organic soils	P	—	—	—	—	G	—	—	G

Source: (EPA 1983b)

[a]Key: E, excellent; G, good; F, fair, P, poor.

TABLE 7-10 Available Products for Surface Seals

Material and Description	Advantages	Disadvantages
Bitumen cements or concretes (AC-40 and AC-20 viscosity grades)	a. Provide tight, impervious barriers covering municipal/hazardous waste. b. Good availability. c. May be used as thick waterproofing layers in flat areas or on slopes.	a. Expensive b. Special heating and storage equipment required for handling. c. Vulnerable to breaking.
Portland cements or concretes (3000 and 5000 psi)	a. Good availability. b. Provides good highly impermeable containers or covers for hazardous waste disposal. Very low water permeability.	a. May crack during curing, allowing potential paths for escaping gases or infiltrating water. b. Leakage from hazardous wastes in liquid form may weaken concrete with time.
Liquid and emulsified asphalts (RC and EC 30, 70, 250, 800, and 3000 liquid asphalts; RS's and CRS's 1 and 2, MS's emulsion)	a. Can be sprayed on soil covers to decrease water and gas permeability. b. Can be mixed with soil to form waterproof layer. c. Penetrate open surfaces, plug voids, then cure. d. Penetrate tight surfaces, plug voids, then cure. e. Provide hard, tight, stable membrane (RC and MC 800 and 3000).	a. Must leave sprayed surface exposed until it either cures (RC's. MC's) or sets (SS's). b. Must be covered for protection. c. Require additional equipment to handle and apply the asphalts. Spraying temperatures range from 75 to 270°F (25 to 130°C). d. Use of RC and MC 800 and 3000 in thick membrane construction may require numerous applications.

Tars
(RT 1, 2, 3, 4, 7, 8, and 9;
RTCB 5 and 6).

a. Can be sprayed on soil surfaces or mixed with particles. Tars mix well with wet aggregate.

b. Penetrate tightly bonded soil surfaces and plug voids (RT 1 and 2).

c. Penetrate loosely bonded fine aggregate surfaces and plug voids (RT 2, 3, and 4).

d. Penetrate loosely bonded coarse aggregate surfaces and plug voids (RT 3 and 4).

e. Low spray on temperatures 60–150°F (15–65°C) for RT 1, 2, 3, and 4 and RTCB 5 and 6.

f. Provide hard, tight, stable surface membrane (RT 7, 8, and 9). May be used in flat areas or on slopes.

g. Provide good penetration, then cure to form hard surface (RTCB 5 and 6).

a. Tar may be removed by traffic if not covered with a protective soil layer.

b. Tars are more susceptible to weathering effects than asphalts. Must be protected from weathering.

c. Require special equipment for handling and application.

d. RT 7, 8, 9 require application temperatures of 150–225°F (65–105°C).

Bituminous fabrics

a. Require minimal special equipment and skill.

b. Resist tearing.

a. Expensive.

Commercial polymeric membranes
Butyl rubber

a. Available in various size sheets.

b. Can be reinforced with fibers for added strength.

c. Can be joined at seams to cover.

d. Good availability

e. Good heat resistance.

f. Very low water permeability.

g. Low vapor transmissivity.

a. Poor resistance to weathering and abrasion.

b. May be damaged by gnawing/burrowing animals if not protected with soil.

c. May be damaged by heavy equipment operating directly on surface and may be punctured by large stones or sharp edges in direct contact.

b. Laps should be sealed.

TABLE 7-10 (*Continued*)

Material and Description	Advantages	Disadvantages
Neoprene rubber (chloroprene rubber)	a. Good resistance to oils, grease, gasoline, acids, and alkalis. b. Good resistance to abrasion, weathering, and flexing. c. Can be joined at seams to cover large areas. d. Can be reinforced with fibers for added strength. e. Very low water permeability.	a. More expensive than other natural and synthetic rubbers. b. Use is limited to special applications because of *a* above. c. May be damaged by gnawing/burrowing animals if not protected with a soil layer. d. May be damaged by heavy equipment operating directly on surface and may be punctured by large stones or sharp edges in direct contact.
Hypalon (chlorinated chlorosulfinated polyethylene)	a. Outstanding resistance to abrasion and weathering. b. Available in various size sheets. c. Can be fiber reinforced for added strength.	a. May be damaged by gnawing, burrowing animals if not protected with a soil layer. b. Does not perform satisfactorily when exposed to amyl acetate, benzene, carbon tetrachloride, creosote oil, cyclohexane, dioctyl phthalate, ethyl acetate, lacquer methylene chloride, naphthalene, nitrobenzene, oleum, toluene, tributyl phosphate, trichloroethylene, turpentine, and xylene. c. For good seam quality, the weather must be at least 50°F (10°C) and sunny. If not, heat has to be applied to seams to develop full early strength.

d. Can be joined at seams to cover large areas. This can be done onsite or at factory.

e. Very low water permeability.

d. May be damaged by heavy equipment operating directly on membrane.

e. May be punctured by large stones or sharp edges in direct contact.

Polyolefin (polyethlene and chlorinated polyethylene)

a. Available in various sizes.

b. Can be joined at seams to cover large areas.

c. Can be fiber reinforced.

d. Chlorinated polyethylene has excellent outdoor durability.

e. Very low water permeability.

a. May be damaged by gnawing/burrowing animals if not protected with a soil layer.

b. May be damaged by heavy equipment operating directly on membrane.

c. May be punctured by large stones or sharp edges in direct contact.

d. Polyethylene has poor durability when exposed.

Elasticized polyolefin (3110)

a. Can be joined at seams to cover large areas. Field bonding of individual sheets is done using heat seaming techniques.

b. Excellent resistance to soil microorganisms, extremes of weather, and ozone attack.

c. Very low water permeability.

a. May be damaged by gnawing/burrowing animals if not protected with a soil layer.

b. May be damaged by heavy equipment operating directly on membrane.

c. May be punctured by large stones or sharp edges in direct contact with membrane.

PVC (polyvinyl chloride)

a. Fair outdoor durability.

b. Available in sheets of various sizes. Factory seaming available.

c. Seams can be bonded in the field with vinyl to vinyl adhesive.

a. May be damaged by gnawing/burrowing animals if not protected with a soil layer.

b. For extended life, this membrane must be covered with soil or other material.

c. May be damaged by heavy equipment.

TABLE 7-10 (*Continued*)

Material and Description	Advantages	Disadvantages
	d. Generally used without reinforcement, however, can be fiber reinforced for special applications. e. Very low water permeability. f. Less permeable to gas than polyethylene.	d. Not as durable as hypalon or chlorinated polyethylene. e. Becomes stiff in cold weather.
EPDM (ethylene-propylene-unsaturated diene terpolymer).	a. Good outdoor durability. Ozone and oxidation resistant. b. Sheets may be bonded to cover large areas. c. Very low water permeability.	a. May be damaged by gnawing/burrowing animals if not protected with a soil layer. b. May be damaged by heavy equipment operating directly on surface and may be punctured by large stones or sharp edges in direct contact.
Sulfur (thermoplastic coating) (molten sulfur)	a. Can be formulated for a wide range of viscosities. b. Can be sprayed on various materials to act as a bonding agent. c. Reduces permeability. d. Resistant to weather extremes (subfreezing to very hot). e. Resistant to acids and salts. f. Can be mixed with fine aggregate to form a type of concrete.	a. Requires high temperatures for workability, 250–300°F (20–150°C). b. Requires special equipment for handling and application. c. May not tolerate much shear deformation. d. If applied to hazardous waste containers prior to land disposal, heat absorption by volatile wastes may cause gas expansion and possible explosion hazards.
Bentonite	a. No special equipment needed. b. Can be mixed with soil.	a. Difficult to handle and spread after wetting. b. Susceptible to shrink–swell.

Source: (EPA 1983b).

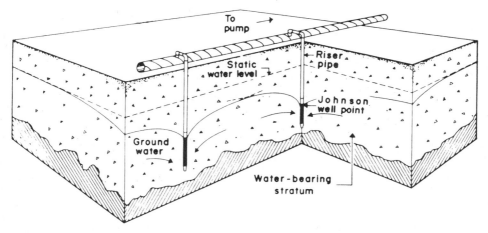

Figure 7-10 Schematic of a well point dewatering system. Source (EPA 1983b).

movement. The wells should also be close enough to permit appropriate drawdown between adjacent wells.

A technique that can be used to collect and divert leachate at a contaminated site involves the emplacement of subsurface drains. This is accomplished by placing tile or perforated pipe in a gradually sloped trench that is then filled with a porous material such as gravel around the tiles, and then backfilling the remainder of the trench with an impermeable material such as clay. The leachate flows through the porous envelope around the drain, and then through the drain or pipe into a collecting point for treatment and disposal. Subsurface drains are conventional components of construction sites. The drain depth and spacing between adjacent drains are functions of site-specific conditions such as soil permeability and the hydraulic capacity of the drain. As a rule, the deeper the

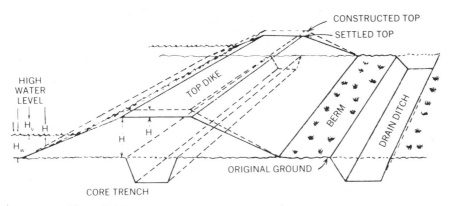

Figure 7-11 Typical dike cross section. Source: (EPA 1983b).

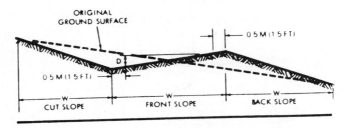

BROADBASE TERRACE CROSS SECTION

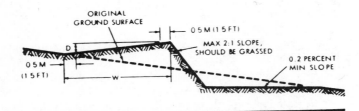

STEEP-BACKSLOPE TERRACE CROSS SECTION

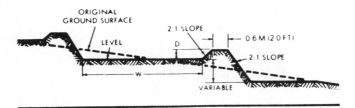

BENCH TERRACE CROSS SECTION

Figure 7-12 Typical terrace cross sections. Source: (EPA 1983b).

392

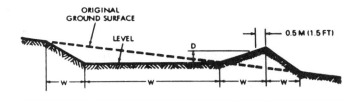

FLAT CHANNEL TERRACE OR ZINGG CONSERVATION BENCH TERRACE CROSS SECTION

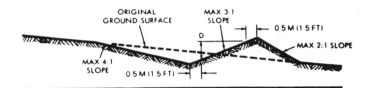

NARROW-BASE TERRACE CROSS SECTION. Slopes are the maximum allowable and should be grassed.

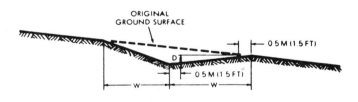

RIDGELESS CHANNEL TERRACE CROSS SECTION

Figure 7-12 *(Continued)*

drain is placed, the wider the spacing between drains, and therefore the fewer number of drains required. Drain materials are selected for compatibility with the soil and leachate conditions, with sufficient strength to withstand loading from equipment movement on the overlying fill. Fired clay is a commonly used drain material.

Other groundwater control technologies such as use of sheet pile cutoff walls or the block displacement method have been developed for use as barriers to contain or divert contaminated groundwater. They have not, however, received significant consideration for use at radiologically contaminated sites and will not be discussed here. Data on these techniques are available in the literature (EPA 1983b).

Surface Water Control Technologies. Surface water contamination results primarily from contact with waste on the surface of the site or with contaminated surface soils. The contaminated surface water can then flow offsite or into the ground. Contamination can also occur from deposition of airborne particulates in surface water bodies. Surface water control technologies are employed to prevent the contact between the surface water and the waste or soil, by preventing run-on from offsite sources or infiltration through the site surface to the waste, to divert, collect, and treat already contaminated surface water before reaching the site boundary, and to control site surface water and wind erosion. Thus, surface water control technologies are designed to prevent water run-on, control

TABLE 7-11 Surface Water Control Technologies

	Primary Function					
Technology	Minimize Run-on	Minimize Infiltration	Reduce Erosion	Protection from Flooding	Collect and Transfer Water	Discharge Water
Flood control dikes				×		
Runoff control dikes	×		×			
Terraces	×		×			
Channels			×		×	
Chutes			×		×	
Downpipes			×		×	
Grading	×		×			
Surface seals		×				
Vegetation	×	×	×			
Seepage basins						×
Seepage ditches						×

Source: (EPA 1983b).

infiltration, prevent erosion, collect and transfer water, store and discharge water, and protect against flooding (EPA 1983b). There are a range of technologies, applied individually or in combination, that are used to accomplish these functions. They are listed in Table 7-11 (EPA 1983b) and the ones of primary applicability to radiological contaminated sites are summarized below. These surface water control technologies may also be incorporated into the design of new LLW disposal sites.

Surface water runoff from precipitation or flooding can be controlled by compacted earthen ridges called dikes. A typical dike cross section is shown in Figure 7-11 (EPA 1983b). The dike may require a core of low-permeability material to eliminate seepage. The design of flood control and runoff control dikes will differ in configuration and construction. Dikes may be used in combination with terraces or other techniques to divert water away from the site. Dike design is governed by state, local, and industry codes.

Terraces are used to intercept and divert surface flow away from a site and

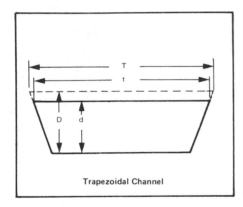

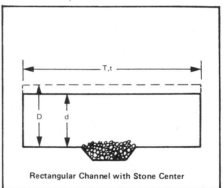

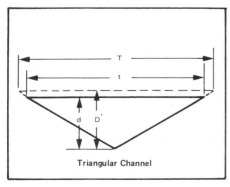

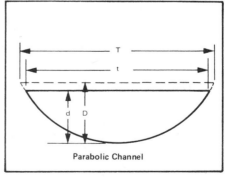

Figure 7-13 Typical channel cross sections. T, total construction top width; t, design top width of water flow; D, total construction depth; d, design depth of flow. Source: (ERA 1983b).

control erosion by reducing slope length. They are constructed as single or multiple embankments and channels across a slope. Typical terrace cross sections are shown in Figure 7-12 (EPA 1983b). The relationship between the terrace cross section dimensions of width and height is a function of slope and other site-specific conditions such as capacity, length, roughness coefficient, and soil erodability (sediment deposits).

Diversion channels, which are wide and shallow excavated ditches, are used to intercept surface water runoff for diversion offsite or to onsite treatment facilities and/or reduce slope length. They will generally be stabilized with vegetation or

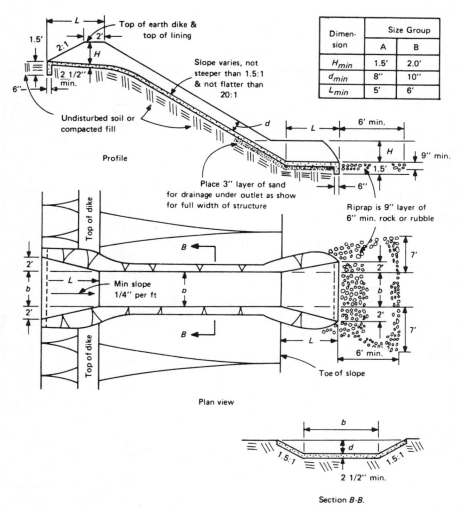

Figure 7-14 Paved chute design. Requirements for chute designs vary according to state regulations. Values given are typical. Source: (EPA 1983b).

rip-rap. Typical diversion channel cross sections are shown in Figure 7-13 (EPA 1983b). The diversion channel is also a conventional, demonstrated technology.

Chutes (flumes) and downpipes, which are used to transfer heavy flows of runoff from one level of a site to a lower level without permitting surface erosion, consist of lined channels with interconnecting drainage pipes typically constructed of piping or flexible tubing. The channels collect the flow at the higher level, generally from a heavy sudden precipitation event, and the downpipes convey the water to stabilized waterways or outlets at the base of the slope. A typical paved chute design in shown in Figure 7-14, and a representative downpipe configuration is shown in Figure 7-15 (EPA 1983b).

A commonly used approach to alter the runoff characteristics of a site is to

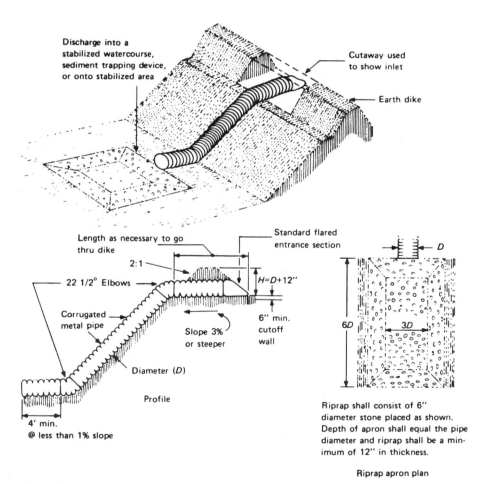

Figure 7-15 Downpipe configuration. Requirements for downpipe design vary according to state regulations. Values given are typical. Source: (EPA 1983b).

TABLE 7-12 Grading Techniques

Technique	Description	Use	Equipment
Excavation	Soil removal	Slope grade construction	Dozer, loader, scraper
Spreading	Soil application smoothing	Slope grade construction	Dozer, loader, grader
Compaction	Compacts soil, increases density	Slope grade construction	Dozer, loader, compactor
Scarification	Roughening technique loosens soil	Preparation for revegetation increases infiltration	Dozer, tractor, harrow
Tracking	Roughening technique grooves soil along contour	Preparation for revegetation increases infiltration	Cleated crawler tractor
Contour furrowing	Roughening technique creates small depressions in soil along contour	Preparation for revegetation increases infiltration	Dozer

Source: (EPA 1983b).

change the natural topography by grading the site to either spread and compact loose soil, roughen and loosen compacted soil, and/or modify the surface gradient. Grading can be used to "optimize the slope at a waste site such that surface runoff increases and infiltration and ponding decrease without significantly increasing erosion" (EPA 1983b), thus increasing the effectiveness of surface soils, and, conversely, to reduce runoff and increase infiltration to prepare a soil base for revegetation. Grading techniques are generally used in conjunction with other surface water control measures. Table 7-12 describes the six basic grading techniques and their uses.

Surface sealing, which is both a surface water control technique and a technology to prevent infiltration and flocculation of leachate, is described under groundwater control technologies. A natural form of surface sealing at a remediated site is accomplished through use of a vegetative cover that serves to stabilize the cover emplaced over the waste by intercepting rainfall, slowing runoff, and holding the soil together. As noted previously, the selection of species is important to minimize uptake of radionuclides from the soil and waste.

REFERENCES

(AEC 1972) U.S. Atomic Energy Commission, "Grand Junction Remedial Action Criteria." 10 CFR Part 12, Washington, D.C., 1972.

(Anderson 1959) Anderson, M. L. *Thorium Recovery from Waste Magnesium–Thorium Sludge.* Physical Research Laboratory, The Dow Chemical Company, Midland, Michigan, June 1959.

(ANS 1982) American Nuclear Society, "The Decommissioning of Nuclear Plants." August, 1982.

(Berlin 1982) Berlin, R. E., J. M. Peterson, and L. Skoski, "Overview of the Formerly Utilized Sites Remedial Action Program (FUSRAP)." In *Methods for Assessing Environmental Impacts of a FUSRAP Property—Cleanup/Interim Storage, Remedial Action.* Argonne National Laboratory, December 1982.

(DOE 1980a) U.S. Department of Energy, "A Background Report for the Formerly Utilized Manhattan Engineer District/Atomic Energy Commission Sites Program." DOE/EV-0097, Washington, D.C., September 1980.

(DOE 1980b) U.S. Department of Energy, "Description of the Formerly Utilized Sites Remedial Action Program." ORO-777, Oak Ridge, Tennessee, September 1980.

(DOE 1981) U.S. Department of Energy, "Background Report for the Uranium Mill Tailings Sites Remedial Action Program." Washington, D.C., April 1981.

(DOE 1982a) U.S. Department of Energy, "The United States Department of Energy Remedial Action Program." 1982.

(DOE 1982b) U.S. Department of Energy, "Commingled Uranium Tailings Study." DOE/DP-0011, Washington, D.C., June 1982.

(DOE 1983) U.S. Department of Energy, Oak Ridge Operations, "Radiological Guidelines for Application to DOE's Formerly Utilized Sites Remedial Action Program." ORO-831, 1983.

(EPA 1982)   U.S. Environmental Protection Agency, "Environmental Radiation Protection Standards for Nuclear Power Operations." 40 CFR Part 190, Washington, D.C., 1982.

(EPA 1983a)   U.S. Environmental Protection Agency, "Standards for Remedial Actions at Inactive Uranium Processing Sites." 40 CFR Part 192, Washington, D.C., 1983.

(EPA 1983b)   U.S. Environmental Protection Agency, "Handbook for Evaluating Remedial Action Technology Plans." EPA-600/2-83-076, August 1983.

(EPA 1986)   U.S. Environmental Protection Agency, "Operable Unit X Feasibility Study." Denver Radium Site, Denver, Colorado, October 1986.

(House 1985)   House of Representatives, Superfund Amendments and Reauthorization Act of 1986, H.R. 2005, Washington, D.C., October 3, 1986.

(NRC 1981)   U.S. Nuclear Regulatory Commission, "Disposal or Onsite Storage of Residual Thorium or Uranium (Either as Natural Ores or without Daughters Present) from Past Operations." SECY 81-576, Washington, D.C., 1981.

(NRC 1982)   U.S. Nuclear Regulatory Commission, "Standards for Protection Against Radiation." 10 CFR Part 20, Washington, D.C., 1982.

APPENDIX A

GLOSSARIES

A-1 GLOSSARY OF TERMS

Actinides   Radioactive elements with atomic number larger than 88.

Activation   The process of making a material radioactive by bombardment with neutrons, protons, or other nuclear particles.

Active maintenance   Any significant remedial activity needed during the period of institutional control to maintain a reasonable assurance that the performance objectives of 10 CFR 61 are met. Such active maintenance includes ongoing activities such as the pumping and treatment of water from a disposal unit or one-time measures such as replacement of a disposal unit cover. Active maintenance does not include custodial activities.

Activity   Radioactivity, the spontaneous emission of radiation, generally α or β particles, often accompanied by γ rays, from the nuclei of an unstable nuclide. As a result of this emission, the radioactive isotope is converted (or decays) into the isotope of a different (daughter) nuclide, which may or may not be radioactive. Ultimately, as a result of one or more stages of radioactive decay, a stable (nonradioactive) nuclide is formed.

Agreement state   A state granted licensing authority over nuclear facilities within its borders by entering into an agreement with the Nuclear Regulatory Commission (formerly, the Atomic Energy Commission) under subsection 274b of the Atomic Energy Act of 1954, as amended by Public Law 86-373 in 1959.

Airborne radioactive material   Any radioactive material in the air in the form of dusts, mists, vapors, or gases.

401

Alpha (α) particle   A positively charged particle emitted by certain radioactive materials. It is made up of two neutrons and two protons bound together, and hence is identical with the nucleus of a helium atom. It is the least penetrating of the three common types of radiation (α, β, γ) emitted by radioactive material, and can be stopped by a sheet of paper. It is not dangerous to plants, animals, or humans unless the α-emitting substance has entered the body.

Aquifer   A water-bearing layer of permeable rock or soil that will yield water in usable quantities to wells.

As low as is reasonably achievable   (ALARA) As low as is reasonably achievable taking into account the state of technology and the economics of improvements in relation to (1) benefits to the public health and safety, (2) other societal and socioeconomic considerations, and (3) the utilization of atomic energy in the public interest.

Atomic energy   All forms of energy released in the course of nuclear fission or nuclear transformation.

Background radiation   The radiation in the natural environment, including cosmic rays and radiation from the naturally radioactive elements. It is also called natural radiation. Levels vary, depending on location.

Barrier   Any material or structure that prevents or substantially delays movement of water or radionuclides.

Beta (β) particle   An elementary particle emitted from a nucleus during radioactive decay, with a single electrical charge and a mass equal to $1/1837$ that of a proton. A negatively charged β particle is identical to an electron. A positively charged β particle is called a positron. β radiation may cause skin burns, and β emitters are harmful if they enter the body.

Biosphere   The part of the earth in which life can exist, including the lithosphere, hydrosphere, and atmosphere; living beings together with their environment.

Biota   The animal and plant life of a region.

Borosilicate glass   A silicate glass containing at least 5% boric acid and used to vitrify calcined waste.

Buffer zone   A portion of the disposal site that is controlled by the licensee and that lies under the disposal units and between the disposal units and the boundary of the site.

Burial grounds   Areas designated for disposal of containers of radioactive wastes by near-surface burial.

By-product material   Any radioactive material (except special nuclear material) yielded in, or made radioactive by exposure to, the radiation incident to the process of producing or utilizing special nuclear material.

Canister   A metal container for radioactive solid waste.

Cask   A container that provides shielding and containment during transportation of radioactive materials.

Containment   The confinement of radioactive waste within a designated boundary.

Crystalline rock An inexact but convenient term designating an igneous or metamorphic rock.

Custodial agency An agency of the government designated to act on behalf of the government owner of the disposal site.

Decommissioning The retirement from active services of nuclear facilities, accompanied by a program to reduce or stabilize radioactive contamination.

Decontamination The removal or cleanup of radiological and chemical contaminants resulting from the onsite process to standards prescribed by the EPA, NRC, and state agencies. Usually decontamination involves cleanup of equipment, buildings, and open lands, and may include restoration of surface and groundwater quality if degradation has occurred.

Discharge In groundwater hydrology, water that issues naturally or is withdrawn from an aquifer.

Disposal facility Engineered and man-made structures and adjuncts at the disposal site for the purpose of waste disposal and supporting activities including disposal units, equipment, security systems, administration facilities, laboratories, and warehouses.

Disposal site That portion of a land disposal facility that is used for disposal of waste. It consists of disposal units and a buffer zone.

Disposal unit A discrete portion of the disposal site into which waste is placed for disposal.

Dome A dome-shaped landform or rock mass, for example, a salt dome.

Dose The quantity of radiation absorbed, per unit of mass, by the body or by any portion of the body. When the regulations specify a dose during a period of time, the dose means the total quantity of radiation absorbed per unit of mass, by the body or by any portion of the body during such period of time.

Engineered barrier A man-made structure or device that is intended to improve the disposal facility's ability to meet the performance objectives of 10 CFR 60 and/or 10 CFR 61.

Environmental assessment A concise public document that serves to (1) briefly provide sufficient evidence and analysis for determining whether to prepare an environmental impact statement or a finding of no significant impact, (2) aid the NRC's compliance with NEPA when no environmental impact statement is necessary, and (3) facilitate preparation of an environmental impact statement when one is necessary.

Environmental impact statement A detailed written statement as required by section 102(2)(C) of NEPA.

Environmental report A document submitted to the Commission by an applicant for a permit, license, or other form of permission, or an amendment to or renewal of a permit, license, or other form of permission, or by a petitioner for rulemaking, in order to aid the Commission in complying with section 102(2) of NEPA.

Exclusive use   Also referred to as "sole use" or "full load"—the sole use of a conveyance by a single consignor and for which all initial, intermediate, and final loading and unloading are carried out in accordance with the direction of the consignor or consignee.

Fault   A fracture or fracture zone along which there has been displacement of the sides relative to one another parallel to the fracture.

Fission product   Any radioactive or stable nuclide produced by fission, including both primary fission fragments and their radioactive decay products.

Food chain   The pathways by which any material passes from the first absorbing organism through plants and animals to humans.

Formerly utilized site   A site once used by or under contract to the Manhattan Engineer District or the Atomic Energy Commission to conduct research with, process, or store uranium or thorium ore or metal derived therefrom.

Fracture   Breaks in rocks caused by intense folding or faulting.

Fuel cycle   Mining, refining, enrichment, and fabrication of fuel elements, use in a reactor, chemical reprocessing of the spent fuel, reenrichment of the fuel material, refabrication of new fuel elements, and management of radioactive wastes.

Fuel reprocessing   The processing of reactor fuel to recover the unused fissionable material.

Gamma (γ) radiation   High-energy, short-wavelength electromagnetic radiation of nuclear origin (radioactive decay). γ Rays are the highest penetrating of the three common types of radioactive decay (α, β, and γ) and are best stopped or shielded against by dense materials, such as lead or depleted uranium.

Geohydrology   The study of the character, source, and mode of occurrence of underground water.

Geologic repository   A system that is intended to be used for, or may be used for, the disposal of radioactive wastes in excavated geologic media. A geologic repository includes (1) the geologic repository operations area, and (2) the portion of the geologic setting that provides isolation of the radioactive waste.

Groundwater   All water that occurs below the land surface.

Grout   A mortar fluid combined with liquid waste to provide a matrix for isolation of the waste.

Hazardous waste   Those wastes designated as hazardous by Environmental Protection Agency regulations in 40 CFR Part 261.

Health physics   The science concerned with recognition, evaluation, and control of health hazards from ionizing radiation.

High-level radioactive waste (HLW)   (1) Irradiated reactor fuel, (2) liquid wastes resulting from the operation of the first cycle solvent extraction system, or equivalent, and the concentrated wastes from subsequent extraction cycles, or equivalent, in a facility for reprocessing irradiated reactor fuel, and (3) solids into which such liquid wastes have been converted.

Host rock   The geologic medium in which the waste is emplaced.

Hot cell   A facility that allows remote viewing and manipulation of radioactive materials.

Hydraulic gradient   The change in static head per unit of lateral distance in a given direction.

Hydrogeologic unit   Any soil or rock unit or zone that by virtue of its porosity or permeability, or lack thereof, has a distinct influence on the storage or movement of groundwater.

Immobilization   Treatment and/or emplacement of the wastes so as to impede their movement.

Interim storage   Storage operations for which (1) monitoring and human controls are provided and (2) subsequent action involving treatment transportation, or final disposition is expected.

Intruder barrier   A sufficient depth of cover over the waste that inhibits contact with waste and helps to ensure that radiation exposures to an inadvertent intruder will meet the performance objectives of 10 CFR 61 or engineered structures that provide equivalent protection to the inadvertent intruder.

Ion exchange   Replacement of ions adsorbed on a solid, or exposed at the surface of a solid, by ions from solution.

Isolation   Inhibiting the transport of radioactive material so that amounts and concentrations of this material entering the accessible environment will be kept within prescribed limits.

Isotope   One of two or more atoms with the same atomic number (the same chemical element) but with different atomic weights. An equivalent statement is that the nuclei of isotopes have the same number of protons but different numbers of neutrons. Isotopes usually have very nearly the same chemical properties but somewhat different physical properties.

Kilowatt-hour (kWh)   Use of electricity for 1 hr at a rate of 1000 W.

Land disposal facility   The land, buildings, and equipment that is intended to be used for the disposal of radioactive wastes into the subsurface of the land. A geologic repository as defined in 10 CFR 60 is not considered a land disposal facility.

Licensed material   Source material, special nuclear material, or by-product material received, possessed, used, or transferred under a general or specific license issued by the Nuclear Regulatory Commission.

License   A license issued under the regulations in Parts 30 through 35, 40, 60, 61, 70, or Part 72 of 10 CFR. Licensee means the holder of such license.

Low-level radiation   Radiation that is of such intensity or concentration that it poses a minimal health hazard (less than 1 Ci per gallon or ft^3).

Low-level waste   Low-level waste has the same meaning as in the Low-Level Waste Policy Act, that is, radioactive waste not classified as high-level radioactive waste, transuranic waste, spent nuclear fuel, or by-product material as defined in Section 11e.(2) of the Atomic Energy Act (uranium or thorium tailings and waste).

Low specific activity material   Any of the following:

(1) Uranium or thorium ores and physical or chemical concentrates of those ores.

(2) Unirradiated natural or depleted uranium or unirradiated natural thorium.

(3) Tritium oxide in aqueous solutions provided the concentration does not exceed 5.0 mCi/mL.

(4) Material in which the radioactivity is essentially uniformly distributed and in which the estimated average concentration per gram of contents does not exceed:

  (i) 0.0001 mCi of radionuclides for which the A_2 quantity is not more than 0.05 Ci;

  (ii) 0.005 mCi of radionuclides for which the A_2 quantity is more than 0.05 Ci, but not more than 1 Ci; or

  (iii) 0.3 mCi of radionuclides for which the A_2 quantity is more than 1 Ci.

(5) Objects of nonradioactive material externally contaminated with radioactive material, provided that the radioactive material is not readily dispersible and the surface contamination, when averaged over an area of 1 m^2, does not exceed 0.0001 mCi (220,000 disintegrations per minute) per cm^2 of radionuclides for which the A_2 quantity is not more than 0.05 Ci, or 0.001 mCi (2,200,000 disintegrations per minute) per cm^2 for other radionuclides.

Mixed LLW   Low-Level radioactive wastes that contain hazardous materials under subpart D of 40 CFR 261 or that manifest hazardous characteristics as described by subpart C of 40 CFR 261.

Multibarrier   A system using the waste form, the container, the overpack, the emplacement medium, and surrounding geologic media as multiple barriers to isolate the waste from the biosphere.

NEPA   The National Environmental Policy Act of 1969, as amended (P.L. 91-190, 83 Stat. 852, 856, as amended by P.L. 94-83, 89 Stat. 424, 42 U.S.C. 4321, et seq.).

Near-surface disposal facility   A land disposal facility in which radioactive waste is disposed of in or within the upper 30 m of the earth's surface.

Nuclear reactor   An apparatus, other than an atomic weapon, designed or used in a self-supporting chain reaction.

Occupational dose   Includes exposure of an individual to radiation (1) in a restricted area, or (2) in the course of employment in which the individual's duties involve exposure to radiation, provided, that "occupational dose" shall not be deemed to include any exposure of an individual to radiation for the purpose of medical diagnosis or medical therapy of such individual.

Package   The waste packaging together with its radioactive contents as presented for transport.

Packaging   The assembly of components necessary to ensure compliance with the packaging requirements of this part. It may consist of one or more receptacles, absorbent materials, spacing structures, thermal insulation, radiation shielding, and devices for cooling or absorbing mechanical shocks. The vehicle, tie-down system, and auxiliary equipment may be designated as part of the packaging.

Partition   To separate one (or more) element(s) from one (or more) other element(s).

Permeability   The relative ease with which a porous medium can transmit a liquid under a hydraulic gradient.

Rad   Radiation adsorbed dose, the basic unit of absorbed dose of ionizing radiation.

Radiation   Any or all of the following: α rays, β rays, γ rays, X rays, neutrons, high-speed electrons, high-speed protons, and other atomic particles, but not sound or radio waves or visible, infrared or ultraviolet light.

Radiation standards   Exposure standards, permissible concentrations, rules for safe handling, regulations for transportation, regulations for industrial control of radiation, and control of radiation exposure by legislative means.

Radioactivity   The spontaneous decay or disintegration of an unstable atomic nucleus, usually accompanied by the emission of ionizing radiation. (Often-shortened to "activity.")

Radiological hazard   A condition in which radiation may cause health effects due to long-term exposure.

Radionuclide   A radioactive nuclide.

Reclamation   The overall final efforts by the site owner that are directed toward site cleanup for possible release or disposal in accordance with regulations.

Rem (roentgen equivalent man)   A quantity used to express the effective dose equivalent for all forms of ionizing radiation.

Remedial action   Action that is necessary for (1) the removal from a remedial action site of residual radioactive material to a disposal site and subsequent control of that material, or (2) stabilization and subsequent control of residual radioactive material at the remedial action site, or (3) both (1) and (2), to comply with standards established under Section 276 of the Atomic Energy Act of 1954, as amended.

Remedial action site   (1) A site at which remedial action is required and that was used under a contract with any predecessor of the Department of Energy, including the Manhattan Engineer District and Atomic Energy Commission, for researching, developing, manufacturing, fabricating, testing, processing, sampling, or storing radioactive material, except a site (a) for which a license (issued by the Nuclear Regulatory Commission or its predecessor agency under the Atomic Energy Act of 1954, or by a State under Section 274 of that Act) for the production or possession at the site of uranium or thorium, or their daughter products, including radium, is in effect on the date of enactment of the Residual Radioactive Material Control Act, or is issued or renewed after that date, or (b) owned or leased by the federal government on or after the date of enactment of the Residual Radioactive Material Control Act and (2) any other location the Secretary of Energy or his designee determines to require remedial action due to contamination with residual radioactive material derived from a site meeting the criteria of part (1) of this definition.

Remotely handled waste   Waste package having surface dose rates greater than 0.2 R/hr requiring shielding and/or remote handling.

Residual radioactive material   Any radioactive material present at a site that results in radiation levels that exceed background levels, including but not limited to waste material, soils, rock, plants, shrubs, personal property, and building materials.

Restricted area   Any area, access to which is controlled by the licensee for purposes of protection of individuals from exposure to radiation and radioactive materials. "Restricted area" shall not include any areas used as residential quarters, although a separate room or rooms in a residential building may be set apart as a restricted area.

Restricted use   A designation following remedial action that requires some control on the activities at a site containing residual radioactive material.

Retrievability   Capability to remove waste from its place in isolation with approximately the same level of effort and exposure as required to emplace it.

Saturated zone   That part of the earth's crust beneath the regional water table in which all voids, large and small, are ideally filled with water under pressure greater than atmospheric.

Seismicity   The phenomenon of earth movements as manifested by earthquakes.

Site closure and stabilization   Those actions that are taken upon completion of operations that prepare the disposal site for custodial care and that ensure that the disposal site will remain stable and will not need ongoing active maintenance.

Slurry   A fluid mixture or suspension of insoluble material.

Solidification   Conversion of liquid radioactive waste to a dry, stable solid.

Source material   (1) Uranium or thorium, or any combination thereof, in any physical or chemical form or (2) ores that contain by weight one-twentieth of one percent (0.05%) or more of (a) uranium, (b) thorium, or (3) any combination thereof. Source material does not include special nuclear material.

Source terms   The quantity of radioactive material (or other pollutant) released to the environment at its point of release (source).

Special nuclear material (1) Plutonium, uranium-233, uranium enriched in the isotope 233 or in the isotope 235, and any other material that the Commission, pursuant to the provisions of section 51 of the Act, determines to be special nuclear material, but does not include source material; or (2) any material artificially enriched by any of the foregoing but does not include source material.

Spent fuel Irradiated nuclear fuel that has undergone at least 1 yr decay since being used as a source of energy in a power reactor. Spent fuel includes the special nuclear material, by-product material, source material, and other radioactive materials associated with fuel assemblies.

Stabilization The activities or measures taken to contain radioactive materials and prevent them from migrating from a site.

Storage Retention of waste in some type of man-made device in a manner permitting retrieval.

Surface contamination Radioactive materials attached to or deposited on a surface.

Surplus facility A potentially radioactive contaminated property under the control of the Department of Energy (including cribs, ponds, trenches, buildings, reactors, and equipment) that has become obsolete in terms of current and future program needs.

Surveillance Observation of the disposal site for purposes of visual detection of need for maintenance, custodial care, evidence of intrusion, and compliance with other license and regulatory requirements.

Tailings The remaining portion of a metal-bearing ore after some or all of such metal, such as uranium, has been extracted. Tailings sometimes are separated into a coarse fraction called sands and a fine fraction known as slimes.

Tectonic Pertaining to the processes causing, and the rock structures resulting from, deformation of the earth's crusts.

Transuranic element (isotope) An element above uranium in the periodic table, that is, with an atomic number greater than 92. All 11 transuranic elements are produced artificially and are radioactive. They are neptunium, plutonium, americium, curium, berkelium, californium, einsteinium, fermium, mendelevium, nobelium, and lawrencium.

Transuranic waste Waste material measured or assumed to contain more than 10 nCi of transuranic elements.

Unrestricted area Any area, access to which is not controlled by the licensee for purposes of protection of individuals from exposure to radiation and radioactive materials, and any area used for residential quarters.

Unsaturated zone The zone between the land surface and the regional water table. Generally, fluid pressure in this zone is less than atmospheric pressure, and some of the voids may contain air or other gases at atmospheric pressure. Beneath flooded areas or in perched water bodies the fluid pressure locally may be greater than atmospheric.

Uplift A structurally high area in the crust, produced by movements that raise the rocks.

Waste form The radioactive waste materials and any encapsulating or stabilizing matrix, and the physical characteristics of the waste, with emphasis on the ability of the waste to be structurally stable in the long term.

Waste management The planning, execution, and surveillance of essential functions related to the control of radioactive (and nonradioactive) waste, including treatment, packaging, transportation, storage, surveillance, and disposal (isolation).

Waste package The waste form and any containers, shielding, packing, and other absorbent materials immediately surrounding an individual waste container.

Water table The upper surface of an unconfined aquifer.

Vicinity property A real property in the vicinity of a radioactive materials processing site that has become contaminated by radioactive materials emanating from the site; also called an associated property.

A-2 GLOSSARY OF ABBREVIATIONS AND ACRONYMS

ACI	American Concrete Institute
ACL	Alternate Concentration Limit
AEC	Atomic Energy Commission
AFR	Away from Reactor
AGNS	Allied Gulf Nuclear Services
AGV	Above-Ground Vault
ALARA	As Low As Reasonably Achievable
ANL	Argonne National Laboratory, Argonne, Illinois
APS	American Physical Society
BEIR	Committee on Biological Effects of Ionizing Radiation of the National Academy of Sciences
BGV	Below-Ground Vault
BMI	Battelle Memorial Institute
BNL	Brookhaven National Laboratory
Bq	Becquerels
BRC	Below Regulatory Concern
BWR	Boiling-Water Reactor
CAA	Clean Air Act
CERCLA	Comprehensive Environmental Response, Compensation and Liability Act
CFR	Code of Federal Regulations
CH	Contact Handled
Ci	Curies
DAW	Dry Active Waste

DEC	New York State Department of Environmental Conservation
D&D	Decontamination and Decommissioning
DEIS	Draft Environmental Impact Statement
DHLW	Defense High-Level Waste
DOE	U.S. Department of Energy
DOT	U.S. Department of Transportation
DWMP	Defense Waste Management Plan
EA	Environmental Assessment
EG&G/ID	EG&G Idaho, Inc.
EIS	Environmental Impact Statement
EMCB	Earth Mounded Concrete Bunker
EPA	Environmental Protection Agency
EPRI	Electric Power Research Institute
ERDA	Energy Research and Development Administration
FCR	Full Core Reserve
FEIS	Final Environmental Impact Statement
FS	Feasibility Study
FRC	Federal Radiation Council
FUSRAP	Formerly Utilized Site Remedial Action Project
GJRAP	Grand Junction Remedial Action Project
HEDL	Hanford Engineering Development Laboratory, Richland, Washington
HIC	High-Integrity Container
HLW	High-Level Waste
HM	Heavy Metal, Generally Uranium and Plutonium
HR	Hanford Reservation
IAEA	International Atomic Energy Agency
ICRP	International Commission on Radiological Protection
IDD	Intermediate Depth Disposal
ILLW	Intermediate-Level Liquid Waste
ILTSF	Intermediate-Level Transuranic Storage Facility
INFCE	International Nuclear Fuel Cycle Evaluation
ISLB	Improved Shallow Land Buriel
JCAE	Joint Committee on Atomic Energy
LANL	Los Alamos National Laboratory, Los Alamos, New Mexico
LASL	Los Alamos Scientific Laboratory
LLRW	Low-Level Radioactive Waste
LLRWPA	Low-Level Radioactive Waste Policy Act
LLW	Low-Level Waste
LSA	Low Specific Activity
LWT	Legal Weight Truck
MCC	Modular Concrete Canister
MCD	Mined Cavity Disposal
MCCD	Modular Concrete Canister Disposal

MED	Manhattan Engineer District (Manhattan Project)
ML	Mound Laboratory
MOU	Memorandum of Understanding
MPC	Maximum Permissible Concentration
MRS	Monitored Retrievable Storage
MTIHM	Metric Tons Initial Heavy Metal
MTU	Metric Tons Uranium
NAS	National Academy of Sciences
NASA	National Aeronautical and Space Administration
NASAP	Nonproliferation Alternative Systems Assessment Program
NCRP	National Council on Radiation Protection
NEPA	National Environmental Policy Act of 1969
NGS	National Geodetic Survey
ORNL	Oak Ridge National Laboratory, Oak Ridge, Tennessee
NPL	National Priority List
NRC	Nuclear Regulatory Commission
NUREG	Nuclear Regulation (Issued by Nuclear Regulatory Commission)
NWPA	Nuclear Waste Policy Act of 1982
NYSERDA	New York State Energy Research and Development Authority
ORIGEN	Oak Ridge Isotope Generation and Depletion Code
OSHA	Occupational Safety and Health Administration
OSSC	Onsite Storage Container
OWT	Overweight Truck
P.L.	Public Law
PNL	Pacific Northwest Laboratory (Battelle), Richland, Washington
PORV	Pilot Operated Relief Valve
PRP	Potentially Responsible Party
PWR	Pressurized Water Reactor
RAP	Remedial Action Project
RCRA	Resource Conservation and Recovery Act
RFP	Rocky Flats Plant, Golden, Colorado
RH	Remote Handled
RI	Remedial Investigation
RMCL	Recommended Maximum Contaminant Level
ROD	Record of Decision
RSO	Radiation Safety Officer
RWMC	Radioactive Waste Management Complex
SAR	Safety Analysis Report
SARA	Superfund Amendments and Reauthorization Act of 1986
SD	Shaft Disposal
SDA	Subsurface Disposal Area
SF	Spent Fuel
SFMP	Surplus Facilities Management Program

SLB	Shallow Land Burial
SNL	Sandia National Laboratory, Albuquerque, New Mexico
SNM	Special Nuclear Material
SRL	Savannah River Laboratory, Aiken, South Carolina
SDWA	Safe Drinking Water Act
SWIMS	Solid Waste Information Management Systems
TRU	Transuranic Waste
TRUPACT	Transuranic Package Transporter (Cask)
TSA	Transuranic Storage Area
TSCA	Toxic Substance Control Act
UMTRAP	Uranium Mill Tailings Remedial Action Project
UNSCEAR	United Nations Scientific Committee on the Effects of Atomic Radiation
USGS	United States Geologic Survey
WIPP	Waste Isolation Pilot Plant, Carlsbad, New Mexico
WVDP	West Valley Demonstration Project, New York (DOE site)

RADIOLOGICAL ASSESSMENT COMPUTER CODES

AIRDOS-EPA   A computer code designed to estimate radionuclide concentrations in air, rates of deposition on ground surfaces, ground surface concentrations, intake rates via inhalation of air and ingestion of meat, milk, and fresh vegetables.

XOQDOQ   A computer program developed for estimating radionuclide concentrations in the air and on the ground from routine or anticipated intermittent releases at nuclear facilities. This program implements NRC Regulatory Guide 1.111.

MILDOS, UDAD   These computer codes provide estimates of potential radiation exposure to individuals and to the general population in the vicinity of uranium processing facilities and mines. The UDAD code has been modified to include thorium series radionuclides and a selected group of fission products and actinides.

PAVAN   A computer code designed to estimate downwind ground-level air concentrations for potential accident releases of radioactive materials from nuclear facilities. This program implements the guidance provided in NRC Regulatory Guide 1.145, "Atmospheric Dispersion Models for Potential Accident Consequence Assessments at Nuclear Power Plants."

GASPAR	A computer code written for the evaluation of radiological impacts due to the release of radioactive material to the atmosphere during normal operation of nuclear power reactors. This program implements the guidance provided in NRC Regulatory Guide 1.145, "Atmospheric Dispersion Models for Potential Accident Consequence Assessments at Nuclear Power Plants."
CRAC2	A computer code developed for the calculation of reactor accident consequences, which include radiation doses and health effects.
LADTAP-II	A computer code written for determining the radiation dose to humans from the pathways in the aquatic environment such as potable water, aquatic foods, shoreline deposits, swimming, boating, and irrigated foods. This program implements the radiological exposure models of NRC Regulatory Guide 1.109. Dose conversion factors used in this code have been updated to include the ICRP-30 recommended values.
LPGS	A computer code designed to calculate the radiological impact resulting from radioactive releases to the hydrosphere. The hydrosphere is represented by estuary, small river, well, lake, and one-dimensional rivers. The code is principally developed to calculate radiation dose to humans as a function of time for the various pathways.
AT1230	This code is a generalized analytical transient, one-, two-, and/or three-dimensional computer code and is developed for estimating the transport of wastes in a groundwater aquifer.
FEMWATER	A finite element model of water flow through saturated–unsaturated porous media. This model is designed for evaluating the moisture content increasing rate within the region of interest.
FEMWASTE	A two-dimensional transient model for estimating the transport of dissolved constituents through porous media. The transport mechanisms considered include convection, hydrodynamic dispersion, chemical sorption, and radioactive decay.
PRESTO-EPA	A computer code developed to evaluate possible radiation effects from shallow land disposal trenches. This model will assess radionuclide transport, radiation exposure, and health impact to a population over the long term.

RADTRANII
A computer code developed to analyze radiation doses to the public from transportation of radioactive material.

QAD
A point kernel code designed to provide an estimate of uncollided γ flux, dose rate, energy deposition, and other quantities that result from a point-by-point representation of a volume-distributed source of radiation.

GGG
An air-scattered radiation dose calculation computer code.

SKYSHINEII
A computer code developed for the analysis of the radiation environment in the vicinity of a building structure containing radiation sources situated within the confines of a nuclear facility.

ANISN
A one-dimensional, discrete ordinate neutron and γ transport computer code.

DOT
A two-dimensional discrete ordinate neutron and γ transport code for reactor physics computation and radiation shielding.

KENO IV
A multigroup Monte Carlo criticality code.

MORSE-CG
A multigroup combinational Monte Carlo neutron and γ transport code.

DOSFACTER
A computer code used to calculate the dose-rate conversion factors for external exposure to photon and electron radiation from radionuclides occurring in routine releases from nuclear fuel cycle facilities. Exposure modes considered are immersion in contaminated air, immersion in contaminated water, and exposure to a contaminated ground surface.

DACRIN
A computer program that permits rapid and consistent estimates of the effective radiation dose to the human respiratory tract and other organs resulting from the inhalation of radioactive aerosols. This program incorporates the lung model proposed by the ICRP Task Group on Lung Dynamics.

PABLM
A computer program that facilitates calculation of internal radiation doses to humans resulting from radioactive materials released to the environment. The program is designed to calculate accumulated radiation doses from the chronic ingestion of food products and from external exposures to the environment.

FOOD
A computer code to calculate radiation doses from deposition of radionuclides on farm or garden soil and crops during either an atmospheric or water

release. Deposition may be either directly from the air or from irrigation water. Fifteen crop or animal product pathways may be chosen.

ARRRG   Radiation doses to humans may be calculated for radionuclides released to bodies of water from which people might obtain fish, other aquatic foods, or drinking water, and in which they might fish, swim, or boat.

SUBDOSA   A computer program that calculates submersion doses from acute release of radionuclides to the atmosphere. Doses are calculated as a function of distance from release point, atmospheric stability, and wind speed for a specified radionuclide inventory. Contributions from both β and γ radiation are included as a function of tissue depth.

CONDOS   A computer program that estimates radiation doses to humans from distribution, use, and disposal of a variety of consumer products that contain radioactive materials. This code utilizes a generalized format in which the life span of a consumer product is divided into five main stages (distribution, transport, use, disposal, and emergencies) that require descriptions of the activities by which humans may be exposed to the products (events) during each stage.

10 CFR 61 CODES   A pair of computer codes (GRWATRR and OPTIONR) that were developed to assist the U.S. NRC in preparation of the FEIS on "Licensing Requirements for Land Disposal of Radioactive Waste."

GRWATRR   A code written to perform an assessment of the impact from groundwater migration of radionuclides with emphasis on waste form and packaging performance parameters, and site selection and design parameters.

OPTIONR   A code that finds the disposal practices for proper interment of radioactive waste. The code calculates both radiological and socioeconomic impact associated with the management, transport, and disposal of such radioactive wastes.

APPENDIX C

UNCONTROLLED HAZARDOUS WASTE SITE RANKING SYSTEM (HAZARD RANKING SYSTEM—HRS)[1]

DOCUMENTATION RECORDS FOR HAZARD RANKING SYSTEM

FACILITY NAME: ....

LOCATION: ....
Cover Sheet

Facility name: ....
Location: ....
EPA Region: ....
Person(s) in charge of the facility: ....
....
....
Name of Reviewer: .... Date: ....
General description of the facility:
(For example: landfill, surface impoundment, pile, container; types of hazar-
  dous substances; location of the facility; contamination route of major
  concern; types of information needed for rating; agency action, etc.)

....
....
....
....
....
Scores: $S_M =$ $(S_{gw} =$ $S_{sw} =$ $S_a =$)
$S_{FE} =$
$S_{DC} =$

---

[1]*Source:* U.S. Environmental Protection Agency 40 CFR Part 300, Appendix, A.

Instructions: The purpose of these records is to provide a convenient way to prepare an auditable record of the data and documentation used to apply the Hazard Ranking System to a given facility. As briefly as possible summarize the information you used to assign the score for each factor (e.g., "Waste quantity = 4,230 drums plus 800 cubic yards of sludges"). The source of information should be provided for each entry and should be a bibliographic-type reference that will make the document used for a given data point easier to find. Include the location of the document and consider appending a copy of the relevant page(s) for ease in review.

Instructions For Completing Ground Water Route Work Sheet

1. Observed Release

Contaminants detected (5 maximum):
Rationale for attributing the contaminants to the facility:

2. Route Characteristics

Depth to Aquifer of Concern

Name/description of aquifer(s) in concern:
Depth(s) from the ground surface to the highest seasonal level of the saturated zone [water table(s)] of the aquifer of concern:
Depth from the ground surface to the lowest point of waste disposal/storage:

Net Precipitation

Mean annual or seasonal precipitation (list months for seasonal):
Mean annual lake or seasonal evaporation (list months for seasonal):
Net precipitation (subtract the above figures):

Permeability of Unsaturated Zone

Soil type in unsaturated zone:
Permeability associated with soil type

Physical State

Physical state of substances at time of disposal (or at present time for generated gases):

Facility Name: _____   Date: _____

Ground Water Route Work Sheet

Rating Factor	Assigned Value (Circle One)	Multi-plier	Score	Max. Score	Ref. (Section)
1 Observed Release	0 45	1		45	3.1
If observed release is given a score of 45, proceed to line 4 .					
If observed release is given a score of 0, proceed to line 2 .					
2 Route Characteristics					3.2
Depth to Aquifer of Concern	0 1 2 3	2		6	
Net Precipitation	0 1 2 3	1		3	
Permeability of the Unsaturated Zone	0 1 2 3	1		3	
Physical State	0 1 2 3	1		3	
Total Route Characteristics Score				15	
3 Containment	0 1 2 3	1		3	3.3
4 Waste Characteristics					3.4
Toxicity/Persistence	0 3 6 9 12 15 18	1		18	
Hazardous Waste Quantity	0 1 2 3 4 5 6 7 8	1		8	
Total Waste Characteristics Score				26	
5 Targets					3.5
Groundwater Use	0 1 2 3	3		9	
Distance to Nearest Well/Population Served	0 4 6 8 10 12 16 18 20 24 30 32 35 40	1		40	
Total Targets Score				49	
6 If line 1 is 45, multiply 1 × 4 × 5 If line 1 is 0, multiply 2 × 3 × 4 × 5				57,330	
7 Divide line 6 by 57,330 and multiply by 100			$S_{gw} =$		

3. *Containment*

Containment

Method(s) of waste or leachate containment evaluated:
Method with highest score:

4. *Waste Characteristics*

Toxicity and Persistence

Compound(s) evaluated:
Compound with highest score:

Hazardous Waste Quantity

Total quantity of hazardous substances at the facility, excluding those with a containment score of 0 (Give a reasonable estimate even if quantity is above maximum):
Basis of estimating and/or computing waste quantity:

5. *Targets*

Groundwater Use

Uses(s) of aquifer(s) of concern within a 3-mile radius of the facility:

Distance to nearest Well

Location of nearest well drawing from *aquifer of concern* or occupied building not served by a public water supply:
Distance to above well or building:

Population Served by Groundwater Wells within a 3-Mile Radius

Identified water-supply well(s) drawing from *aquifer(s) of concern* within a 3-mile radius and populations served by each:
Computation of land area irrigated by supply well(s) drawing from aquifer(s) of concern within a 3-mile radius, and conversion to population (1.5 people per acre):
Total population served by ground water within a 3-mile radius:

Surface Water Route Work Sheet

Facility Name: _____   Date: _____

Surface Water Route Work Sheet

Rating Factor	Assigned Value (Circle One)	Multiplier	Score	Max. Score	Ref. (Section)
[1] Observed Release	0 45	1		45	4.1
If observed release is given a value of 45, proceed to line [4].					
If observed release is given a value of 0, proceed to line [2].					
[2] Route Characteristics					4.2
Facility Slope and Intervening Terrain	0 1 2 3	1		3	
1-yr 24-hr Rainfall	0 1 2 3	1		3	
Distance to Nearest Surface Water	0 1 2 3	2		6	
Physical State	0 1 2 3	1		3	
Total Route Characteristics Score				15	
[3] Containment	0 1 2 3	1		3	4.3
[4] Waste Characteristics					4.4
Toxicity/Persistence	0 3 6 9 12 15 18	1		18	
Hazardous Waste Quantity	0 1 2 3 4 5 6 7 8	1		8	
Total Waste Characteristics Score				26	
[5] Targets					4.5
Surface Water Use	0 1 2 3	3		9	
Distance to a Sensitive Environment	0 1 2 3	2		6	
Population Served/ Distance to Water Intake Downstream	0 4 6 8 10 12 16 18 20 24 30 32 35 40	1		40	
Total Targets Score				55	
[6] If line [1] is 45, multiply [1] $\times$ [4] $\times$ [5] If line [1] is 0, multiply [2] $\times$ [3] $\times$ [4] $\times$ [5]				64,350	
[7] Divide line [6] by 64,350 and multiply by 100			$S_{sw} =$		

422

Instructions For Completing Surface Water Route Work Sheet

1. Observed Release

Contaminants detected in surface water at the facility or downhill from it (5 maximum):

Rationale for attributing the contaminants to the facility:

2. Route Characteristics

Facility Slope and Intervening Terrain

Average slope of facility in percent:

Name/description of nearest downslope surface water:

Average slope of terrain between facility and above-cited surface water body in percent:

Is the facility located either totally or partially in surface water?

Is the facility completely surrounded by areas of higher elevation?

1-yr 24-hr Rainfall in Inches

Distance to Nearest Downslope Surface Water

Physical State of Wastes

3. Containment

Containment

Method(s) of waste or leachate containment evaluated:

Method with the highest score:

4. Waste Characteristics

Toxicity and Persistence

Compound(s) evaluated:

Compound with the highest score:

Hazardous Waste Quantity

Total quantity of hazardous substances at the facility, excluding those with a containment score of 0 (Give a reasonable estimate even if quantity is above maximum):

Basis of estimating and/or computing waste quantity:

5. Targets

Surface Water Use

Use(s) of surface water within 3 miles downstream of the hazardous substance:
Is there tidal influence?

Distance to a Sensitive Environment

Distance to 5-acre (minimum) coastal wetland, if 2 miles or less:
Distance to 5-acre (minimum) fresh-water wetland, if 1 mile or less:
Distance to critical habitat of an endangered species or national wildlife refuge, if 1 mile or less:

Population Served by Surface Water

Location(s) of water-supply intake(s) within 3 miles (free-flowing bodies) or 1 mile (static water bodies) downstream of the hazardous substance and population served by each intake:
Computation of land area by above-cited intake(s) and conversion to population (1.5 people per acre):
Total population served:
Name/description of nearest of above water bodies:
Distance to above-cited intakes, measured in stream miles:

Instruction for Completing Air Route Work Sheet

1. Observed Release

Contaminants detected:
Date and location of detection of contaminants
Methods used to detect the contaminants:
Rationale for attributing the contaminants to the site:

2. Waste Characteristics

Reactivity and Incompatibility

Most reactive compound:
Most incompatible pair of compounds:

Toxicity

Most toxic compound:

Air Route Work Sheet

Facility Name: _____   Date: _____

Air Route Work Sheet

Rating Factor	Assigned Value (Circle One)	Multi-plier	Score	Max. Score	Ref. (Section)
1 Observed Release	0 45	1	0	45	5.1
Date and Location:					
Sampling Protocol:					
If line 1 is 0, the $S_a = 0$. Enter on line 5. If line 1 is 45, then proceed to line 2.					
2 Waste Characteristics					5.2
Reactivity and Incompatibility	0 1 2 3	1		3	
Toxicity	0 1 2 3	3		9	
Hazardous Waste	0 1 2 3 4 5 6 7 8	1		8	
Total Waste Characteristics Score				20	
3 Targets					5.3
Population Within 4-Mile Radius	0 9 12 15 18 21 24 27 30	1		30	
Distance to Sensitive Environment	0 1 2 3	2		6	
Land Use	0 1 2 3	1		3	
Total Targets Score				39	
4 Multiply 1 × 2 × 3				35,100	
5 Divide line 4 by 35,100 and multiply by 100			$S_a = 0$		

Hazardous Waste Quantity

Total quantity of hazardous waste:
Basis of estimating and/or computing waste quantity:

3. Targets

Population Within 4-Mile Radius

Circle radius used, give population, and indicate how determined:
0 to 4 mi   0 to 1 mi   0 to 1/2 mi   0 to 1/4 mi

Distance to a Sensitive Environment

Distance to 5-acre (minimum) coastal wetland, if 2 miles or less:
Distance to 5-acre (minimum) fresh-water wetland, if 1 mile or less:
Distance to critical habitat of an endangered species, if 1 mile or less:

Land Use

Distance to commercial/industrial area, if 1 mile or less:
Distance to national or state park, forest, or wildlife reserve, if 2 miles or less:
Distance to residential area, if 2 miles or less:
Distance to agricultural land in production within past 5 yr, if 1 mile or less:
Distance to prime agricultural land in production within past 5 yr, if 2 miles or less:
Is a historic or landmark site (National Register of Historic Places and National Natural Landmarks) within view of the site?

Work Sheet for Computing S_M

Facility Name: .... Date: ....

Worksheet for Computing S_M

	S	S^2
Groundwater Route Score (S_{gw})		
Surface Water Route Score (S_{sw})		
Air Route Score (S_a)		
$S_{gw}^2 + S_{sw}^2 + S_a^2$	/////////	
$\sqrt{S_{gw}^2 + S_{sw}^2 + S_a^2}$	/////////	
$\sqrt{S_{gw}^2 + S_{sw}^2 + S_a^2}/1.73 = S_M =$	/////////	

Fire And Explosion Work Sheet

Facility Name: _____   Date: _____

Fire and Explosion Work Sheet

Rating Factor	Assigned Value (Circle One)	Multiplier	Score	Max. Score	Ref. (Section)
1 Containment	1 3	1		3	7.1
2 Waste Characteristics					7.2
Direct Evidence	0 1 2 3	1		3	
Ignitability	0 1 2 3	1		3	
Reactivity	0 1 2 3	1		3	
Incompatibility	0 1 2 3	1		3	
Hazardous Waste Quantity	0 1 2 3 4 5 6 7 8	1		8	
Total Waste Characteristics Score				20	
3 Targets					7.3
Distance to Nearest Population	0 1 2 3 4 5	1		5	
Distance to Nearest Building	0 1 2 3	1		3	
Distance to Sensitive Environment	0 1 2 3	1		3	
Land Use	0 1 2 3	1		3	
Population within 2-Mile Radius	0 1 2 3 4 5	1		5	
Buildings within 2-Mile Radius	0 1 2 3 4 5	1		5	
Total Targets Score				24	
4 multiply $\boxed{1}$ × $\boxed{2}$ × $\boxed{3}$				1,440	
5 Divide line $\boxed{4}$ by 1,440 and multiply by 100				$S_{FE} = 0$	

Direct Contact Work Sheet

Facility Name: _____ Date: _____

Direct Contact Work Sheet

Rating Factor	Assigned Value (Circle One)	Multi-plier	Score	Max. Score	Ref. (Section)
1 Observed Incident	0 45	1		45	8.1
If line 1 is 45, proceed to line 4 . If line 1 is 0, proceed to line 2 .					
2 Accessibility	0 1 2 3	1		3	8.2
3 Containment	0 15	1			8.3
4 Waste Characteristics Toxicity	0 1 2 3	5		15	8.4
5 Targets Population within 1-Mile Radius Distance to a Critical Habitat	0 1 2 3 4 5 0 1 2 3	4 4		20 12	8.5
Total Targets Score				32	
6 If line 1 is 45, multiply 1 × 4 × 5 If line 1 is 0, multiply 2 × 3 × 4 × 5				21,600	
7 Divide line 6 by 21,600 and multiply by 100					

$S_{DC} =$

CURRENT STATUS OF INTERSTATE COMPACTS AND INDEPENDENT LARGE-VOLUME-GENERATING STATES

Interstate agreements for cooperative provision of low-level radioactive waste (LLW) management and disposal capability (known as compacts) were encouraged by the Low-Level Radioactive Waste Policy Act of 1980 (the Act — P.L. 96–573). The Act required that any compact receive Congressional ratification before it would be authorized to exclude wastes from nonmember states from disposal at a regional facility. As described in Chapter 3, the Act provided that such exclusion could be invoked as of January 1, 1986. New disposal capacity was not available by that time and extension of the exclusion date to January 1, 1993 was one of the agreements reached in the Low-Level Radioactive Waste Policy Amendments Act of 1985 (the Amendments Act — P.L. 99–240). The Amendments Act also served as the ratification vehicle for those compacts that had been submitted for Congressional approval. There are several other interstate agreements in various stages of completion although the compacts have not yet received Congressional ratification. As discussed below there have been realignments in some compact memberships and it is possible that additional changes will evolve in the next several years. Other states have certified their intention to proceed on a "go-it-alone" or nonaligned basis to provide disposal capability necessary for the LLW produced within their borders.

The specific agreements negotiated differ among the several compacts. These differences reflect particular concerns among the member states and include such things as procedures for selecting a host state for a new disposal facility, criteria for selecting technologies for treatment and/or disposal, the responsibility for facility siting, development, and regulation, and fee structures. The current compact membership and some significant features of the compact agreements are described in the following paragraphs. Similar information is also provided for several of the large volume nonaligned state programs.

As noted, there is a wide range of approaches to meeting the January 1, 1993 deadline in the Amendments Act by which each state or compact must be able to provide for the disposal of LLW produced within its borders. The "sited states" (South Carolina, Nevada, Washington) in which disposal facilities are currently operating and the Department of Energy each has responsibility for reviewing the progress being made toward reaching that goal as discussed in Chapter 3. The sited states determine which compacts or states are in compliance and eligible for continued access or subject to penalty surcharges. DOE determines the compact or state's eligibility for the 25% surcharge rebate for milestone compliance. Both the sited states and DOE provided criteria for demonstrating compliance with the January 1, 1988 milestone, which requires that the host state for the regional disposal facility has been identified, a siting plan has been prepared, and authority delegated to accomplish the actions identified in the siting plan. Such actions include site selection, license application, facility development and operation, and licensing authority. Only those states that did not submit information regarding milestone compliance were found by DOE not to be eligible for surcharge rebates as of January 1, 1988. These states were New Hampshire, North Dakota (which is negotiating with the Northwest Compact for disposal at the Richland site), Puerto Rico, and Vermont. Similar criteria will likely be provided for determining compliance with subsequent milestones.

The currently ratified compacts, and other selected compacts or large volume generating states are

Compact Name:	Northwest Interstate Compact on Low-level Radioactive Waste Management
Party States:	Alaska, Hawaii, Idaho, Montana, Oregon, Utah, Washington
Host State:	Washington

Notes: Richland, Washington is the site of an operating regional disposal facility. The site is located on part of the Federal Hanford Reservation leased by the State of Washington from the Department of Energy. It is operated by U.S. Ecology and is expected to continue to function as the regional disposal site for an indefinite period after January 1, 1993. In 1987 the Richland facility accepted 555,192 ft^3 of waste for disposal (30% of the national total). The compact generated 7% of the national waste volume in 1987. The compact legislation includes provisions for party states to require that waste produced within their borders be packaged and transported in compliance with the host state's regulations. There is also explicit agreement in the compact that cooperative waste management is not limited to LLW. Article IV(5) states that "in consideration of the State of Washington allowing access to its low-level waste disposal facility by generators in other party states, party states such as Oregon and Idaho which host hazardous chemical waste disposal facilities will allow access to such facilities by generators within other party states."

Current issues that the state of Washington is addressing include the adequacy of liability coverage, alternative instruments for providing the desired level of coverage both for onsite cleanup and for transportation, baseline data on waste already in place, disposal of mixed waste and the status of lead and lead shielding, and procedures for site closure and long-term care.

Compact Name:	Central Interstate Low-level Radioactive Waste Compact
Party States:	Arkansas, Kansas, Louisiana, Nebraska, Oklahoma
Host State:	Nebraska

Notes. In addition to the five party states, Iowa, Minnesota, Missouri, and North Dakota were eligible to join the Central Interstate Compact. The Commission has selected U.S. Ecology to develop a disposal site for the region. The five compact party states generated 9% of the national waste volume in 1987. Rates for disposal at the regional facility are subject to the approval of the host state based upon criteria established by the Commission. The host state is further empowered to establish fees in addition to the rates based on cost of service and rate of return to the operator. These fees are to cover any costs incurred by the host state with regard to the regional facility.

Compact Name:	Southeast Interstate Low-Level Radioactive Waste Management Compact
Party States:	Alabama, Florida, Georgia, Mississippi, North Carolina, South Carolina, Tennessee, Virginia
Host State:	North Carolina

Notes. The disposal facility operated by Chem-Nuclear Services, Inc. in Barnwell, South Carolina is identified as the initial regional facility in the compact legislation. In 1987 the Barnwell site accepted 957,335 ft^3 of waste for disposal (52% of the national total). The eight states that are Party to the Southeast Compact generated 30% of the national waste volume in 1987. Article 4(E)(6) requires the Commission to identify a host state for a second facility that must be in operation "in no event later than 1991." The schedule for this region is therefore somewhat more stringent than that imposed by the Amendments Act. Following a host state selection process in which all party states were involved and that included development, weighting, and application of technical criteria, the Commission designated North Carolina to host the subsequent disposal facility. North Carolina has developed siting criteria and is in the process of soliciting proposals to develop and operate the site. Compact provisions require that each party state serve as a host state before one state is requested to host a second facility. The costs of the Commission's operations are funded by a fee imposed on waste volumes disposed at the regional facility.

Compact Name:	Central Midwest Interstate Low-Level Radioactive Waste Compact
Party States:	Illinois, Kentucky
Host State:	Illinois

Notes. Ilinois is the host state for the compact in accordance with provisions in Article II that "a party state with a total volume of waste recorded on low-level radioactive waste manifests for any year that is less than 10 percent of the total volume recorded on such manifests for the region during the same year shall not be designated a host state." The compact generated 10% of the national waste volume in 1987. Kentucky is a relatively small-volume generating state. The waste existing at the Maxey Flats disposal facility in Kentucky is specifically excluded from the regional disposal facility by Article V(f) of the compact. In its siting program, Illinois developed plans to characterize four sites in areas identified by progressive screening studies. The state also encouraged volunteer proposals to host the disposal facility and essentially provided a local veto power over site selection. Current efforts are focused on selecting the four areas to be characterized.

Compact Name:	Midwest Interstate Low-Level Radioactive Waste Management Compact
Party States:	Iowa, Indiana, Michigan, Minnesota, Missouri, Ohio, Wisconsin
Host State:	Michigan

Notes. The compact generated 8% of the national waste volume in 1987. Michigan has created a Low-Level Radioactive Waste Management Authority with a single member directly responsible to the Governor. The Authority member is charged with developing siting criteria, selecting a disposal method, choosing candidate sites, characterizing those sites, and selecting the site that will be developed as the regional disposal facility. Within 30 days after such selection, the State Legislature may disagree with the site so identified and select among one of the other sites already characterized. There are several issues that Michigan is seeking to resolve with the other party states in the compact: provisions for shared liability for the disposal site, commitment to remain within the compact regardless of changes in federal status related to Congressional authority to review compacts every 5 yr, and agreement on funding the site operation, closure, and postclosure care. With regard to the last issue of financial stability, the Authority is being funded in the short term by loans from regional generators. The Authority is seeking clarification on whether long-term revenue is pledged by the generators or by the party states.

Compact Name:	Rocky Mountain Low-level Radioactive Waste Compact
Party States:	Colorado, Nevada, New Mexico, Wyoming
Host State:	Colorado

Notes. The compact legislation provides that the currently existing regional disposal facility in Beatty, Nevada will be phased out and a new regional facility will be developed. Colorado will host the next facility. The compact provides that each party state must host a facility before any one state is subject to hosting a second facility. Article II(c) provides that only those party states that generate 20% or more of the region's volume are obligated to host a disposal facility. States party to the compact generated less than 1% of the national waste volume in 1987.

Compact Name: Northeast Interstate Low-Level Radioactive Waste
                         Management Compact
Party States: Connecticut, New Jersey
Host State: Dual Host States

Notes. The Commission designated each party state to plan to site a disposal facility capable of safely managing an equitable portion of the region's waste. The compact generated 7% of the national waste volume in 1987. Further evaluation is being conducted to support a choice among alternatives such as disposal facilities in each state that would accept all types of waste, separation by waste classification with disposal of Class A waste in one state and Class B–C waste in the other state, treatment in one state and disposal in the other, or disposal of mixed hazardous and radioactive waste in one state and the rest of the waste in the other state. Siting is the responsibility of the Connecticut Hazardous Waste Management Service in Connecticut and of the New Jersey Low-Level Radioactive Waste Disposal Facility Siting Board in that state.

Compact Name: Appalachian Low-Level Radioactive Waste Management
                         Compact
Party States: Pennsylvania, West Virginia, Maryland, Delaware
Host State: Pennsylvania

Notes. Pennsylvania negotiated the Appalachian Compact with West Virginia and identified itself as the host state. It provided eligibility status for states with contiguous borders and low volumes relative to those generated in Pennsylvania. The Department of Environmental Resources is responsible for regulation of the facility and has developed siting and design criteria that must be complied with by any developer/operator. The compact generated 11% of the national waste volume in 1987. Site selection, facility construction, and operation are to be performed by a commercial contractor selected in response to a Request for Proposals. The Appalachian Compact has been submitted for Congressional ratification

Compact: Western
Party States: Arizona, South Dakota (North Dakota is eligible)
Host state: Arizona

Notes. The Western Compact has been introduced for Congressional ratific-ation. The compact generated 1% of the national waste volume in 1987. The party states to this compact are also eligible to join the proposed Southwest Compact with California. Negotiations are continuing among the several states.

State: California

Notes. California has investigated agreement with several other states, including Arizona, North Dakota, and South Dakota. The state has not yet submitted legislation seeking Congressional ratification of any compact status. Following development of siting criteria and procedures, the state solicited proposals from commercial developer/operators to conduct the final site selection and construct and operate the site. U.S. Ecology was selected to develop a disposal facility in a desert area in southeastern California.

State: Texas

Notes. Texas established an independent Authority to select, develop, and run the state's disposal facility for state-generated waste only. Siting studies have considered both state-owned and non-state-owned areas and characterization is scheduled to begin in late spring 1988 on a site in Hudspeth County. Texas has completed a "de minimis" rule making in accordance with which short-lived radioactive materials are eligible to be disposed of as nonradioactive.

State: New York

Notes. Responsibility for selecting a site and disposal methodology for a facility to accept only waste generated in New York State is vested in a five-member Siting Commission. The Commission is composed of a medical doctor, a health physicist, a geologist, a professional engineer, and a knowledgeable private citizen who serves as the Chairman. Development and operation of the facility are the responsibility of the State's Energy Research and Development Authority. Licensing responsibility for the site itself rests with the Department of Environ-mental Conservation. Site operation will be licensed by the Labor Department. Other agencies and private groups are involved through participation in an Advisory Committee to the Siting Commission as well as through the several decision-making and rule-making proceedings conducted by the primary agencies. Most of the development costs, for example, the costs of the Siting Commission and the Department of Environmental Conservation regulations development, are being funded by annual appropriations on the electric utilities producing power with nuclear units. Eligible costs are prorated by the number of fully operational nuclear power reactors in the state.

REFERENCES

(DOE 1987) "1986 Annual Report on Low-Level Radioactive Waste Management progress," Report to Congress in Response to Public Law 99–240. U.S. Department of Energy, Washington, D.C., June 1987.

(NRC 1987) "Current Status of Each State in Providing Disposal of Low-Level Radioactive Waste—August 18, 1987." Memorandum from Spiros Droggitis, U.S. Nuclear Regulatory Commission Office of Governmental and Public Affairs.

(EG&G 1987) "State Development of New Low-Level Radioactive Waste Disposal Capacity: Progress and Issues." Presented by T. D. Tait, EG&G Idaho, Inc. at the World Nuclear Fuel Market International Conference on Nuclear Energy, San Diego, California, October 20, 1987.

(EG&G 1988a) "Progress in Developing New Commercial LLRW Disposal Facilities and DOE Assistance." Presented by T. D. Tait and S. T. Hinschberger at Waste Management '88, Tucson, Arizona, March 1988.

(EG&G 1988b) "U.S. Department of Energy Low-Level Waste Management Program National Status Report as of 2/29/88." Provided courtesy of Paul Smith, EG&G Idaho, Inc.

INDEX

Above–ground vaults (AGV), 233, 274–278
Absorption, 180–182, 374
Accelerator targets, 48
Accident(s), 12–17, 155–158
  analysis, 155–158
  Chernobyl, 12, 13
  Goiania, 12
  Khyshtym, 12, 13
  safety analysis report (SAR), 157–158, 251
  Three Mile Island, 12, 13
  transportation, 157
Acid digestion, 187, 190, 192, 194, 196
Actinides, 301, 302, 401
Activation products, 31
Active maintenance, 10, 111, 261, 263, 276, 401
Adsorption, 373–374
Agitated hearth incinerator, 190, 192, 194, 196
Agreement states, 63, 65, 86, 401
Airborne releases, 15, 401
  mechanisms, 126, 375–381
Airborne surveys, 13, 151
Allied Gulf Nuclear Services (AGNS), 68
Antarctic Treaty of 1959, 301
As low as reasonably achievable (ALARA), 41, 83, 110, 111, 132, 141–144, 148, 259, 302, 336, 402
Asphalt (bitumen), 179–180
Atomic Energy Act of 1954, as amended, 2, 63, 65, 83, 86, 306
Atomic Energy Commission(AEC), 63, 306–308

Audit, 149–150
Away from reactor (AFR) storage, 211

Baling, 186
Barnwell, South Carolina, 70, 76, 222–223, 358
Barrier:
  engineered, 4, 265, 266, 270, 375–390, 402–403
  intruder, 14, 111, 112, 272
  multi–, 286, 291, 294, 300, 406
Baseline monitoring, 145
Bathtub effect, 70, 75
Beatty, Nevada, 69, 76, 358–359
Below–ground vault (BGV), 233, 270–272
Below regulatory concern (BRC), 3, 83, 117. *See also* "de minimis"
Benefit–cost analysis, 3, 136–138
Bioassay, 151–152
Biosphere, 402
Borosilicate glass, 37
Brine influx, 108, 291. *See also* Waste Isolation Pilot Plant, (WIPP)
Bruce Nuclear Generating Station, Canada, 279
Buffer zone, 4, 253, 258, 402
Byproduct material, 239, 402

Calcination, 5
Carbon–14, 50, 75, 76, 85
Casks:
  onsite storage, 106, 207–211
  spent fuel storage, 104–106, 207–211